2 0 1 0

STATE OF THE WORLD

Transforming Cultures

From Consumerism to Sustainability

Other Norton/Worldwatch Books

State of the World 1984 through *2009*
(an annual report on progress toward a sustainable society)

Vital Signs 1992 through *2003* and *2005* through *2007*
(a report on the trends that are shaping our future)

Saving the Planet
Lester R. Brown
Christopher Flavin
Sandra Postel

How Much Is Enough?
Alan Thein Durning

Last Oasis
Sandra Postel

Full House
Lester R. Brown
Hal Kane

Power Surge
Christopher Flavin
Nicholas Lenssen

Who Will Feed China?
Lester R. Brown

Tough Choices
Lester R. Brown

Fighting for Survival
Michael Renner

The Natural Wealth of Nations
David Malin Roodman

Life Out of Bounds
Chris Bright

Beyond Malthus
Lester R. Brown
Gary Gardner
Brian Halweil

Pillar of Sand
Sandra Postel

Vanishing Borders
Hilary French

Eat Here
Brian Halweil

Inspiring Progress
Gary T. Gardner

2 0 1 0

STATE OF THE WORLD

Transforming Cultures

From Consumerism to Sustainability

A Worldwatch Institute Report on
Progress Toward a Sustainable Society

Erik Assadourian, *Project Director*

Mona Amodeo
Robin Andersen
Ray Anderson
Cecile Andrews
Judi Aubel
Albert Bates
Walter Bortz
Robert Costanza
Cormac Cullinan
Jonathan Dawson
John de Graaf
Robert Engelman
Joshua Farley
Susan Finkelpearl
Kate Ganly
Gary Gardner
Amy Han
Jim Hartzfeld
Toby Hemenway
Yoshie Kaga
Ida Kubiszewski
Susan Linn
Johanna Mair
Michael Maniates
Pamela Miller
Kevin Morgan
Peter Newman
David W. Orr
Michael Renner
Jonah Sachs
Ingrid Samuelsson
Juliet Schor
Michael H. Shuman
Roberta Sonnino
Wanda Urbanska

Linda Starke and Lisa Mastny, *Editors*

W · W · NORTON & COMPANY
NEW YORK LONDON

1776 Massachusetts Avenue, N.W.
Suite 800
Washington, DC 20036
www.worldwatch.org

Printed in the United States of America.

The text of this book is composed in Galliard, with the display set in ScalaSans. Book design, cover design, and composition by Lyle Rosbotham; manufacturing by Victor Graphics.

First Edition
ISBN 978-0-393-33726-6

W. W. Norton & Company, Inc., 500 Fifth Avenue, New York, N.Y. 10110
www.wwnorton.com

W. W. Norton & Company Ltd., Castle House, 75/76 Wells Street, London W1T 3QT

1 2 3 4 5 6 7 8 9 0

♻ This book is printed on recycled paper.

Worldwatch Institute Board of Directors

Acknowledgments

This book was conceived in the fall of 2008 during a conversation over dinner with former Worldwatch Board Chair Øystein Dahle. Over pasta in Oslo, the two of us discussed how much consumer cultures will need to change for the human species to truly thrive. Upon returning to Washington, I proposed the idea of confronting this issue head-on in *State of the World 2010*. Somewhat to my surprise, the Worldwatch staff and Board of Directors gave me the green light to proceed. First, a thank you to all of them for trusting that such a topic would be a valuable theme for our flagship publication, even if it proves controversial. Thanks especially to Worldwatch President Christopher Flavin for trusting me to run with this idea.

After that brief moment of elation, the long process of building this book began. A *State of the World* committee was formed and the counsel of its members proved essential throughout. Many thanks to all of you for the hours spent discussing the newest ideas as they developed, for suggesting authors and topics, and for helping the project move forward.

Much of last spring was devoted to recruiting the highly talented group of authors who are listed on the Contents page. I want to especially thank these individuals—all of whom agreed to freely share their knowledge and insights with *State of the World* readers. Without their generosity, this book would not have been possible.

This year we have a number of short Boxes as well, which complement the longer articles and add more voices and views to the report. Many thanks to these authors as well for their time and thoughtful contributions: Yann Arthus-Bertrand, Eduardo Athayde, Almut Beringer, Michael Braungart, Raj Chengappa, Patrick Curry, Øystein Dahle, Anne H. Ehrlich, Paul R. Ehrlich, Gregory C. Farrington, Satish Kumar, Serge LaTouche, William McDonough, Julie Ozanne, Lucie Ozanne, and Alexander Rose.

Those who helped with the research over the past year deserve special attention here. I am grateful for the ideas shared and the assistance given in making this book possible. Thanks to Franny Armstrong, Diane Assadourian, Andrew Balmford, Mark Beam, Guy P. Brasseur, Gene Brockhoff, Brian Burke, Tony Carr, Robert Corell, Joel Cowan, Scott Denman, Nancy Durkee, Duane Elgin, Hilary French, Jim Freund, Nina Frisak, Marcin Gerwin, Alex Hallatt, Harry Halloran, Jody Heymann, Yeşne Iren, Chris Jung, Hayrettin Karaca, William Kilbourne, Lynne LaCarrubba, Shawna Larson, Kalle Lasn, Annie Leonard, Ling Li, Lisa Lucero, Jan Lundberg, Mia MacDonald, Michael Maniates, Susanne Martikke, Marc Matthieu, Jim McDonough, Krystal McKay, Bill McKibben, Olivier Milhomme, Molly O'Meara Sheehan, Pete Palmer, Nadina Perera, Barbara Petruzzi, Andrea Prothero, Paul Reitan, Joan Roberts, Regina Rowland, Peter Sawtell, Vernon Scarborough, Blair Shane, David Stoesz, Robert Welsch, and, at UNESCO, Aline Bory-Adams, Bernard Combes, Hans d'Orville, Mark Richmond, and Ariana Stahmer. I am grateful to all of you!

I'd also like to thank Muhammad Yunus, who kindly shared his wisdom and his story with us in this year's Foreword. His support for our humble book is quite an honor.

One other special contributor I'd like to acknowledge is artist Chris Jordan, whose beautiful image graces our cover. *Gyre* is one of the most striking portrayals of the threat of consumerism and the possibility of this moment to change course that I have seen, and we are very happy to have the chance to display it on the cover.

Behind the scenes, there were three special people without whom this project would not have succeeded. First and foremost, Linda Starke, editor extraordinaire and *State of the World* elder, who was a joy to work with and selflessly put this project over herself time and again. She was an exemplar of calm throughout the stressful concluding months and, considering the challenges, was truly a role model for me. Thank you, Linda.

A large thanks also to Gary Gardner for helping to improve several of the contributions, including my own. While surely a thankless job at times, I appreciate the many hours of the summer that Gary sacrificed to get this book from manuscript to finished product. Thanks as well to Lisa Mastny, who also spent several weeks this summer helping to polish several articles and produce a fascinating chronology of the environmental events of the past year.

Even further behind the scenes, but without whom this book would have been significantly weaker, are the eight project interns who this year pursued hard-to-find data, examples, and ideas, helped recruit authors, and even contributed several interesting Boxes and an article. In order of their appearance, let me express my appreciation for each one.

Helene Gallis—intern of many countries, candid, and creative—started when this project was still formless and played an important early role in molding it. She then continued to prove herself invaluable as she helped out to the very end, researching, reviewing, and writing.

Eddie Kasner, when not studying public health and farmers' use of pesticides in China, helped recruit several authors and research sustainable dietary norms and health care. Amy Han was truly a "Jill-of-all-trades," cheerfully becoming the creator and Webmaster of the Transforming Cultures Web site, doing research, writing blog posts, and finishing off with a captivating article on music's role in building sustainable societies.

Valentina Agostinelli of Italy also enthusiastically assisted with research and assiduously monitored the year of environmental events, helping to produce this year's timeline. Kevin Green was a data-finding machine, which at a place like Worldwatch is a compliment of the highest order. Without Kevin's diligent research, the overview chapter would not be so chockful of useful information.

Mami Shijo, coming from our partner organization Worldwatch Japan, played an important role in finding several bits of data and helping to explore the Japanese blogosphere—a project intern Emiko Akaishi expanded further in her month here. The message of *State of the World 2010* is one that will need to be heard in Japan—a leading consumer culture—as much as in North America and Europe, so thanks to both of you for helping start that conversation.

And finally, in the last month of production, Fulbright Fellow Stefanie Bowles swooped in and helped finalize the book, expanded its content, and kept me on task. Good timing, Stefanie!

My wife, Aynabat Yaylymova, deserves a special note for putting up with me these last several months, as I increasingly lived in the office—and all the ideas of sustainable living discussed in these pages started to fade from practice and even memory.

At Worldwatch, I'd like to acknowledge the staff and the many ways they contributed to this book. First, thanks to Robert Engelman, Gary Gardner, and Michael Renner for sharing their expertise in their articles and expanding the breadth and depth of this volume. A special thanks to Alice McKeown, who reviewed many articles and raised the bar in every case. Juliane Diamond, Brian Halweil, Danielle Nierenberg, Thomas Prugh, Molly Theobald, and senior fellow Zoë Chafe also helped with reviewing—thank you.

While the words of this book come from the authors, its beauty comes from Worldwatch Art Director Lyle Rosbotham. Lyle designed this book from cover to cover and found the gorgeous pictures to grace many of its pages. If a picture is worth a thousand words, then he singlehandedly added a hundred rich pages to *State of the World 2010*, all without destroying a single extra tree!

Thanks to Patricia Shyne for all her work with our partners around the world to ensure that the ideas and examples of *State of the World* are dispersed far and wide. And to our communications team, Darcey Rakestraw and Julia Tier, for spreading this message even further, both through press outreach and through helping to coordinate the new Transforming Cultures blog.

A big bow of gratitude goes as well to the development team—Courtney Berner, Trudy Loo, Meghan Nicholson, and Mary C. Redfern—for helping ensure the support needed to make this book a success.

Thanks also to Ben Block, Amanda Chiu, Anna da Costa, Yingling Liu, and Janet Sawin for suggesting topics and authors. And to John Mulrow and summer intern Ben Gonin for help in analyzing just how little we can consume before hitting unsustainable levels. Finally, thanks to Barbara Fallin for ensuring the smooth administration of this project and to Corey Perkins for keeping Worldwatch's electronic infrastructure humming.

Beyond the Institute, I'd like to extend my gratitude to our many publishing partners. First, in the United States, we appreciate the dedication of W. W. Norton & Company, which has published *State of the World* in all 27 years of its existence. Thank you Amy Cherry, Erica Stern, and Devon Zahn for your work in producing the book and ensuring that it gets distributed broadly in bookstores and university classrooms across the United States.

Without our strong network of international publishing partners, we would have a limited international audience and lessened effect. We very much appreciate the work that all of them do to get Worldwatch's findings translated and disseminated as quickly and as widely as possible. We give special thanks to Eduardo Athayde of the Universidade Mata Atlântica in Brazil; Sylvia Shao of Environment Science Press in China; Tuomas Seppa of Gaudeamus & Otatieto in Finland; Klaus Milke and colleagues at Germanwatch, Ralf Fuechs and colleagues at the Heinrich Böll Foundation, and Jacob Radloff of OEKOM in Germany; Yiannis Sakiotis and Michalis Probonas of the Evonymos Ecological Library in Greece; Zsuzsa Foltanyi of Earth Day Foundation in Hungary; Kartikeya Sarabhai and Kiran Chhokar of the Centre for Environment Education in India; Anna Bruno Ventre and Gianfranco Bologna of WWF Italy; Soki Oda of Worldwatch Japan; Melanie Gabriel Camacho and Cecilia Geiger of Africam Safari and Diana Isabel Jaramillo and Fabiola Escalante of UDLAP in Mexico; Marcin Gerwin of Earth Conservation in Poland; Monica Di Donato of Area Sostenibilidad CIP Ecosocial and Anna Monjo of Icaria Editorial for the Castilian version and Helena Cots of the Centre UNESCO de Catalunya for the Catalan version in Spain; Sang-ik Kim of the Korean Federation of Environmental Movement in South Korea; Øystein Dahle, Hans Lundberg, and Ivana Kildsgaard of Worldwatch Norden in Norway and Sweden; George Cheng of Taiwan Watch Institute in Taiwan; Yeşim Erkan of TEMA in Turkey; Professor Marfenin and Anna Ignatieva of Center of Theoretical Analysis of Environmental Problems at the International Independent University of Environmental and Political Sciences in Russia; and Jonathan Sinclair Wilson, Michael Fell, Gudrun Freese, and Alison Kuznets of Earthscan in the United Kingdom.

Our great customer service team at Direct Answer, Inc. also helps ensure that our readers are effectively served and their questions are swiftly answered. We are grateful to Katie Rogers, Katie Gilroy, Lolita Harris, Cheryl Marshall, Valerie Proctor, Ronnie Hergett, Marta Augustyn, Heather Cranford, Colleen Curtis, Sharon Hackett, and Karen Piontkowski for providing first-rate customer service.

We want to express our deep appreciation to the many foundations and institutions whose support over the past year has made *State of the World 2010* and Worldwatch's many other projects possible: The Heinrich Böll Foundation; The Casten Family Foundation of the Chicago Community Trust; the Compton Foundation, Inc.; the Del Mar Global Trust; the Bill & Melinda Gates Foundation; the Goldman Environmental Prize; the Richard and Rhoda Goldman Fund; the Good Energies Foundation; the Hitz Foundation; the W. K. Kellogg Foundation; the Steven C. Leuthold Family Foundation; the Marianists Sharing Fund of the USA; the Netherlands Environment Ministry; the V. Kann Rasmussen Foundation; the Royal Norwegian Ministry of Foreign Affairs; the Shared Earth Foundation; The Renewable Energy & Energy Efficiency Partnership; the Shenandoah Foundation; Stonyfield Farm; the TAUPO Fund; the United Nations Environment Programme; the United Nations Population Fund; the UN Foundation; the Wallace Genetic Foundation, Inc.; the Wallace Global Fund; the Johanette Wallerstein Institute; the Winslow Foundation; and the World Wildlife Fund–Europe.

State of the World 2010 would not exist without the generous contributions of the many individuals who support the Institute as Friends of Worldwatch. These gifts make up nearly one third of the Institute's annual operating budget and are indispensable to our work. We are profoundly grateful to all the Friends of Worldwatch for their commitment to the Institute and its vision for a sustainable world. And thanks to the many Worldwatch supporters who invested directly in this year's report when they learned about it through a fundraising appeal this spring. Your generosity—even at a time when this project was still just a concept—is much appreciated.

Finally, saving the most important acknowledgment for last, I want to thank you. If you are reading this, I can assume that you are interested in digging deeply into this topic—as who else would plow through four pages of names? The goal of this book is to help get human cultures back on track before we undermine the ecological systems that we as a species depend on. Your help in changing cultures is essential. As the book indicates, there are countless ways to get involved. Many more will be discussed on our Web site, at blogs.worldwatch.org/transformingcultures. And while visiting the Web site, consider starting a discussion group about the report or mobilizing your own network to bring about the change you want to see. This is how new cultures start!

Erik Assadourian
Project Director

Worldwatch Institute
1776 Massachusetts Ave., NW
Washington, DC 20036
www.worldwatch.org
blogs.worldwatch.org/transformingcultures
eassadourian@worldwatch.org

Contents

BOXES

TABLES

FIGURES

Foreword

Muhammad Yunus
Founder, Grameen Bank, and 2006 Nobel Peace Prize Laureate

I am pleased that the Worldwatch Institute has chosen to tackle the difficult issue of cultural change in *State of the World 2010*. Over the past three decades, at the heart of my work with microfinance, I had to challenge the centuries-old belief that poor, illiterate women cannot be agents of their own prosperity. Microfinance rejects this fundamental cultural misconception.

Culturally rooted fallacies are difficult to slay. My early requests to established bankers to lend to poor women were met with clear and strong objections. "Poor people are not bankable. They're not creditworthy," a local banker insisted, adding for good measure, "You can say goodbye to your money." The initial experiment was highly encouraging—our borrowers turned out to be excellent customers who repaid their debts on time. The conventional bankers were unimpressed, calling the results a fluke. When we were successful in multiple villages, they shrugged their shoulders.

I realized that their cultural presumptions about the poor would not budge easily, no matter how many successes we earned. Their minds were made up—*Poor people are not creditworthy!* My job, I realized, was to sow the seeds of a new financial culture by turning this false notion on its head: the truth is not that the poor are not creditworthy, but that conventional banks are not people-worthy.

So we set out to create a different kind of bank, one geared to serve the poor. Conventional banks are built around the principle that "the more you have, the more you can get." We reversed that principle to the less you have, the higher your priority for receiving a loan. Thus began a new culture of finance and poverty alleviation, in which the poorest are served first and a fistful of capital could turn abject poverty into a livelihood.

After years of careful cultivation, these ideals became Grameen Bank, which today lends a billion dollars annually to 8 million borrowers. Our average loan is $360, and 99 percent of funds are paid back on time. Programs now include lending to beggars, micro-savings accounts, and micro-insurance policies. And we are proud to note that microcredit has expanded worldwide.

A financial industry for impoverished people, mostly women. That is a cultural change.

Now I know that cultural assumptions, even well-established ones, can be overturned, which is why I am excited about *State of the World 2010*. It calls for one of the greatest cultural shifts imaginable: from cultures of consumerism to cultures of sustainability. The book goes well beyond standard prescriptions for clean technologies and enlightened policies. It advocates rethinking the foundations of modern consumerism—the practices and values regarded as "natural," which paradoxically undermine nature and jeopardize human prosperity.

Worldwatch has taken on an ambitious agenda in this volume. No generation in history has achieved a cultural transformation as sweeping as the one called for here. The book's many articles demonstrate that such a shift is possible by reexamining core assumptions of modern life, from how businesses are run and what is taught in classrooms to how weddings are celebrated and the way cities are organized. Readers may not agree with every idea presented here. But it is hard not to be impressed with the book's boldness: its initial assumption is that wholesale cultural transformation is possible. I believe this is possible after having lived through the cultural transformation of women in Bangladesh. Culture, after all, is for making it easy for people to unleash their potential, not for standing there as a wall to stop them from moving forward. Culture that does not let people grow is a dead culture. Dead culture should be in the museum, not in human society.

Preface

Christopher Flavin
President, Worldwatch Institute

The past five years have witnessed an unprecedented mobilization of efforts to combat the world's accelerating ecological crisis. Since 2005, thousands of new government policies have been enacted, hundreds of billions of dollars have been invested in green businesses and infrastructure, scientists and engineers have greatly accelerated development of a new generation of "green" technologies, and the mass media have turned environmental problems into a mainstream concern.

Amid this flurry of activity, one dimension of our environmental dilemma remains largely neglected: its cultural roots. As consumerism has taken root in culture upon culture over the past half-century, it has become a powerful driver of the inexorable increase in demand for resources and production of waste that marks our age. Of course, environmental impacts on this scale would not be possible without an unprecedented population explosion, rising affluence, and breakthroughs in science and technology. But consumer cultures support—and exaggerate—the other forces that have allowed human societies to outgrow their environmental support systems.

Human cultures are numerous and diverse—and in many cases have deep and ancient roots. They allow people to make sense of their lives and to manage their relationships with other people and the natural world. Strikingly, anthropologists report that many traditional cultures have at their core respect for and protection of the natural systems that support human societies. Unfortunately, many of these cultures have already been lost, along with the languages and skills they nurtured, pushed aside by a global consumer culture that first took hold in Europe and North America and is now pressing to the far corners of the world. This new cultural orientation is not only seductive but powerful. Economists believe that it has played a big role in spurring economic growth and reducing poverty in recent decades.

Even if these arguments are accepted, there can be no doubt that consumer cultures are behind what Gus Speth has called the "Great Collision" between a finite planet and the seemingly infinite demands of human society. More than 6.8 billion human beings are now demanding ever greater quantities of material resources, decimating the world's richest ecosystems, and dumping billions of tons of heat-trapping gases into the atmosphere each year. Despite a 30-percent increase in resource efficiency, global resource use has expanded 50 percent over the past three decades. And those numbers could continue to soar for decades to come as more than 5 billion people who currently consume one tenth as many resources per person as the average European try to follow the trail blazed by the world's affluent.

State of the World has touched on the cultural dimensions of sustainability in the past—particularly in *State of the World 2004*, which focused on consumption. But these discussions have been brief and superficial. Early last year, my colleague Erik Assadourian convinced me that the elephant in the room could no longer be ignored. At Worldwatch, no good idea goes unpunished, and Erik became the Project Director for this year's book.

While shifting a culture—particularly one that is global in scope—sounds daunting if not impossible, the chapters that follow will convince you otherwise. They contain scores of examples of cultural pioneers—from business leaders and government officials to elementary school teachers and Buddhist monks. These pioneers are convincing their customers, constituents, and peers of the advantages of cultures based on nurturing the natural world and ensuring that future generations live as well or better than the current one.

Religious values can be revitalized, business models can be transformed, and educational paradigms can be elevated. Even advertisers, lawyers, and musicians can make cultural shifts that allow them to contribute to sustainability rather than undermine it.

While the destructive power of modern cultures is a reality that many government and business decisionmakers continue to willfully ignore, it is keenly felt by a new generation of environmentalists who are growing up in an era of global limits. Young people are always a potent cultural force—and often a leading indicator of where the culture is headed. From modern Chinese who draw on the ancient philosophy of Taoism to Indians who cite the work of Mahatma Gandhi, from Americans who follow the teachings of the new *Green Bible* to Europeans who draw on the scientific principles of ecology, *State of the World 2010* documents that the renaissance of cultures of sustainability is already well under way.

To ensure that this renaissance succeeds, we will need to make living sustainably as natural tomorrow as consumerism is today. This volume shows that this is beginning to happen. In Italy, school menus are being reformulated, using healthy, local, and environmentally sound foods, transforming children's dietary norms in the process. In suburbs like Vauban, Germany, bike paths, wind turbines, and farmers' markets are not only making it easy to live sustainably, they are making it hard not to. At the Interface Corporation in the United States, CEO Ray Anderson radicalized a business culture by setting the goal of taking nothing from Earth that cannot be replaced by Earth. And in Ecuador, rights for the planet have even entered into the Constitution—providing a strong impetus to safeguard the country's ecological systems and ensure the long-term flourishing of its people.

While sustainability pioneers are still few in number, their voices are growing louder, and at a moment of profound economic and ecological crisis, they are being heard. As the world struggles to recover from the most serious global economic crisis since the Great Depression, we have an unprecedented opportunity to turn away from consumerism.

Forced deprivation is causing many to rethink the benefits of ever-greater levels of consumption—and its accompanying debt, stress, and chronic health problems. In early 2009, *Time Magazine* proclaimed the "end of excess" and called for Americans to push the "reset" button on their cultural values. In fact, many people are already questioning the cowboy culture, buying smaller cars, moving into less grandiose homes, and questioning the suburban sprawl that has characterized the postwar era. And in poor countries around the globe, the disadvantages of the "American model" are being discussed openly. In *Blessed Unrest*, Paul Hawken has documented the recent rise of a plethora of diverse non-governmental movements that are working to redefine human beings' relationships to the planet and each other.

While consumerism remains powerful and entrenched, it cannot possibly prove as durable as most people assume. Our cultures are in fact already sowing the seeds of their own destruction. In the end, the human instinct for survival must triumph over the urge to consume at any cost.

Christopher Flavin

State of the World: A Year in Review

Compiled by Lisa Mastny with Valentina Agostinelli

This timeline covers some significant announcements and reports from October 2008 through September 2009. It is a mix of progress, setbacks, and missed steps around the world that are affecting environmental quality and social welfare.

Timeline events were selected to increase awareness of the connections between people and the environmental systems on which they depend.

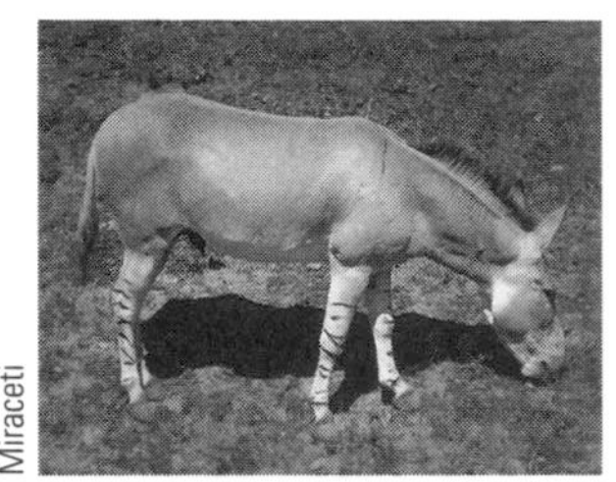
Miraceti
Critically endangered African Wild Ass

BIODIVERSITY
IUCN warns that an "extinction crisis" is under way, with one in four mammals at risk of disappearing because of habitat loss, hunting, and climate change.

MARINE SYSTEMS
Study reports that carbon dioxide is raising ocean acidity at least 10 times faster than was previously thought, with negative effects on shellfish species.

WATER
UN draft treaty calls on countries with shared aquifers to cooperate to protect these waters and to prevent and control their pollution.

CLIMATE
UNEP reports that "brown clouds" of soot, smog, and toxic chemicals are absorbing sunlight and heating the air, aggravating impacts of climate change.

Photodisc

FORESTS
Brazil initiates a crackdown on illegal timber businesses in the Amazon after loggers ransack government offices and steal contraband wood.

OCTOBER NOVEMBER

2008 STATE OF THE WORLD: A YEAR IN REVIEW

2 4 6 8 10 12 14 16 18 20 22 24 26 28 30 2 4 6 8 10 12 14 16 18 20 22 24 26 28

RELIGION
U.S. publisher releases *Green Bible* to spread the message of Creation Care to both religious and non-religious readers.

FORESTS
Provincial governors agree to protect Sumatra's endangered forests, a move that could help cut greenhouse gases from Indonesia, the third-largest emitter.

CLIMATE
Study says ice loss at the poles can best be explained by the human-caused buildup of greenhouse gases, rather than by natural shifts.

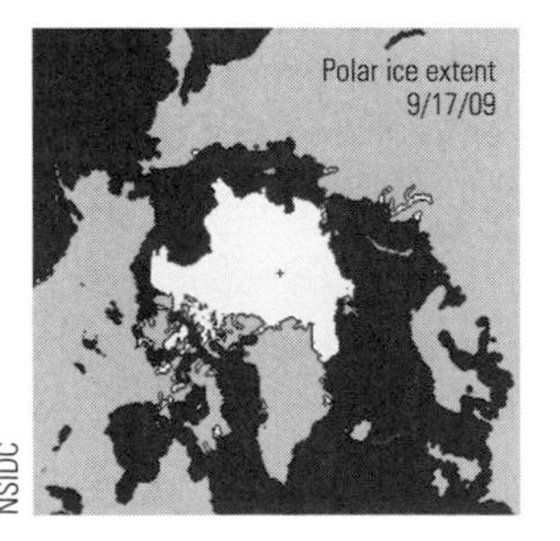

NSIDC

MARINE SYSTEMS
Study warns that fisheries targeting small-to-medium sized "forage fish" to feed farmed fish, pigs, and poultry are affecting both marine ecosystems and human food security.

ENERGY
The Vatican activates a solar energy system to power key buildings and commits to using renewable energy to meet 20 percent of its needs by 2020.

Jessica Flavin

NATURAL DISASTERS
Venice, Italy, suffers its worst flooding in 22 years as high waters reach 1.6 meters, submerging most of the city before gradually receding.

CLIMATE
EU approves package to reduce greenhouse gas emissions by 20 percent, improve energy efficiency by 20 percent, and achieve a 20-percent share for renewables by 2020.

CLIMATE
Researchers say ice loss from Greenland in summer 2008 was nearly three times greater than in 2007, spanning the area of two Manhattans.

BIODIVERSITY
Report says some 15,000 of the world's 50,000 medicinal plant species now face extinction, threatened by habitat loss, overharvesting, and pollution.

Forest & Kim Starr

Hibiscus kokio

DECEMBER — 2009 — JANUARY

MARINE SYSTEMS
Report says nearly a fifth of world coral reefs are dead and the rest may be lost in 20–40 years because of rising water temperatures, ocean acidification, and other threats.

Camilla Sharkey

CLIMATE
California commits to the first comprehensive U.S. plan to slash greenhouse gases, pledging to cut emissions to 1990 levels by 2020.

NATURAL DISASTERS
Munich Re reports that in 2008, the number of devastating weather-related natural disasters increased to 40—a record high—and killed more than 220,000 people.

China Earthquake Administration

Sichuan earthquake of 12 May 2008

CLIMATE
Japan launches the world's first satellite dedicated to monitoring greenhouse gas emissions.

POLLUTION
Study reports that "light pollution" from skyscrapers, cars, and other reflective surfaces is disrupting wildlife behavior and ecosystems.

NATURAL DISASTERS
Officials say more than 4 million people and 2 million cattle in northern China face drinking water shortages due to the worst drought in half a century.

NASA

CLIMATE
NASA's Orbiting Carbon Observatory satellite, intended to help track the sources and sinks of carbon dioxide, crashes into the Pacific Ocean.

FOOD
U.S. White House breaks ground on its first vegetable garden since World War II, in an effort to highlight healthy eating.

Samantha Appleton

CONSERVATION
U.S. President Barack Obama signs law protecting some 8,000 square kilometers in nine states as wilderness areas, off limits to resource exploitation and development.

FEBRUARY MARCH

2009 STATE OF THE WORLD: A YEAR IN REVIEW

2 4 6 8 10 12 14 16 18 20 22 24 26 28 | 2 4 6 8 10 12 14 16 18 20 22 24 26 28 30

CLIMATE
Study warns that climate change due to increased atmospheric carbon dioxide is largely irreversible for 1,000 years after emissions stop.

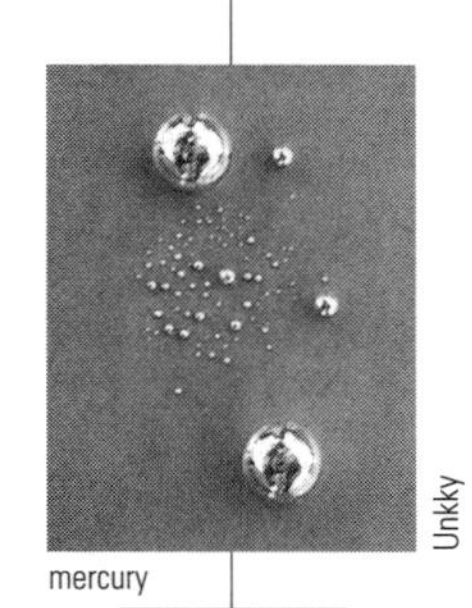
mercury
Unkky

GOVERNANCE
Environment ministers from more than 140 countries agree to create a new international treaty to address emissions and discharge of mercury.

CLIMATE
NASA reports that black carbon is responsible for 50 percent of the increase in Arctic warming from 1890 to 2007.

NOAA, Harley D. Nygren

NATURAL DISASTERS
Southern Africa is hit by the worst floods since 1965, killing more than 100 people and displacing thousands.

NATURAL DISASTERS
Magnitude 6.3 earthquake tears through L'Aquila in central Italy, devastating historic mountain towns and killing nearly 300 people.

HEALTH
Report reveals that 6 out of 10 Americans, or 186 million people, live in areas where air pollution endangers lives.

Massimo Catarinella
Los Angeles in smog

WATER
Scientists report that in roughly a third of the world's largest rivers, decreases in water flow outnumber increases by 2.5 to 1.

FORESTS
Sierra Leone and Liberia announce a new Transboundary Peace Park to protect the shared Gola forests, one of the largest intact rainforests in West Africa.

World Resources Institute

APRIL 2 4 6 8 10 12 14 16 18 20 22 24 26 28 30
MAY 2 4 6 8 10 12 14 16 18 20 22 24 26 28 30

HEALTH
Swine flu virus outbreak claims up to 68 lives in Mexico, leading WHO to declare a "public health emergency of international concern."

ENERGY
World's largest commercial solar power tower plant begins operations near Seville, Spain, with a capacity of 20 megawatts.

CONSUMPTION
Reports says U.S. sales of organic products reached $24.6 billion in 2008, a 17-percent increase over 2007 despite tough economic times.

MARINE SYSTEMS
Six Asia-Pacific countries launch an initiative to protect the 5.7 million-square-kilometer Coral Triangle, home to 76 percent of known coral species.

Robert Young

CLIMATE
Study reports that some 160 villages in northern Syria were deserted in 2007–08 because of severe drought, hinting at future climate change impacts in the Middle East.

Jared Tarbell

FOOD
FAO says world hunger will reach a historic high in 2009, with 1.02 billion people—about one seventh of humanity—going hungry every day.

FORESTS
Indigenous residents protesting oil and gas exploration on their lands battle police in Peru's Amazon, with reports of 9 police and 25 protester deaths.

Harmen Piekema

BIODIVERSITY
Study says the global trade in frogs for pets and food is spreading two severe diseases blamed for driving amphibians toward extinction.

WILDLIFE
EU ministers approve a regulation that forbids the marketing of seal products in the European Union, in response to animal welfare concerns about seal hunting.

JUNE JULY

2009 STATE OF THE WORLD: A YEAR IN REVIEW

2 4 6 8 10 12 14 16 18 20 22 24 26 28 30 2 4 6 8 10 12 14 16 18 20 22 24 26 28 30

TOXICS
Survey finds that thousands of everyday products and materials in the United States and worldwide contain harmful radioactive metals.

CLIMATE
Study says that hydrofluorocarbons, long hailed as a substitute for ozone-depleting gases, are a growing greenhouse threat, given their large atmospheric-warming capacity.

FOOD
San Francisco adopts landmark food policy to improve access to healthy food while supporting local agriculture and reducing shipping-related greenhouse gas emissions.

Barb Howe
San Francisco farmers' market

MARINE SYSTEMS
Study finds that fish stocks are beginning to rebuild in 5 of 10 large marine ecosystems under intensive management, suggesting that efforts to curb overfishing are succeeding.

FOOD
Researchers in Japan identify two genes that make rice plants grow longer stems and survive floods, potentially enabling farmers to grow high-yielding species in flood-prone areas.

USDA NRCS

AGRICULTURE
Study reveals that nearly half the world's farmlands have at least 10 percent tree cover—more than 10 million square kilometers in total—suggesting widespread use of agroforestry.

CLIMATE
U.S. agency reports that the world ocean surface temperature was the warmest on record for any June–August season since 1880.

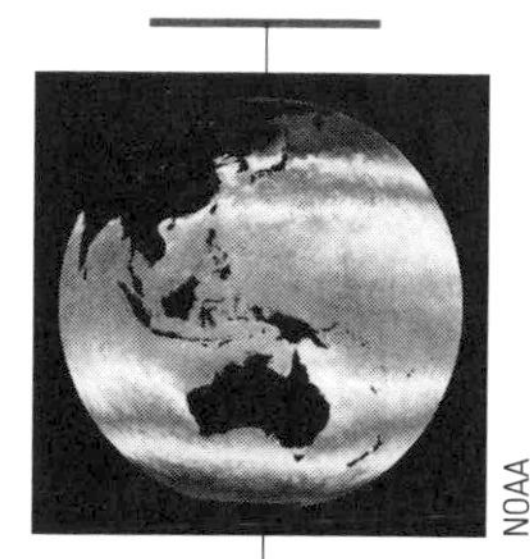
NOAA

SECURITY
U.N. Security Council unanimously commits to the goal of a world without nuclear weapons, as both the United States and Russia pledge to slash arsenals.

AUGUST

SEPTEMBER

See page 191 for sources.

2 4 6 8 10 12 14 16 18 20 22 24 26 28 30 2 4 6 8 10 12 14 16 18 20 22 24 26 28

CLIMATE
Study says Australia has surpassed the United States as the world's biggest per capita producer of carbon emissions.

ogwen

Australian road train

GOVERNANCE
FAO countries agree on the first-ever global treaty that aims to close fishing ports to vessels involved in illegal, unreported, and unregulated fishing.

BIODIVERSITY
WWF reports the discovery of 163 new species in the Greater Mekong region in 2008, many of which are at risk of extinction due to climate change.

ENERGY
G-20 leaders meeting in Pittsburgh, Pennsylvania, commit to phasing out nearly $300 billion in fossil fuel subsidies while providing targeted support for the world's poorest households.

2010

STATE OF THE WORLD

Transforming Cultures

From Consumerism to Sustainability

The Rise and Fall of Consumer Cultures

Erik Assadourian

In the 2009 documentary *The Age of Stupid*, a fictional historian who is possibly the last man on Earth looks at archival film footage from 2008 and contemplates the last years in which humanity could have saved itself from global ecological collapse. As he reflects on the lives of several individuals—an Indian businessman building a new low-cost airline, a British community group concerned about climate change but fighting a new wind turbine development in the area, a Nigerian student striving to live the American dream, and an American oilman who sees no contradiction between his work and his love of the outdoors—the historian wonders, "Why didn't we save ourselves when we had the chance?" Were we just being stupid? Or was it that "on some level we weren't sure that we were worth saving?" The answer has little to do with humans being stupid or self-destructive but everything to do with culture.[1]

Human beings are embedded in cultural systems, are shaped and constrained by their cultures, and for the most part act only within the cultural realities of their lives. The cultural norms, symbols, values, and traditions a person grows up with become "natural." Thus, asking people who live in consumer cultures to curb consumption is akin to asking them to stop breathing—they can do it for a moment, but then, gasping, they will inhale again. Driving cars, flying in planes, having large homes, using air conditioning...these are not decadent choices but simply natural parts of life—at least according to the cultural norms present in a growing number of consumer cultures in the world. Yet while they seem natural to people who are part of those cultural realities, these patterns are neither sustainable nor innate manifestations of human nature. They have developed over several centuries and today are actively being reinforced and spread to millions of people in developing countries.

Preventing the collapse of human civilization requires nothing less than a wholesale transformation of dominant cultural patterns. This transformation would reject consumerism—the cultural orientation that leads people to find meaning, contentment, and acceptance through what they consume—as taboo and establish in its place a new cultural framework centered on sustainability. In the process, a revamped understanding of "natural" would emerge: it would mean individual

Erik Assadourian is a Senior Researcher at the Worldwatch Institute and Project Director of *State of the World 2010*.

and societal choices that cause minimal ecological damage or, better yet, that restore Earth's ecological systems to health. Such a shift—something more fundamental than the adoption of new technologies or government policies, which are often regarded as the key drivers of a shift to sustainable societies—would radically reshape the way people understand and act in the world.

Transforming cultures is of course no small task. It will require decades of effort in which cultural pioneers—those who can step out of their cultural realities enough to critically examine them—work tirelessly to redirect key culture-shaping institutions: education, business, government, and the media, as well as social movements and long-standing human traditions. Harnessing these drivers of cultural change will be critical if humanity is to survive and thrive for centuries and millennia to come and prove that we are, indeed, "worth saving."

The Unsustainability of Current Consumption Patterns

In 2006, people around the world spent $30.5 trillion on goods and services (in 2008 dollars). These expenditures included basic necessities like food and shelter, but as discretionary incomes rose, people spent more on consumer goods—from richer foods and larger homes to televisions, cars, computers, and air travel. In 2008 alone, people around the world purchased 68 million vehicles, 85 million refrigerators, 297 million computers, and 1.2 billion mobile (cell) phones.[2]

Consumption has grown dramatically over the past five decades, up 28 percent from the $23.9 trillion spent in 1996 and up sixfold from the $4.9 trillion spent in 1960 (in 2008 dollars). Some of this increase comes from the growth in population, but human numbers only grew by a factor of 2.2 between 1960 and 2006. Thus consumption expenditures per person still almost tripled.[3]

As consumption has risen, more fossil fuels, minerals, and metals have been mined from the earth, more trees have been cut down, and more land has been plowed to grow food (often to feed livestock as people at higher income levels started to eat more meat). Between 1950 and 2005, for example, metals production grew sixfold, oil consumption eightfold, and natural gas consumption 14-fold. In total, 60 billion tons of resources are now extracted annually—about 50 percent more than just 30 years ago. Today, the average European uses 43 kilograms of resources daily, and the average American uses 88 kilograms. All in all, the world extracts the equivalent of 112 Empire State Buildings from the earth every single day.[4]

The exploitation of these resources to maintain ever higher levels of consumption has put increasing pressure on Earth's systems and in the process has dramatically disrupted the ecological systems on which humanity and countless other species depend.

The Ecological Footprint Indicator, which compares humanity's ecological impact with the amount of productive land and sea area available to supply key ecosystem services, shows that humanity now uses the resources and services of 1.3 Earths. (See Figure 1.) In other words, people are using about a third more of Earth's capacity than is available, undermining the resilience of the very ecosystems on which humanity depends.[5]

In 2005 the Millennium Ecosystem Assessment (MA), a comprehensive review of scientific research that involved 1,360 experts from 95 countries, reinforced these findings. It found that some 60 percent of ecosystem services—climate regulation, the provision of fresh water, waste treatment, food from fisheries, and many other services—were being degraded or used unsustainably. The findings were so unsettling that the MA Board warned that "human activity is putting such strain on the natural functions of Earth that the ability of the planet's ecosystems to sustain future generations can no longer be taken for granted."[6]

The shifts in one particular ecosystem service—climate regulation—are especially disturbing. After remaining at stable levels for the past 1,000 years at about 280 parts per million, atmospheric concentrations of carbon dioxide (CO_2) are now at 385 parts per million, driven by a growing human population consuming ever more fossil fuels, eating more meat, and converting more land to agriculture and urban areas. The Intergovernmental Panel on Climate Change found that climate change due to human activities is causing major disruptions in Earth's systems. If greenhouse gas emissions are not curbed, disastrous changes will occur in the next century.[7]

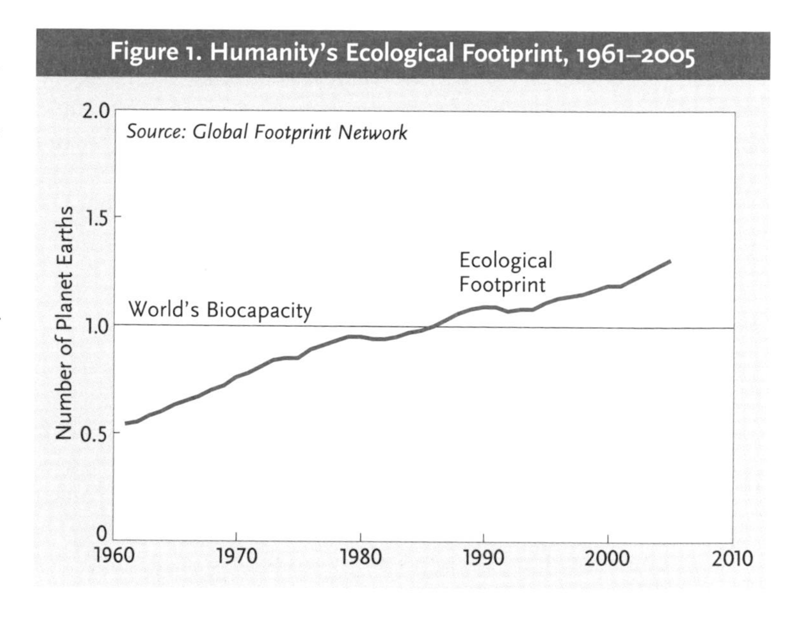

A May 2009 study that used the Integrated Global Systems Model of the Massachusetts Institute of Technology found that unless significant action is taken soon, median temperature increases would be 5.1 degrees Celsius by 2100, more than twice as much as the model had projected in 2003. A September 2009 study reinforced that finding, stating that business as usual would lead to a 4.5 degree Celsius increase by 2100, and that even if all countries stuck to their most ambitious proposals to reduce greenhouse gas emissions, temperatures would still go up by 3.5 degrees Celsius. In other words, policy alone will not be enough. A dramatic shift in the very design of human societies will be essential.[8]

These projected levels of temperature change mean the odds would be great that ocean levels would increase by two or more meters due to the partial melting of Greenland or Western Antarctica ice sheets, which in turn would cause massive coastal flooding and potentially submerge entire island nations. The one sixth of the world who depend on glacier- or snowmelt-fed rivers for water would face extreme water scarcity. Vast swaths of the Amazon forest would become savanna, coral reefs would die, and many of the world's most vulnerable fisheries would collapse. All of this would translate into major political and social disruptions—with environmental refugees projected to reach up to 1 billion by 2050.[9]

And climate change is just one of the many symptoms of excessive consumption levels. Air pollution, the average loss of 7 million hectares of forests per year, soil erosion, the annual production of over 100 million tons of hazardous waste, abusive labor practices driven by the desire to produce more and cheaper consumer goods, obesity, increasing time stress—the list could go on and on. All these problems are often treated separately, even as many of their roots trace back to current consumption patterns.[10]

In addition to being excessive overall, modern consumption levels are highly skewed, leading to disproportionate responsibility for modern environmental ills among the rich.

According to a study by Princeton ecologist Stephen Pacala, the world's richest 500 million people (roughly 7 percent of the world's population) are currently responsible for 50 percent of the world's carbon dioxide emissions, while the poorest 3 billion are responsible for just 6 percent. These numbers should not be surprising, for it is the rich who have the largest homes, drive cars, jet around the world, use large amounts of electricity, eat more meat and processed foods, and buy more stuff—all of which has significant ecological impact. Granted, higher incomes do not always equate with increased consumption, but where consumerism is the cultural norm, the odds of consuming more go up when people have more money, even for ecologically conscious consumers.[11]

In 2006, the 65 high-income countries where consumerism is most dominant accounted for 78 percent of consumption expenditures but just 16 percent of world population. People in the United States alone spent $9.7 trillion on consumption that year—about $32,400 per person—accounting for 32 percent of global expenditures with only 5 percent of global population. It is these countries that most urgently need to redirect their consumption patterns, as the planet cannot handle such high levels of consumption. Indeed, if everyone lived like Americans, Earth could sustain only 1.4 billion people. At slightly lower consumption levels, though still high, the planet could support 2.1 billion people. But even at middle-income levels—the equivalent of what people in Jordan and Thailand earn on average today—Earth can sustain fewer people than are alive today. (See Table 1.) These numbers convey a reality that few want to confront: in today's world of 6.8 billion, modern consumption patterns—even at relatively basic levels—are not sustainable.[12]

A 2009 analysis of consumption patterns across socioeconomic classes in India made this particularly clear. Consumer goods are broadly accessible in India today. Even at annual income levels of about $2,500 per person in purchasing power parity (PPP), many households have access to basic lighting and a fan. As incomes reach about $5,000 per year PPP, access to television becomes standard and access to hot water heaters grows. By $8,000 a year PPP, most people have an array of consumer goods, from washing machines and DVD players to kitchen appliances and computers. As incomes rise further, air conditioning and air travel become common.[13]

Not surprisingly, the richest 1 percent of Indians (10 million people), who earn more than $24,500 PPP a year, are now each respon-

Table 1. Sustainable World Population at Different Consumption Levels

Consumption Level	Per Capita Income, 2005	Biocapacity Used Per Person, 2005	Sustainable Population at this Level
	(GNI, PPP, 2008 dollars)	(global hectares)	(billion)
Low-income	1,230	1.0	13.6
Middle-income	5,100	2.2	6.2
High-income	35,690	6.4	2.1
United States	45,580	9.4	1.4
Global average	9,460	2.7	5.0

Source: See endnote 12.

sible for more than 5 tons of CO_2 emissions annually—still just a fifth of American per capita emissions but twice the average level of 2.5 tons per person needed to keep temperatures under 2 degrees Celsius. Even the 151 million Indians earning more than $6,500 per person PPP are living above the threshold of 2.5 tons per person, while the 156 million Indians earning $5,000 are nearing it, producing 2.2 tons per person.[14]

As the Ecological Footprint Indicator and Indian survey demonstrate, even at income levels that most observers would think of as subsistence—about $5,000–6,000 PPP per person a year—people are already consuming at unsustainable levels. And today, more than a third of the world's people live above this threshold.[15]

The adoption of sustainable technologies should enable basic levels of consumption to remain ecologically viable. From Earth's perspective, however, the American or even the European way of life is simply not viable. A recent analysis found that in order to produce enough energy over the next 25 years to replace most of what is supplied by fossil fuels, the world would need to build 200 square meters of solar photovoltaic panels every second plus 100 square meters of solar thermal every second plus 24 3-megawatt wind turbines every hour nonstop for the next 25 years. All of this would take tremendous energy and materials—ironically frontloading carbon emissions just when they most need to be reduced—and expand humanity's total ecological impact significantly in the short term.[16]

Add to this the fact that population is projected to grow by another 2.3 billion by 2050 and even with effective strategies to curb growth will probably still grow by at least another 1.1 billion before peaking. Thus it becomes clear that while shifting technologies and stabilizing population will be essential in creating sustainable societies, neither will succeed without considerable changes in consumption patterns, including reducing and even eliminating the use of certain goods, such as cars and airplanes, that have become important parts of life today for many. Habits that are firmly set—from where people live to what they eat—will all need to be altered and in many cases simplified or minimized. These, however, are not changes that people will want to make, as their current patterns are comfortable and feel "natural," in large part because of sustained and methodical efforts to make them feel just that way.[17]

In considering how societies can be put on paths toward a sustainable future, it is important to recognize that human behaviors that are so central to modern cultural identities and economic systems are not choices that are fully in consumers' control. They are systematically reinforced by an increasingly dominant cultural paradigm: consumerism.

Consumerism Across Cultures

To understand what consumerism is, first it is necessary to understand what culture is. Culture is not simply the arts, or values, or belief systems. It is not a distinct institution functioning alongside economic or political systems. Rather, it is all of these elements—values, beliefs, customs, traditions, symbols, norms, and institutions—combining to create the overarching frames that shape how humans perceive reality. Because of individual cultural systems, one person can interpret an action as insulting that another would find friendly—such as making a "thumbs up" sign, which is an exceptionally vulgar gesture in some cultures. Culture leads some people to believe that social roles are designated by birth, determines where people's eyes focus when they talk to others, and even dictates what forms of sexual relationships (such as monogamy, polyandry, or polygamy) are acceptable.[18]

Cultures, as broader systems, arise out of the complex interactions of many different elements of social behaviors and guide humans at

Hegariz

Grub to go: Sago grubs, a gourmet delicacy in New Guinea.

an almost invisible level. They are, in the words of anthropologists Robert Welsch and Luis Vivanco, the sum of all "social processes that make the artificial (or human constructed) seem natural." It is these social processes—from direct interaction with other people and with cultural artifacts or "stuff" to exposure to the media, laws, religions, and economic systems—that shape people's realities.[19]

Most of what seems "natural" to people is actually cultural. Take eating, for example. All humans eat, but what, how, and even when they eat is determined by cultural systems. Few Europeans would eat insects because these creatures are intrinsically repulsive to them due to cultural conditioning, though many of them would eat shrimp or snails. Yet in other cultures, bugs are an important part of cuisine, and in some cases—like the Sago grub for the Korowai people of New Guinea—bugs are delicacies.[20]

Ultimately, while human behavior is rooted in evolution and physiology, it is guided primarily by the cultural systems people are born into. As with all systems, there are dominant paradigms that guide cultures—shared ideas and assumptions that, over generations, are shaped and reinforced by leading cultural actors and institutions and by the participants in the cultures themselves. Today the cultural paradigm that is dominant in many parts of the world and across many cultural systems is consumerism.[21]

British economist Paul Ekins describes consumerism as a cultural orientation in which "the possession and use of an increasing number and variety of goods and services is the principal cultural aspiration and the surest perceived route to personal happiness, social status, and national success." Put more simply: consumerism is a cultural pattern that leads people to find meaning, contentment, and acceptance primarily through the consumption of goods and services. While this takes different forms in different cultures, consumerism leads people everywhere to associate high consumption levels with well-being and success. Ironically though, research shows that consuming more does not necessarily mean a better individual quality of life. (See Box 1.)[22]

Consumerism has now so fully worked its way into human cultures that it is sometimes hard to even recognize it as a cultural construction. It simply seems to be natural. But in fact the elements of cultures—language and symbols, norms and traditions, values and institutions—have been profoundly transformed by consumerism in societies around the world. Indeed, "consumer" is now often used interchangeably with person in the 10 most commonly used languages of the world, and most likely in many more.[23]

Consider symbols—what anthropologist Leslie White once described as "the origin and basis of human behavior." In most countries today people are exposed to hundreds if not thousands of consumerist symbols every day. Logos, jingles, slogans, spokespersons, mascots—all these symbols of different brands routinely bombard people, influencing behavior even at unconscious levels. Many people today recognize these consumerist symbols more easily than they do common wildlife

Box 1. Do High Consumption Levels Improve Human Well-being?

Ultimately, whether high consumption levels make people better off is irrelevant if they lead to the degradation of Earth's systems, as ecological decline will undermine human well-being for the majority of society in the long term. But even assuming this threat were not looming, there is strong evidence that higher levels of consumption do not significantly increase the quality of life beyond a certain point, and they may even reduce it.

First, psychological evidence suggests that it is close relationships, a meaningful life, economic security, and health that contribute most to well-being. While there are marked improvements in happiness when people at low levels of income earn more (as their economic security improves and their range of opportunities grows), as incomes increase this extra earning power converts less effectively into increased happiness. In part, this may stem from people's tendency to habituate to the consumption level they are exposed to. Goods that were once perceived as luxuries can over time be seen as entitlements or even necessities.

By the 1960s, for instance, the Japanese already viewed a fan, a washing machine, and electric rice cookers as essential goods for a satisfactory living standard. In due course, a car, an air conditioner, and a color television were added to the list of "essentials." And in the United States, 83 percent of people saw clothes dryers as a necessity in 2006. Even products around only a short time quickly become viewed as necessities. Half of Americans now think they must have a mobile phone, and one third of them see a high-speed Internet connection as essential.

A high-consumption lifestyle can also have many side effects that do not improve well-being, from increased work stress and debt to more illness and a greater risk of death. Each year roughly half of all deaths worldwide are caused by cancers, cardiovascular and lung diseases, diabetes, and auto accidents. Many of these deaths are caused or at least largely influenced by individual consumption choices such as smoking, being sedentary, eating too few fruits and vegetables, and being overweight. Today 1.6 billion people around the world are overweight or obese, lowering their quality of life and shortening their lives, for the obese, by 3 to 10 years on average.

Source: See endnote 22.

species, birdsong, animal calls, or other elements of nature. One study in 2002 found that British children could identify more Pokémon characters (a brand of toy) than common wildlife species. And logos are recognized by children as young as two years old. One investigation of American two-year-olds found that although they could not identify the letter M, many could identify McDonald's M-shaped golden arches.[24]

Cultural norms—how people spend their leisure time, how regularly they upgrade their wardrobes, even how they raise their children—are now increasingly oriented around purchasing goods or services. One norm of particular interest is diet. It now seems natural to eat highly sweetened, highly processed foods. Children from a very early age are exposed to candy, sweetened cereals, and other unhealthy but highly profitable and highly advertised foods—a shift that has had a dramatic impact on global obesity rates. Today, fast-food vendors and soda machines are found even in schools, shaping children's dietary norms from a young age and in turn reinforcing and perpetuating these norms throughout societies. According to a study by the U.S. Centers for Disease Control and Prevention, nearly two thirds of U.S. school districts earn a percentage of the revenue from vending machine sales, and a third receive financial awards from soda companies when a certain amount of their product is sold.[25]

Traditions—the most ritualized and deeply rooted aspects of cultures—are also now shaped by consumerism. From weddings that cost an average $22,000 in the United States to funeral norms that pressure grieving loved ones to purchase elaborate coffins, headstones, and other expensive symbolic goods, consumerism is deeply embedded in how people observe rituals. Choosing to celebrate rituals in a simple manner can be a difficult choice to make, whether because of norms, family pressure, or advertising influence.[26]

Christmas demonstrates this point well. While for Christians this day marks the birth of Jesus, for many people the holiday is more oriented around Santa Claus, gift giving, and feasting. A 2008 survey on Christmas spending in 18 countries found that individuals spent hundreds of dollars on gifts and hundreds more on socializing and food. In Ireland, the United Kingdom, and the United States—the three with the largest expenditures—individuals on average spent $942, $721, and $581 on gifts, respectively. Increasingly, even many non-Christians celebrate Christmas as a time to exchange gifts. In Japan, Christmas is a big holiday, even though only 2 percent of the population is Christian. As Reverend Billy of the tongue-in-cheek consumer education effort The Church of Stop Shopping notes: "We think we are consumers at Christmas time. No! We are being consumed at Christmastime."[27]

Consumerism is also affecting peoples' values. The belief that more wealth and more material possessions are essential to achieving the good life has grown noticeably across many countries in the past several decades. One annual survey of first-year college students in the United States has investigated students' life priorities for more than 35 years. Over this time the importance of being well-off financially has grown while the importance of developing a meaningful life philosophy has fallen. (See Figure 2.) And this is not just an American phenomenon. A study by psychologists Güliz Ger and Russell Belk found high levels of materialism in two thirds of the 12 countries they surveyed, including several transitional economies.[28]

Figure 2. Aspirations of First-Year College Students in the United States, 1971–2008

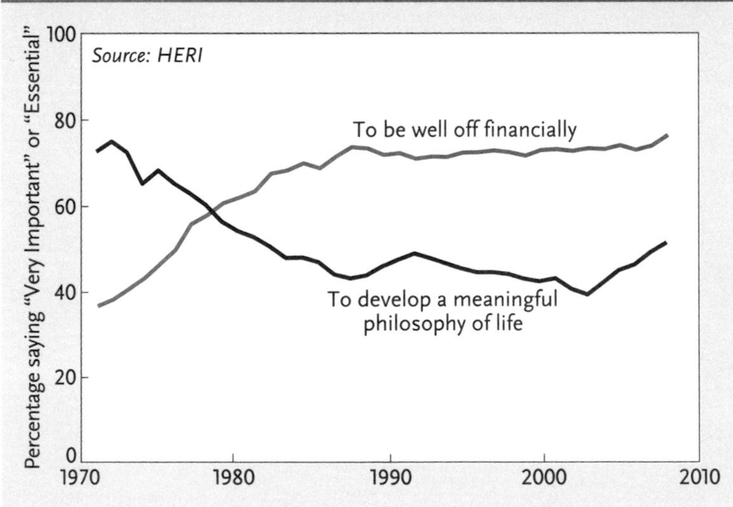

While consumerism is now found in nearly all cultures, it is not without consequences. On this finite planet, defining success and happiness through how much a person consumes is not sustainable. Moreover, it is abundantly clear that this cultural orientation did not just happen to appear as a byproduct of growing incomes. It was engineered over several centuries. Today, since consumerism has been internalized by many societies, it is self-perpetuating to some extent, yet institutions within society—

including businesses, the media, governments, and educational facilities—continue to prop up this cultural orientation. These institutions also are actively working to expand markets around the world for new consumer goods and services. Understanding the role of these institutional drivers will be essential in order to cultivate new cultures of sustainability.

Institutional Roots of Consumerism

As long ago as the late 1600s, societal shifts in Europe began to lay the groundwork for the emergence of consumerism. Expanding populations and a fixed base of land, combined with a weakening of traditional sources of authority such as the church and community social structures, meant that a young person's customary path of social advancement—inheriting the family plot or apprenticing in a father's trade—could no longer be taken for granted. People sought new avenues for identity and self-fulfillment, and the acquisition and use of goods became popular substitutes.[29]

Meanwhile, entrepreneurs were quick to capitalize on these shifts to stimulate purchase of their new wares, using new types of advertising, endorsements by prominent people, creation of shop displays, "loss-leaders" (selling a popular item at a loss as a way to pull customers into a store), creative financing options, even consumer research and the stoking of new fads. For example, one eighteenth-century British pottery manufacturer, Josiah Wedgwood, had salespeople drum up excitement for new pottery designs, creating demand for newer lines of products even from customers who already had a perfectly good, but now seemingly outdated, set of pottery.[30]

Still, traditional social mores blocked the rapid advance of a consumerist mindset. Peasants with extra income traditionally would increase landholdings or support community works rather than buy new fashions or home furnishings—two of the earliest consumer goods. Workers whose increased productivity resulted in greater pay tended to favor more leisure time rather than the wealth that a full day at increased pay might have brought them.[31]

But over time the emerging consumerist orientation was internalized by a growing share of the populace—with the continued help of merchants and traders—redefining what was understood as natural. The universe of "basic necessities" grew, so that by the French Revolution, Parisian workers were demanding candles, coffee, soap, and sugar as "goods of prime necessity" even though all but the candles had been luxury items less than 100 years earlier.[32]

By the early 1900s, a consumerist orientation had become increasingly embedded in many of the dominant societal institutions of many cultures—from businesses and governments to the media and education. And in the latter half of the century, new innovations like television, sophisticated advertising techniques, transnational corporations, franchises, and the Internet helped institutions to spread consumerism across the planet.

Arguably, the strongest driver of this cultural shift has been business interests. On a diverse set of fronts, businesses found ways to coax more consumption out of people. Credit was liberalized, for instance, with installment payments, and the credit card was promoted heavily in the United States, which led to an almost 11-fold increase in consumer credit between 1945 and 1960. Products were designed to have short lives or to go out of style quickly (strategies called, respectively, physical and psychological obsolescence). And workers were encouraged to take pay raises rather than more time off, increasing their disposable incomes.[33]

Perhaps the biggest business tool for stoking consumption is marketing. Global advertising expenditures hit $643 billion in 2008, and in countries like China and India they are growing at 10 percent or more per year. In the United States, the average "consumer"

Cereal content: a comic book ad from 1964.

sees or hears hundreds of advertisements every day and from an early age learns to associate products with positive imagery and messages. Clearly, if advertising were not effective, businesses would not spend 1 percent of the gross world product to sell their wares, as they do. And they are right: studies have demonstrated that advertising indeed encourages certain behaviors and that children, who have difficulty distinguishing between advertising and content, are particularly susceptible. As one U.S. National Academy of Sciences panel found, "food and beverage marketing influences the preferences and purchase requests of children, influences consumption at least in the short term, is a likely contributor to less healthful diets, and may contribute to negative diet-related health outcomes and risks among children and youth."[34]

In addition to direct advertising, product placement—intentionally showing products in television programs or movies so that they are positively associated with characters—is a growing practice. Companies spent $3.5 billion placing their products strategically in 2004 in the United States, four times the amount spent 15 years earlier. And, like advertising, product placements influence choices. Research has found, for example, a causal relationship between cigarette smoking in the movies and the initiation of this behavior in young adults in a "dose-response" manner, meaning that the more that teenagers are exposed to cigarette smoking in the movies, the more likely they are to start smoking.[35]

Other clever marketing efforts are also increasingly common tools. In "word of mouth" marketing, people who are acting as unpaid "brand agents" push products on unsuspecting friends or acquaintances. In 2008, U.S. businesses spent $1.5 billion on this kind of marketing, a number expected to grow to $1.9 billion by 2010. One company, BzzAgent, currently has 600,000 of these brand agents volunteering in its network; they help to spread the good word about new products—from the latest fragrance or fashion accessory to the newest juice beverage or coffee drink—by talking about them to their friends, completing surveys, rating Web sites, writing blogs, and so on. In Tokyo, Sample Lab Ltd. recently brought this idea to a new level with a "marketing café" specifically created to expose consumers to samples of new products. Companies now even harness anthropologists to figure out what drives consumers' choices, as Disney did in 2009 in order to better target male teens, one of their weaker customer bases.[36]

Any of these marketing strategies, taken alone, stimulates interest in a single good or service. Together these diverse initiatives stimulate an overall culture of consumerism. As economist and marketing analyst Victor Lebow explained in the *Journal of Retailing* over 50 years ago, "A specific advertising and promotional campaign, for a particular product at a particular time, has no automatic guarantee of success, yet it may contribute to the general pressure by which wants are stimulated and maintained. Thus its very failure may serve to fertilize this soil, as does so much else that seems to go down the drain." Industries, even as they pursue limited agendas of expanding sales for their products, play a significant role in stimulating consumerism. And whether intentionally or not, they transform cultural norms in the process. (See Table 2.)[37]

The media are a second major societal institution that plays a driving role in stimulating consumerism, and not just as a vehicle for marketing. The media are a powerful tool for transmitting cultural symbols, norms, customs, myths, and stories. As Duane Elgin, author and media activist, explains: "To control a society, you don't need to control its courts, you don't need to control its armies, all you need to do is control its stories. And it's television and Madison Avenue that is telling us most of the stories most of the time to most of the people."[38]

Between television, movies, and increasingly the Internet, the media are a dominant form of leisure time activity. In 2006, some 83 percent of the world's population had access to television and 21 percent had access to the Internet. (See Table 3.) In countries that belong to the Organisation for Economic Co-operation and Development, 95 percent of households have at least one television, and people watch about three to four hours a day on average. Add to this the two to three hours spent online each day, plus radio broadcasts, newspapers, magazines, and the 8 billion movie tickets sold in 2006 worldwide, and it becomes clear that media exposure consumes anywhere from a third to half of people's waking day in large parts of the world.[39]

During those hours, much of media output reinforces consumer norms and promotes materialistic aspirations, whether directly by extolling the high-consumption lives of celebrities and the wealthy or more subtly through stories that reinforce the belief that happiness comes from being better off financially, from buying the newest consumer gadget or fashion accessory, and so on. There is clear evidence that media exposure has an impact on norms, values, and preferences. Social modeling studies have found connections between such exposure and violence, smoking, reproductive norms, and various unhealthy behaviors. One study found that for every additional hour of television people watched each week, they spent an additional $208 a year on stuff (even though they had less time in a day to spend it).[40]

Government is another institution that often reinforces the consumerist orientation. Promoting consumer behavior happens in myriad ways—perhaps most famously in 2001 when U.S. President George W. Bush, U.K. Prime Minister Tony Blair, and several other western leaders encouraged their citizens to go out and shop after the terrorist attacks of September 11th. But it also happens more systemically. Subsidies for particular industries—especially in the transportation and energy sectors, where cheap oil or electricity has ripple effects throughout the economy—also work to stoke consumption. And to the extent that manufacturers are not required to internalize the environmental and social costs of production—when pollution of air or water is unregulated, for example—the cost of goods is artificially low, stimulating their use. Between these subsidies and externalities, total support of polluting business interests was pegged at $1.9 trillion in 2001.[41]

Some of these government actions are driven by "regulatory capture," when special interests wield undue influence over regulators. In 2008, that influence could be observed in the United States through the $3.9 billion spent on campaign donations by business

Table 2. How Industries Have Shifted Cultural Norms

Industry	Shift
Bottled water	This $60-billion industry sold 241 billion liters of water in 2008, more than double the amount sold in 2000. Through its global advertising efforts, the industry has helped create the impression that bottled water is healthier, tastier, and more fashionable than publicly supplied water, even as studies have found some bottled water brands to be less safe than public tap water and to cost 240 to 10,000 times as much.
Fast food	Fast food is now a $120-billion industry in the United States, with about 200,000 restaurants in operation. Among major restaurant chains, half are now hamburger joints. In the early 1900s, the hamburger was scorned in the United States as a dirty "food for the poor," but by the 1960s the hamburger had become a loved meal. By spending an annual $1.2 billion in advertising, promoting convenience and value, and providing play places for children, McDonald's in particular has helped transform dietary norms. It now serves 58 million people every day in its 32,000 restaurants spread across 118 countries.
Disposable paper products	From paper towels and plates to diapers and facial tissue, the disposable paper product industry has cultivated the belief that these products provide convenience and hygiene. In China, the market for these goods hit $14.6 billion in 2008, up 11 percent from the previous year. For many around the world, use of these products is today seen as a necessity, although this is a belief actively cultivated over many years by the industry. In China, when the disposable diaper industry entered the market it worked aggressively to make the use of "split-pants" taboo and instead to have disposable diapers be a symbol of affluence and sophistication.
Vehicles	Car companies are the second largest advertiser in the United States. They spent $15.6 billion on ads in 2008 and actively pushed the image of cars as sexy, exciting, and liberating. Since the 1920s, car companies have played an aggressive role in shifting the American culture to be car-centric, lobbying for increased road support, supporting organizations that fought against regulating car usage, even buying up several public trolley systems and dismantling them. Today car companies everywhere continue to promote auto-centric societies. In 2008, they spent $67 million on lobbying and $19 million on campaign contributions in the United States alone.
Pet industry	Views of specific animal species are primarily determined by cultures. The pet industry, which earns $42 billion globally each year on pet food alone, is a driving force in making it seem natural to view dogs, cats, and several other animals as friends and even members of the family. The "humanization" of these animals is a stated strategy of the industry and in 2005 was backed by over $300 million in advertising in the United States. As these pets are increasingly humanized, consumers become more willing to spend greater sums on expensive foods, veterinary services, clothing, and toys. Pets, however, consume considerable ecological resources. For example, two pet German Shepherds use more resources in a year than the average Bangladeshi does.

Source: See endnote 37.

interests (71 percent of total contributions) and the $2.8 billion spent by business interests to lobby policymakers (86 percent of total lobbying dollars).[42]

A clear example of official stimulation of consumption came in the 1940s when governments started to actively promote consumption as a vehicle for development. For

Table 3. Media Access by Global Income Group, 2006

Income Group	Population	Household Consumption Expenditure Per Capita	Households with Television	Internet Users
	(million)	(PPP 2008 dollars)	(percent)	(per 100 people)
World	6,538	5,360	83	21
High-income	1,053	21,350	98	59
Upper-middle-income	933	6,090	93	22
Lower-middle-income	3,619	1,770	80	11
Low-income	933	780	16	4

Source: See endnote 39.

example, the United States, which came out of World War II relatively unscathed, had mobilized a massive war-time economy—one that was poised to recede now that the war was over. Intentionally stimulating high levels of consumption was seen as a good solution to address this (especially with the memory of the Great Depression still raw). As Victor Lebow explained in 1955, "our enormously productive economy demands that we make consumption our way of life, that we convert the buying and use of goods into rituals, that we seek our spiritual satisfactions, our ego satisfactions, in consumption."[43]

Today, this same attitude toward consumption has spread far beyond the United States and is the leading policy of many of the world's governments. As the global economic recession accelerated in 2009, wealthy countries did not see this as an opportunity to shift to a sustainable "no-growth" economy—essential if they are to rein in carbon emissions, which is also on the global agenda—but instead primed national economies with $2.8 trillion of new government stimulus packages, only a small percentage of which focused on green initiatives.[44]

Finally, education plays a powerful role in cultivating consumerism. As with governments, in part this is because education seems to be increasingly susceptible to business influence. Today schools accept classroom materials sponsored by business interests, like the "bias-balanced" energy education materials by groups representing oil companies in Canada. And *Channel One News*, a 12-minute daily "news" program with 2 minutes of commercials and some segments sponsored by products or companies, is now shown in 8,000 middle and high schools across the United States, exposing 6 million students—nearly a quarter of all American teens—to marketing and product placements with the tacit support of educators.[45]

Perhaps the greatest critique of schools is that they represent a huge missed opportunity to combat consumerism and to educate students about its effects on people and the environment. Few schools teach media literacy to help students critically interpret marketing; few teach or model proper nutrition, even while providing access to unhealthy or unsustainable consumer products; and few teach a basic understanding of the ecological sciences—specifically that the human species is not unique but in fact just as dependent on a functioning Earth system for its survival as every other species. The lack of integration of this basic knowledge into the school curriculum, coupled with repeated exposure to consumer goods and advertising and with leisure time focused in large part on television, helps reinforce the unrealistic idea that humans are separate from Earth and the illusion that perpetual increases in consumption are ecologically possible and even valuable.

Cultivating Cultures of Sustainability

Considering the social and ecological costs that come with consumerism, it makes sense to intentionally shift to a cultural paradigm where the norms, symbols, values, and traditions encourage just enough consumption to satisfy human well-being while directing more human energy toward practices that help to restore planetary well-being.

In a 2006 interview, Catholic priest and ecological philosopher Thomas Berry noted that "we might summarize our present human situation by the simple statement: In the 20th century, the glory of the human has become the desolation of the Earth. And now, the desolation of the Earth is becoming the destiny of the human. From here on, the primary judgment of all human institutions, professions, and programs and activities will be determined by the extent to which they inhibit, ignore, or foster a mutually enhancing human-Earth relationship." Berry made it clear that a tremendous shift is necessary in society's institutions, in its very cultures, if humans are to thrive as a species long into the future. Institutions will have to be fundamentally oriented on sustainability.[46]

How can this be done? In an analysis on places to intervene in a system, environmental scientist and systems analyst Donella Meadows explained that the most effective leverage point for changing a system is to change the paradigm of the system—that is to say, the shared ideas or basic assumptions around which the system functions. In the case of the consumerism paradigm, the assumptions that need to change include that more stuff makes people happier, that perpetual growth is good, that humans are separate from nature, and that nature is a stock of resources to be exploited for human purposes.[47]

Although paradigms are difficult to change and societies will resist efforts to do so, the result of such a change can be a dramatic transformation of the system. Yes, altering a system's rules (with legislation, for instance) or its flow rates (with taxes or subsidies) can change a system too, but not as fundamentally. These will typically produce only incremental changes. Today more systemic change is needed.[48]

Cultural systems vary widely, as noted earlier, and so too would sustainable cultures. Some may use norms, taboos, rituals, and other social tools to reinforce sustainable life choices; others may lean more on institutions, laws, and technologies. But regardless of which tools are used, and the specific result, there would be common themes across sustainable cultures. Just as a consumerism paradigm encourages people to define their well-being through their consumption patterns, a sustainability paradigm would work to find an alternative set of aspirations and reinforce this through cultural institutions and drivers.

Ecological restoration would be a leading theme. It should become "natural" to find value and meaning in life through how much a person helps restore the planet rather than how much that individual earns, how large a home is, or how many gadgets someone has.

Equity would also be a strong theme. As it is the richest who have some of the largest ecological impacts, and the very poorest who often by necessity are forced into unsustainable behaviors like deforestation in a search for fuelwood, more equitable distribution of resources within society could help to curb some of the worst ecological impacts. Recent research also shows that societies that are more equitable have less violence, better health, higher literacy levels, lower incarceration rates, less obesity, and lower levels of teen pregnancy—all substantial bonus dividends that would come with cultivating this value.[49]

More concretely, the role of consumption and the acceptability of different types of consumption could be altered culturally as well. Again, while the exact vision of this will vary across cultural systems, three simple goals should hold true universally.

First, consumption that actively undermines well-being needs to be actively discouraged. The examples in this category are many: consuming excessive processed and junk foods, tobacco use, disposable goods, and giant houses that lead to sprawl and car dependency and to such social ills as obesity, social isolation, long commutes, and increased resource use. Through strategies such as government regulation of choices available to consumers, social pressures, education, and social marketing, certain behaviors and consumption choices can be made taboo. At the same time, creating easy access to healthier alternatives is important—such as offering affordable, easily accessible fruits and vegetables to replace unhealthy foods.[50]

Back for more: Factory farm freedom fighters from The Meatrix II.

Second, it will be important to replace the private consumption of goods with public consumption, the consumption of services, or even minimal or no consumption when possible. By increasing support of public parks, libraries, transit systems, and community gardens, much of the unsustainable consumption choices today could be replaced by sustainable alternatives—from borrowing books and traveling by bus instead of by car to growing food in shared gardens and spending time in parks.

The clearest example of this is transportation. Reorganizing infrastructure to support walkable neighborhoods and public transit could lead to a dramatic reduction in road transportation—which pollutes locally, contributes about 17 percent to total greenhouse gas emissions, and leads to 1.3 million deaths from accidents each year. The centrality of cars is a cultural norm, not a natural fact—cultivated over decades by car interests. But this can once again be redirected, extracting cars from cities, as Masdar in Abu Dhabi, Curitiba in Brazil, Perth in Australia, and Hasselt in Belgium have already started to demonstrate. For example, the Hasselt city council, facing rapid growth in car usage and budget shortfalls, decided in the mid-1990s to bolster the city's public transit system and make it free for all residents instead of building another expensive ring road. In the 10 years since then, bus ridership has jumped 10-fold, while traffic has lessened and city revenues have increased from an enlivened city center.[51]

Third, goods that do remain necessary should be designed to last a long time and be "cradle to cradle"—that is, products need to eliminate waste, use renewable resources, and be completely recyclable at the end of their useful lives. As Charles Moore, who has followed the routes of plastic waste through oceans, explains, "Only we humans make waste that nature can't digest," a practice that will have to stop. The cultivation of both psychological and physical obsolescence will need to be discouraged so that, for example, a computer will stay functional, upgradable, and fashionable for a decade rather than a year. Rather than gaining praise from friends for owning the newest phone or camera, having an "old faithful" that has lasted a dozen years will be celebrated.[52]

Having a vision of what values, norms, and behaviors should be seen as natural will be essential in guiding the reorientation of cultures toward sustainability. Of course, this cultural transformation will not be easy. Shifting cultural systems is a long process measured in decades, not years. Even consumerism, with sophisticated technological advances and many devoted resources, took centuries to become dominant. The shift to a culture of sustainability will depend on powerful networks of cultural pioneers who initiate, champion, and drive forward this new, urgently needed paradigm. (See Box 2.)[53]

As the spread of consumerism also demonstrates, leading cultural institutions can be harnessed by specific actors and can play a central role in redirecting cultural norms—whether government, the media, or education.

The good news is that this process has already started, as discussed in the 25 articles that follow this chapter. Significant efforts are being undertaken to redirect societies' cultural orientation by harnessing six powerful institutions: education, business, government, and the media, which have played such powerful roles in driving consumerism, plus social movements and sustainable traditions, both old and new.

In the realm of education, there are early signs that every aspect is being transformed—from preschool to the university, from the museum to the school lunch menu. The very act of walking to and from school is being used to teach children to live sustainably, as "walking buses" in Italy, New Zealand, and elsewhere demonstrate. In Lecco, Italy, for example, 450 elementary school students walk with a "driver" and volunteering parents along 17 routes to 10 different schools each day. There are no school buses in the city. Since their creation in 2003, these "piedibuses" have prevented over 160,000 kilometers of driving and thus have reduced carbon emissions and other auto pollutants. Along with reducing the ecological impact of children's commutes, the piedibuses teach road safety (in a supervised setting), provide exercise, and help children connect with nature on the way to school.[54]

The basic role of business is also starting to be readdressed. Social enterprises are challenging the assumption that profit is the primary or even sole purpose of business. More businesses—from the Grameen Bank in Bangladesh to a restaurant chain in Thailand called Cabbages and Condoms—are putting their social missions front and center, helping people while being financially successful as well. New corporate charters—like the B Corporation (the B stands for Benefit)—are even being designed to ensure that businesses over time are legally bound to consider the well-being of Earth, workers, customers, and other stakeholders as they make business decisions.[55]

In government, some innovative shifts are taking place. A long-standing government role known as "choice editing," in which governments encourage good choices while discouraging bad ones, is being harnessed to reinforce sustainable choices—everything from questioning perverse subsidies to outright bans of unsustainable technologies like the incandescent lightbulb. And more than that, entire ideas are being reassessed, from security to law. New concepts like Earth jurisprudence, in which the Earth community has fundamental rights that human laws must incorporate, are starting to take hold. In September

Box 2. The Essential Role of Cultural Pioneers

Considering that consumerism is such a powerful force and that the majority of resources and wealth are still overwhelmingly being used to stimulate it, how realistic is it to think that the pattern can shift? James Davison Hunter's analysis of how cultures change is instructive. As Hunter, the Director of the Institute for Advanced Studies in Culture at the University of Virginia, explains, cultural change can best be understood not through the Great Man approach (whereby heroic individuals redirect the course of history), but through the Great Network approach. "The key actor in history is not individual genius but rather the network."

When networks come together, they can change history. But not always. Change depends on "overlapping networks of leaders" of similar orientation and with complementary resources (whether cultural clout, money, political power, or other assets) acting "in common purpose." Networks can spread many ideas, whether consumption patterns, habits, political views, or even a new cultural paradigm.

But as Hunter notes, as culture is driven by institutions, success will depend on pulling ideas of sustainability into the center of these institutions, not allowing them to remain on the periphery. This means that as individuals internalize new norms and values personally, they also need to actively spread these ideas along their networks. They need to bring these ideas directly to the center of leading human institutions—spreading them through all available vehicles—so that others adopt this orientation and use their own leadership capacities to spread it even further. Like brand agents who now volunteer to surreptitiously promote the newest consumer product, individuals who recognize the dangerous ecological and social disruptions arising from unsustainable consumerism need to mobilize their networks to help spread a new paradigm. These networks, tapping whatever resources they have—financial, cultural, political, or familial—will play essential roles in pioneering a new cultural orientation.

The story of the documentary *The Age of Stupid* illustrates this point. The filmmakers raised funds from small investments by friends and supporters, and they marketed the film and organized 600 showings in over 60 countries by tapping into a global network of concerned individuals. They then channeled the momentum of the film to build a climate change campaign. This campaign, 10:10, encourages people to commit to reduce their carbon emissions by 10 percent in 2010 and to mobilize policymakers to do the same. By October 2009, some 900 businesses, 220 schools, 330 organizations, and 21,000 individuals had signed the 10:10 pledge.

And if all these networks of pioneers fail? As scientist James Lovelock notes, "Civilization in its present form hasn't got long." Consumerism—due to its ecological impossibility—cannot continue much longer. The more seeds sown by cultural pioneers now, the higher the probability that the political, social, and cultural vacuum created by the decline of consumerism will be filled with ideas of sustainability as opposed to other less humanistic ideologies.

Source: See endnote 53.

2008, Ecuador even incorporated this into its new constitution, declaring that "Nature or Mother Earth, where life is reproduced and exists, has the right to exist, persist, maintain and regenerate its vital cycles, structures, functions and its evolutionary processes" and that "every person, community, and nation will be able to demand the recognition of nature's rights before public institutions."[56]

Film, the arts, music, and other forms of

media are all starting to draw more attention to sustainability. Even a segment of the marketing community is mobilizing to use the knowledge of the industry to persuade people to live sustainably. These "social marketers" are creating ads, videos for the Internet, and campaigns to drive awareness about issues as diverse as the dangers of smoking, the importance of family planning, and the problems associated with factory farming. One social marketing campaign by Free Range Studios, *The Meatrix*, spoofed the global blockbuster movie *The Matrix* by following a group of farm animals as they rebel against factory farms and the ecological and social ills these operations cause. This generally unpalatable message, treated in a humorous way, spread virally across the Internet. It has reached an estimated 20 million viewers to date while costing only $50,000, a tiny fraction of what a 30-second TV ad would have cost to reach an audience of the same size.[57]

A host of social movements are starting to form that directly or indirectly tackle issues of sustainability. Hundreds of thousands of organizations are working, often quietly on their own and unknown to each other, on the many essential aspects of building sustainable cultures—such as social and environmental justice, corporate responsibility, restoration of ecosystems, and government reform. "This unnamed movement is the most diverse movement the world has ever seen," explains environmentalist Paul Hawken. "The very word movement I think is too small to describe it." Together these have the power to redirect the momentum of consumerism and provide a vision of a sustainable future that appeals to everyone. Efforts to promote working less and living more simply, the Slow Food movement, Transition Towns, and ecovillages are all inspiring and empowering people to redirect both their own lives and broader society toward sustainability.[58]

Finally, cultural traditions are starting to be reoriented toward sustainability. New eco-friendly ways to celebrate rituals are being established, for instance, and are becoming socially acceptable. Family size norms are starting to shift. Lost traditions like the wise guidance of elders are being rediscovered and used to support the shift to sustainability. And religious organizations are starting to use their mighty influence to tackle environmental issues—printing *Green Bibles*, encouraging their congregations to conserve energy, investing institution funds responsibly, and taking a stance against abuses of Creation, such as razing forests and blowing up mountaintops for coal.[59]

Perhaps in a century or two, extensive efforts to pioneer a new cultural orientation will no longer be needed as people will have internalized many of these new ideas, seeing sustainability—rather than consumerism—as "natural." Until then, networks of cultural pioneers will be needed to push institutions to proactively and intentionally accelerate this shift. Anthropologist Margaret Mead is often quoted as saying: "Never doubt that a small group of thoughtful, committed citizens can change the world. Indeed, it's the only thing that ever has." With many interconnected citizens energized, organized, and committed to spreading a sustainable way of life, a new cultural paradigm can take hold—one that will allow humanity to live better lives today and long into the future.[60]

Traditions Old and New

Countless choices in human lives are reinforced, driven by, or stem from traditions, whether religious traditions, rituals, cultural taboos, or what people learn from elders and their families. Taking advantage of these traditions and in some cases reorienting them to reinforce sustainable ways of life could help make human societies a restorative element of broader ecological systems. As many cultures throughout history have found, traditional ways can often help enhance rather than undermine sustainable life choices.

This section considers several important traditions in people's lives and in society. Gary Gardner of Worldwatch suggests that religious organizations, which cultivate many of humanity's deepest held beliefs, could play a central role in cultivating sustainability and deterring consumerism. Considering the financial resources of these bodies, their moral authority, and the fact that 86 percent of the people in the world say they belong to an organized religion, getting religions involved in spreading cultures of sustainability will unquestionably be essential.[1]

Rituals and taboos play an important role in human lives and help reinforce norms, behaviors, and relationships. So Gary Gardner also looks at rites of passage, holidays, political rituals, and even daily actions that can be redirected from moments that stimulate consumption to those that reconnect people with the planet and remind them of their dependence on Earth for continued well-being.

Traditions shape not just day-to-day activities but major life choices, such as how many children to have. Tapping into traditions—families' influence, religious teachings, and social pressures—to shift family size norms to more sustainable levels will be essential in global efforts to stabilize population growth. Robert Engelman of Worldwatch points out that the prerequisite of this will be to ensure that women have the ability to control their reproductive choices and that their families and governments let them make these choices in ways that respect their decisions.

Another important and unfortunately diminishing force for sustainability is the wisdom of elders. Through their long lives and breadth of experience, elders traditionally held a place of respect in communities and served as knowledge keepers, religious leaders, and shapers of community norms. These roles, however, have weakened as consumerism and its subsequent celebration of youth and rejection of tradition have spread across the planet. Recognizing the power of elders and taking advantage of all they know, as Judi Aubel of the

Grandmother Project describes, can be an important tool in cultivating traditions that reinforce sustainable practices.

Finally, one long-lived tradition that has been dramatically altered in the past several generations is farming. Albert Bates of The Farm and Toby Hemenway of Pacific University describe how sustainable societies will depend on sustainable agricultural practices—systems in which farming methods no longer deplete soils and pollute the planet but actually help to replenish soils and heal scarred landscapes while providing healthy food and livelihoods.

Several Boxes in these articles also discuss important traditions, including the need for ethical systems to internalize humanity's dependence on Earth's systems, the value of rekindling an understanding of geologic-scale time, and the importance of reorienting dietary norms to encourage healthy and sustainable food choices.

These are just some of the many traditions that need to be critically examined and recalibrated to reflect a changing reality—one in which 6.8 billion people live on Earth, another 2.3 billion are projected to join by 2050, and the ecological systems on which humanity depends are under serious strain. Cultures in the past have also faced ecological crises. Some, like the Rapanui of Easter Island, failed to alter their traditions. The Rapanui continued, for example, to dedicate too many resources to their ritual building of Moai statues—until their society buckled under the strain and Easter Island's population collapsed. Others have been more like the Tikopians, who live on a small island in the southwestern Pacific Ocean. When they saw the dangers they faced as ecological systems became strained, they made dramatic changes in social roles, family planning strategies, and even their diet. Recognizing the resource-intensive nature of raising pigs, for instance, they stopped raising them altogether. As a result, Tikopia's population stayed stable and continues to thrive today.[2]

—*Erik Assadourian*

Engaging Religions to Shape Worldviews

Gary Gardner

When Pan Yue, Vice-Minister of China's Ministry of Environmental Protection, wants to advance environmentalism these days, he often reaches for an unusual tool: China's spiritual heritage. Confucianism, Taoism, and Buddhism, says Pan, can be powerful weapons in "preventing an environmental crisis" because of each tradition's respect for nature. Mary Evelyn Tucker, a Confucian scholar at Yale University, elaborates: "Pan realizes that the ecological crisis is also a crisis of culture and of the human spirit. It is a moment of re-conceptualizing the role of the human in nature."[1]

Religious groups have responded with interest to Pan's overtures. In October 2008, a group of Taoist masters met to formulate a formal response to climate change, with initiatives ranging from solar-powered temples to a Taoist environmental network. Inspiration came from the Taoist concept of yin and yang, the interplay of opposites to create a balanced whole, which infuses the climate crisis with transcendent meaning. "The carbon balance between Earth and Sky is off-kilter," explains a U.N. official who attended the meeting, interpreting the Taoist view. "It is...significant that the current masters of Taoism in China have started to communicate precisely through this ancient yet new vocabulary."[2]

The Chinese Taoists are not alone in their activism. Bahá'ís, Christians, Hindus, Jews, and Muslims—encouraged by a partnership of the United Nations and the Alliance for Religions and Conservation (a U.K. nonprofit)—developed seven-year climate and environment plans that were announced in November 2009, just before the start of the U.N. climate conference in Copenhagen. The plans are the latest religious efforts to address the sustainability crises of our time, including climate change, deforestation, water scarcity, and species loss. By greening their activities and uncovering or re-emphasizing the green dimensions of sacred texts, religious and spiritual groups are helping to create sustainable cultures.[3]

How influential such efforts will be is unclear—in most faiths, environmental activism generally involves a small minority. But in principle, religious people—four out of every five people alive today identify themselves as this—could become a major factor in forging new cultures of sustainability. There is plenty of precedent. The anti-apartheid and U.S. civil rights movements, the Sandinista revolution in Nicaragua, the Jubilee 2000

Gary Gardner is a senior researcher at the Worldwatch Institute who focuses on sustainable economies.

debt-reduction initiative, the nuclear-freeze initiative in the United States in the 1980s—all these featured significant input and support from religious people and institutions. And indigenous peoples, drawing on an intimate and reciprocal relationship with nature, help people of all cultures to reconnect, often in a spiritual way, with the natural world that supports all human activity.[4]

The Greening of Religion

Over the past two decades, the indicators of engagement on environmental issues by religions and spiritual traditions have grown markedly. And opinion polls reveal increased interest in such developments. The World Values Survey, a poll of people in dozens of countries undertaken five times since the early 1980s, reports that some 62 percent of people worldwide feel it is appropriate for religious leaders to speak up about environmental issues, suggesting broad latitude for religious activism.[5]

More specific data from the United States suggest that faith communities are potentially an influential gateway to discussions about environmental protection. A 2009 poll found that 72 percent of Americans say that religious beliefs play at least a "somewhat important" role in their thinking about the stewardship of the environment and climate change.[6]

Another marker of the cultural influence of religious and spiritual traditions is the emergence of major reference works on religion and sustainability, giving the topic added legitimacy. Over the past decade, an encyclopedia, two journals, and a major research project on the environmental dimensions of 10 world religions have documented the growth of religions in the environmental field. (See Table 4.) Dozens of universities now offer courses on the religion/sustainability nexus, and the 2009 Parliament of the World's Religions had major panels on the topic.[7]

Table 4. Reference Works on Religion and Nature

Initiative	Date Appeared	Description
"Religions of the World and Ecology" Project	1995–2005	A Harvard-based research project that produced 10 volumes, each devoted to the relationship between a major world religion and the environment
Encyclopedia of Religion and Nature	2005	A 1,000-entry reference work that explores relationships among humans, the environment, and religious dimensions of life
The Spirit of Sustainability	2009	One volume in the 10-volume *Berkshire Encyclopedia of Sustainability*, examining the values dimension of sustainability through the lens of religions
Green Bible	2008	The New Revised Standard Version, with environmentally oriented verses in green and with essays from religious leaders about environmental topics; printed on recycled paper using soy-based ink
Worldviews: Global Religions, Culture, and Ecology and *Journal for the Study of Religion, Nature, and Culture*	1995, 1996	Journals devoted to the linkages among the spheres of nature, spirit, and culture

Source: See endnote 7.

Religious activism on behalf of the environment is now common—in some cases, to the point of becoming widespread, organized, and institutionalized. Three examples from the realms of water conservation, forest conservation, and energy and climate illustrate this broad-based impact.

First, His All Holiness, Patriarch Bartholomew, ecumenical leader of more than 300 million Orthodox Christians, founded Religion, Science and the Environment (RSE) in 1995 to advance religious and scientific dialogue around the environmental problems of major rivers and seas. RSE has organized shipboard symposia for scientists, religious leaders, scholars, journalists, and policymakers to study the problems of the Aegean, Black, Adriatic, and Baltic Seas; the Danube, Amazon, and Mississippi Rivers; and the Arctic Ocean.[8]

In addition to raising awareness about the problems of specific waterways, the symposia have generated initiatives for education, cooperation, and network-building among local communities and policymakers. Sponsors have included the Prince of Wales; attendees include policymakers from the United Nations and World Bank; and collaborators have included Pope John Paul II, who signed a joint declaration with Patriarch Bartholomew on humanity's need to protect the planet.[9]

Second, "ecology monks"—Buddhist advocates for the environment in Thailand—have taken stands against deforestation, shrimp farming, and the cultivation of cash crops. In several cases they have used a Buddhist ordination ritual to "ordain" a tree in an endangered forest, giving it sacred status in the eyes of villagers and spawning a forest conservation effort. One monk involved in tree ordinations has created a nongovernmental organization to leverage the monks' efforts by coordinating environmental activities of local village groups, government agencies, and other interested organizations.[10]

Third, Interfaith Power and Light (IPL), an initiative of the San Francisco–based Regeneration Project, helps U.S. faith communities green their buildings, conserve energy, educate about energy and climate, and advocate for climate and energy policies at the state and federal level. Led by Reverend Sally Bingham, an Episcopal priest, IPL is now active in 29 states and works with 10,000 congregations. It has developed a range of innovative programs to help faith communities green their work and worship, including Cool Congregations, which features an online carbon calculator and which in 2008 awarded $5,000 prizes to both the congregation with the lowest emissions per congregant and the congregation that reduced emissions by the greatest amount.[11]

These and other institutionalized initiatives, along with the thousands of individual grassroots religious projects at congregations worldwide—from Bahá'í environmental and solar technology education among rural women in India to Appalachian faith groups' efforts to stop mountaintop mining and the varied environmental efforts of "Green nuns"—suggest that religious and spiritual traditions are ready partners, and often leaders, in the effort to build sustainable cultures.[12]

Silence on False Gods?

In contrast to their active involvement in environmental matters, the world's religious traditions seem to hold a paradoxical position on consumerism: while they are well equipped to address the issue, and their help is sorely needed, religious involvement in consumerism is largely limited to occasional statements from religious leaders.

Religious warnings about excess and about excessive attachment to the material world are legion and date back millennia. (See Table 5.) Wealth and possessiveness—key features of a consumer society—have long been linked by religious traditions to greed, corruption, selfishness, and other character flaws. Moreover,

Table 5. Selected Religious Perspectives on Consumption

Faith	Perspective
Bahá'í Faith	"In all matters moderation is desirable. If a thing is carried to excess, it will prove a source of evil." (Bahá'u'lláh, *Tablets of Bahá'u'lláh*)
Buddhism	"Whoever in this world overcomes his selfish cravings, his sorrow fall away from him, like drops of water from a lotus flower." (*Dhammapada*, 336)
Christianity	"No one can be the slave of two masters....You cannot be the slave both of God and money." (Matthew, 6:24)
Confucianism	"Excess and deficiency are equally at fault." (*Confucius*, XI.15)
Hinduism	"That person who lives completely free from desires, without longing...attains peace." (*Bhagavad Gita*, II.71)
Islam	"Eat and drink, but waste not by excess: He loves not the excessive." (*Qur'an*, 7.31)
Judaism	"Give me neither poverty nor riches." (Proverbs, 30:8)
Taoism	"He who knows he has enough is rich." (*Tao Te Ching*)

Source: See endnote 13.

faith groups have spiritual and moral tools that can address the spiritual roots of consumerism—including moral suasion, sacred writings, ritual, and liturgical practices—in addition to the environmental arguments used by secular groups. And local congregations, temples, parishes, and ashrams are often tight-knit communities that are potential models and support groups for members interested in changing their consumption patterns.[13]

Moreover, of the three drivers of environmental impact—population, affluence, and technology—affluence, a proxy for consumption, is the arena in which secular institutions have been least successful in promoting restraint. Personal consumption continues upward even in wealthy countries, and consumer lifestyles are spreading rapidly to newly prospering nations. Few institutions exist in most societies to promote simpler living, and those that do have little influence. So sustainability advocates have looked to religions for help, such as in the landmark 1990 statement "Preserving and Cherishing the Earth: An Appeal for Joint Commitment in Science and Religion" led by Carl Sagan and signed by 32 Nobel Laureates.[14]

Despite the logic for engagement, religious intervention on this issue is sporadic and rhetorical rather than sustained and programmatic. It is difficult to find religious initiatives that promote simpler living or that help congregants challenge the consumerist orientation of most modern economies. (Indeed, an extreme counterexample, the "gospel of prosperity," encourages Christians to see great wealth and consumption as signs of God's favor.) Simplicity and anti-consumerism are largely limited to teachings that get little sustained attention, such as Pope Benedict's July 2009 encyclical, *Charity in Truth*, a strong statement on the inequities engendered by capitalism and the harm inflicted on both people and the planet. Or simplicity is practiced by those who have taken religious vows, whose commitment to this lifestyle—while often respected by other people—is rarely put forth as a model for followers.[15]

Advocating a mindful approach to consumption could well alienate some of the faithful in many traditions. But it would also address directly one of the greatest modern threats to religions and to spiritual health: the insidious message that the purpose of human life is to consume and that consumption is the path to happiness. Tackling these heresies could nudge many faiths back to their spiritual and scriptural roots—their true source of power and legitimacy—and arguably could attract more followers over the long run.

Contributions to a Culture of Sustainability

Most religious and spiritual traditions have a great deal to offer in creating cultures of sustainability.

Educate about the environment. As religious traditions embrace the importance of the natural environment, it makes sense to include ecological instruction in religious education—just as many Sunday Schools include a social justice dimension in their curricula. Teaching nature as "the book of Creation," and environmental degradation as a sin, for example—positions adopted by various denominations in recent years—is key to moving people beyond an instrumentalist understanding of the natural world.[16]

Educate about consumption. In an increasingly "full world" in which human numbers and appetites press against natural limits, introducing an ethic of limited consumption is an urgent task. Religions can make a difference here: University of Vermont scholar Stephanie Kaza reports, for example, that some 43 percent of Buddhists surveyed at Buddhist retreat centers were vegetarians, compared with 3 percent of Americans overall. Such ethical influence over consumption, extended to all wisdom traditions and over multiple realms in addition to food, could be pivotal in creating cultures of sustainability. (See Box 3.)[17]

Educate about investments. Many religious institutions avoid investments in weapons, cigarettes, or alcohol. Why not also steer funds toward sustainability initiatives, such as solar power and microfinance (the *via positiva*, in the words of the Archbishop of Canterbury)? This is what the International Interfaith Investment Group seeks to do with institutional religious investments. In addition, why not stress the need for personal portfolios (not just institutional ones) to be guided ethically as well? In the United States alone the value of investment portfolios under professional management was more than $24 trillion in 2007, only 11 percent of which was socially responsible investment.[18]

Express the sacredness of the natural world in liturgies and rituals. The most important assets of a faith tradition are arguably the intangible ones. Rituals, customs, and liturgical expressions speak to the heart in a profound way that cognitive knowledge cannot. Consider the power of the Taoist yin and yang framing of climate change, or of Christian "carbon fasts" at Lent, or of the Buddhist, Hindu, and Jain understanding of *ahimsa* (non-harming) as a rationale for vegetarianism. How else might religious and spiritual traditions express sustainability concerns ritualistically and liturgically?

Reclaim forgotten assets. Religious traditions have a long list of little-emphasized economic teachings that could be helpful for building sustainable economies. These include prohibitions against the overuse of farmland and pursuit of wealth as an end in itself, advocacy of broad risk-sharing, critiques of consumption, and economies designed to serve the common good. (See Table 6.) Much of this wisdom would be especially helpful now, as economies are being restructured and as people seem open to new rules of economic action and a new understanding of ecological economics.[19]

Coming Home

Often painted as conservative and unchanging institutions, many religions are in fact rapidly embracing the modern cause of environmen-

Box 3. A Global Ecological Ethic

The modern global ecocrisis is a strong signal that "environmentally at least, all established ethics are inadequate," in the words of ethicists Richard Sylvan and David Bennett. Most ethical systems today are indifferent to the steady degradation of natural systems and need to be reformed or replaced. Ecological ethics is a complementary ethical system that gives the natural world a voice in ethical discourse.

A specifically ecological ethic is "ecocentric" (perceiving and protecting value in all of nature), not "anthropocentric" (restricting value to humanity alone). It recognizes that humans are only a part of life on Earth, that humans need the rest of the planet and its inhabitants vastly more than they need humans, and that there is an ethical dimension to all human relationships with the planet. Indeed, a truly ecocentric ethic recognizes that in certain situations, the needs or rights of Earth or its other inhabitants take precedence over purely or narrowly human ones.

An ecological ethic is distinct from ethics rooted in enlightened human self-interest, the basis for virtually all ethical philosophies until now. Anthropocentric ethics encourages rather than counters the human inclination toward short-termism, greed, and limited sympathies. It also denies any responsibility for the effects of human behavior on the millions of other species and living individuals on this planet.

Suppose, for example, that a company wants to cut down a forest of old-growth hardwood trees and convert them into paper products. Company officials argue that local jobs depend on the logging, that the public needs the logs for paper and wood products, that the old-growth trees can be replaced by purpose-grown ones that are just as good, and so on. This is anthropocentric ethics at work.

An argument based on ecological ethics would assert that undisturbed trees are more useful to society because of their ecological value—they stabilize the climate, air, and soil upon which people ultimately depend. Furthermore, it would show that an old-growth forest is vastly richer (in terms of biodiversity) than a planted monoculture and can never, as such, be replaced; that it has value in and of itself regardless of its use-value to humans; and that its conversion into, say, cardboard and toilet paper would be despicable or even mad. When this full toolbox of arguments is given standing, the ecological point of view has a decent chance of prevailing. The paradox is that ecological ethics, though infused with nonhuman dimensions, greatly increases the likelihood of humanity's survival.

The prospects for institutionalizing ecological ethics may be growing as humanity recognizes its radical dependence on the environment. To advance the cause will require work on many fronts. To begin, it will be necessary to replace the sense of self as consumer with a sense of self as green citizen. This implies developing some limits to consumption—fewer disposable items, for example.

It will also require appreciating and adopting many of the principles emerging from "traditional ecological knowledge"—local or bioregional ecological wisdom, spiritual values, ritual practices, and ethics—that has sustained traditional peoples for millennia. Where such knowledge survives, it must be protected and encouraged; where it does not, it must be rediscovered and re-embodied in "invented traditions" that re-root humans in the natural world.

Finally, developing an ecological ethics will require the help of the world's spiritual and religious traditions, which are highly influential in shaping the ethical sensibilities of a large share of humanity.

—Patrick Curry
University of Kent, Canterbury, U.K.
Source: See endnote 17.

Table 6. Economic Precepts of Selected Religious and Spiritual Traditions

Economic Teaching or Principle	Description
Buddhist economics	Whereas market economies aim to produce the highest levels of production and consumption, "Buddhist economics" as espoused by E. F. Schumacher focuses on a spiritual goal: to achieve enlightenment. This requires freedom from desire, a core driver of consumerist economies but for Buddhists the source of all suffering. From this perspective, consumption for its own sake is irrational. In fact, the rational person aims to achieve the highest level of well-being with the least consumption. In this view, collecting material goods, generating mountains of refuse, and designing goods to wear out—all characteristics of a consumer economy—are absurd inefficiencies.
Catholic economic teachings	At least a half-dozen papal encyclicals and countless bishops' documents argue that economies should be designed to serve the common good and are critical of unrestrained capitalism that emphasizes profit at any cost. The July 2009 encyclical *Charity in Truth* is a good recent example.
Indigenous economic practices	Because indigenous peoples' interactions with nature are relational rather than instrumental, resource use is something done with the world rather than to the world. So indigenous economic activities are typically characterized by interdependence, reciprocity, and responsibility. For example, the Tlingit people of southern Alaska, before harvesting the bark of cedar trees (a key economic resource), make a ritual apology to the spirits of the trees and promise to use only as much as needed. This approach creates a mindful and minimalist ethic of resource consumption.
Islamic finance	Islamic finance is guided by rules designed to promote the social good. Because money is intrinsically unproductive, Islamic finance deems it ethically wrong to earn money from money (that is, to charge interest), which places greater economic emphasis on the "real" economy of goods and services. Islamic finance reduces investment risk—and promotes financial stability—by pooling risk broadly and sharing rewards broadly. And it prohibits investment in casinos, pornography, and weapons of mass destruction.
Sabbath economics	The biblical books of Deuteronomy and Exodus declare that every seventh ("Sabbath") year, debts are to be forgiven, prisoners set free, and cropland fallowed as a way to give a fresh start to the poor and the imprisoned and to depleted land. Underlying these economic, social, and environmental obligations are three principles: extremes of consumption should be avoided; surplus wealth should circulate, not concentrate; and believers should rest regularly and thank God for their blessings.

Source: See endnote 19.

tal protection. Yet consumerism—the opposite side of the environmental coin, and traditionally an area of religious strength—has received relatively little attention thus far. Ironically, the greatest contribution the world's religions could make to the sustainability challenge may be to take seriously their own ancient wisdom on materialism. Their special gift—the millennia-old paradoxical insight that happiness is found in self-emptying, that satisfaction is found more in relationships than in things, and that simplicity can lead to a fuller life—is urgently needed today. Combined with the newfound passion of many religions for healing the environment, this ancient wisdom could help create new and sustainable civilizations.

Ritual and Taboo as Ecological Guardians

Gary Gardner

"Keeping kosher," the ancient Jewish practice of observing dietary laws, has great practical and symbolic value for many Jews. It promotes awareness of the abundant generosity of the divine and prescribes a particular, respectful relationship with the fruits of God's creation. Some observant Jews are now working to establish an "eco-kosher" tradition: right eating and right consumption to preserve environmental health. Eco-kosher would infuse Jewish commandments with modern meaning: *Bal Tashchit*, the injunction not to waste, might apply to excessive or non-recycled food packaging; *Tzaar Baalei Chayyim*, the commandment to avoid cruelty to animals, could speak to confined livestock operations; and *Shmirat Haguf*, the requirement that people take care of their bodies, might prohibit foods that have been sprayed with pesticides. The environmental framing of ancient kosher rituals and prohibitions adds a powerful transcendent dimension to environmental protection.[1]

Transforming cultures of consumerism into cultures of sustainability will require a broad set of tools, including, perhaps surprisingly, ritual and taboo. Rituals—defined here as formal acts, repeated regularly, that have deep meaning for a community of people—help people to internalize and communicate deep-seated values. And taboos—the cultural prohibition of specific acts and products—might also help to proscribe human activities in an environmentally degraded world.[2]

Although commonly associated with spiritual practices, rituals and taboos are as much a secular as a religious phenomenon. A prime minister or president singing the national anthem, hand over heart, is engaged in a powerful ritualistic behavior that speaks deeply to compatriots, for example. And disrespecting a flag or other national symbol is a common taboo in many countries.

Whether secular or religious, political or personal, rituals and taboos in a consumer culture often reinforce that culture and the environmental problems it brings. But increasingly these practices are being used to bring mindfulness to modern habits of consumption, as the example of eco-kosher suggests. Ritual and taboo could become powerful, if largely intangible, tools for building cultures of sustainability.

The Power of Ritual

Ritual communication has long had an important role in protecting the natural environ-

Gary Gardner is a senior researcher at the Worldwatch Institute who focuses on sustainable economies.

ment. Cultural ecologist E. N. Anderson observes that in indigenous societies that have managed resources well for sustained periods, the credit often goes to "religious or ritual representation of resource management." This is in part because of the nature of ritual. Anthropologist Roy Rappaport and others suggest that ritual is a more powerful form of communication than even language and that this advantage is useful for environmental protection, especially in cultures like indigenous ones that are deeply embedded in the natural environment. Rituals express deep, culturally accepted truths in ways that language, which is easily manipulated and often used in service of falsehoods, cannot.[3]

As an example of the power of ritual, Swedish historian of religions Anne-Christine Hornborg cites the effort by the Mi'kmaq people on Cape Breton Island, Nova Scotia, to stop development of a quarry proposed for a Mi'kmaq sacred mountain in the early 1990s. While a range of groups, including environmentalists, stepped up to oppose the project, most of them used data, analysis, and rhetoric to highlight environmental and other impacts of the quarry. The quarry company easily parried these arguments with its own statistics and analyses.[4]

The Mi'kmaq, however, took a different approach, relying on ritual, including a sweat lodge, drumming, and pow-wows, as their "argument," and documenting that the mountain was a traditional Mi'kmaq sacred site. The company had a difficult time countering the Mi'kmaq rituals because, as Hornborg explains it, rituals are "immune to bureaucratic control." Or, as another scholar eloquently summarized it, "You cannot argue with a song." In the end, the company dropped its bid. While many reasons are given by different parties for the company's decision, the Mi'kmaq rituals, says Hornborg, were a powerful and possibly decisive influence.[5]

Rappaport and other scholars cite many examples of cultures that use ritual and taboo for environmental protection. The Tsembaga people of New Guinea, for instance, use elaborate pig festivals that include ritual slaughters and pig-eating rituals to achieve ecological balance. Ritual pig slaughtering, which occurs when pig populations have grown too large, lowers ecological pressures, redistributes land and pigs among people, and ensures that the neediest are the first to receive limited supplies of pork.[6]

Ethnographers tell similar stories. In Ghana, the traditional beliefs and taboos of the Ningo people protect turtles, which are viewed as gods, and mollusks, whose habitat is found in a sacred lagoon. Harvesting each species is forbidden, but no such taboos exist in neighboring Ghanaian coastal cultures. As a result, some 80 percent of turtle nesting areas along the Ghanaian coast are found in Ningo protected areas, and mollusks are up to seven times more prevalent in areas protected by taboo than in neighboring areas.[7]

These examples are not isolated cases of conservation. A 1997 analysis of species-specific taboos found strong overlap between taboos and official assessments of species endangerment: some 62 percent of reptiles and 44 percent of mammals protected by indigenous rituals and taboos were also identified as threatened in the World Conservation Union's *Red List* of endangered species, suggesting that indigenous peoples are skilled monitors of species endangerment. And as the examples just cited suggest, indigenous peoples have also developed strategies for protecting species, perhaps through co-evolutionary processes whereby human practices, including taboos, change in step with threats to the well-being of various species.[8]

Rituals of Consumerism

Rituals in consumer cultures may be powerful carriers of meaning, just as they are in indigenous cultures, but many also help to spread

McKay Savage

Less toxic than most: In Chennai, India, a statue of Ganesh is made almost entirely of fruits and vegetables.

consumerist values. Consider modern rites of passage—weddings, funerals, bar/bat mitzvahs, and *quinceañeras*, for example—which in many cases have become events marked by heavy consumption, compared with their old-fashioned predecessors.

The Wedding Report, a market research firm, says that weddings are a $60-billion industry in the United States, with the average celebration in 2008 costing nearly $22,000. The expenditures cover a range of goods and services—invitations, gifts, meals, paper goods, flowers, rings, guest travel, and attire, to name a few—each with its own ecological footprint. Guests who fly in for the event, for instance, have an extraordinary carbon footprint. The reception can have a large impact as well, especially if meat is served and if the food was not grown locally. And the two new gold rings that the celebrants exchange required the removal of tons of ore and earth, along with toxic flows of chemicals to extract the gold.[9]

Modern funerals, too, can carry an unnecessary ecological footprint. Today funerals in western countries typically involve an elaborate casket, embalming, flowers, and a cemetery plot with a concrete liner and marble headstone. The materials requirements for funerals in the United States—some 1.5 million tons of concrete and 14,000 tons of steel for funeral vaults, and 90,000 tons of steel and nearly 3,000 tons of copper and bronze for caskets—is not huge as a share of all concrete and metals used in the country. But many of the features of modern funerals are recent innovations that are entirely unnecessary. After all, just a few generations ago, even in industrial countries, the body of the deceased was prepared at home—wrapped in a shroud or placed in a simple wooden box. And in some cultures today the ritual has scarcely any environmental impact: in the Tibetan "sky burial" the body of the deceased, believed to be an empty vessel now devoid of a soul, is cut up and left for vultures to feed on. However unpalatable to the western mind, this ritual is environmentally restorative and does not spread consumerist values.[10]

Traditional holidays and feasts can be occasions of heavy consumption and environmental impact. Christmas is a commonly cited example, but other holidays make the point as well. In India the festival of Ganesh Chathauri—which honors Ganesh, the god that is half-elephant, half-man—typically involves the use of thousands of large idols painted in bright colors. At the festival's end, these are immersed in rivers, lakes, and the sea, where the paints and other materials contaminate the water. In the Bangalore area, where an estimated 25,000–30,000 idols have been used in festivals in recent years, a test of four lakes found increased acidification, a doubling of dissolved solids, a tenfold increase in iron content, and a 200–300 percent increase in copper in sediments. Many observers have called for alternative ways of marking Ganesh Chathauri—using biodegradable materials for the idols, for example, or ritually sprinkling them in lieu of immersing them in water bodies.[11]

Shopping itself has become a major ritual around some holidays. In the United States, "Black Friday"—the day after Thanksgiving and a non-working day for most people—is a shopping extravaganza that marks the opening of the Christmas shopping season. A Web site promoting Black Friday deals is up months before the day arrives, and people line up outside of malls and major stores, many of which open their doors before dawn. Black Friday has become a popular shopping ritual in itself, with extensive media coverage. And it now stands as a symbol of excess, with some stores experiencing violence, injuries, and even death as shoppers rush the doors at opening time.[12]

Rituals and Taboos for Sustainable Consumption

Modern rituals for sustainability can be developed out of virtually any dimension of the human experience. "Green funerals" are increasingly common, in which families can choose an environmentally benign end-of-life ritual that foregoes embalming, uses a simple wooden box or even a shroud for the deceased, avoids use of a burial vault, and in some cases marks the grave with shrubs, trees, or a stone native to the area, leaving the burial field or forest in an entirely natural state. According to the Centre for Green Burial in the United Kingdom, green burials are now available in Australia, Canada, Europe, and the United States.[13]

Holidays are another opportunity to green common rituals. New Year's Day, for instance, is celebrated in many cultures, whether on the Gregorian, Chinese, Hebrew, Islamic, or other calendar. For many people, entering a new year is foremost about marking the passage of time. And in this era of civilizational transition—an epoch akin to the shift from hunter-gatherers to farmers, or from agrarian to industrial societies—the new year may be a time to reflect in a long-term sense. (See Box 4.)[14]

But New Year's Day is also a time to set a new direction. In Peru and other Latin American countries, for example, people make effigies to represent all that was bad in the year past, then burn them at midnight. In Japan, *Bonenkai* or "forget-the-year parties" are held in December to prepare for the new year by bidding farewell to the concerns of the past year. Would annual cleansing rituals be an appropriate time to review personal and community failures to respect and preserve the natural world—and to vow to do better in the new year?

Earth Day is a relatively new calendar-based ritual that was established specifically to promote environmental awareness and care for the planet. Since its founding in 1970, Earth Day has become a global celebration, with more than a billion people participating, according to the Earth Day Network. The group claims to work with more than 15,000 organizations in 174 countries to create "the only event celebrated simultaneously around the globe by people of all backgrounds, faiths and nationalities." Such a global platform could become a powerful place from which to lead the entire human family in ritual appreciation of the planet.[15]

Fasting, a ritual discipline practiced in many religions, is being used by many people to raise consciousness about personal practices that might be used for a more sustainable world. In 2009 the bishops of Liverpool and London called on Christians to undertake a

Box 4. Deepening Perceptions of Time

The Long Now Foundation was founded in 01996 to help change long-term thinking from being difficult and rare to common and easy. (The foundation uses five-digit dates; the extra zero is to solve the deca-millennium bug that will come into effect in about 8,000 years.) It started with an idea from Danny Hillis, who pioneered the massive parallel logic of today's fastest super-computers. Hillis wanted to build an all-mechanical 10,000-year clock as an icon to long-term thinking.

Hillis was inspired by a story relayed to him by *Whole Earth Catalog* editor Stewart Brand: "I think of the oak beams in the ceiling of College Hall at New College, Oxford. Last century, when the beams needed replacing, carpenters used oak trees that had been planted in 1386 when the dining hall was first built. The 14th-century builder had planted the trees in anticipation of the time, hundreds of years in the future, when the beams would need replacing."

Over the last 14 years, several prototypes and material studies have been completed of the clock, and the monument-scale version is now being built. It will be located at one of the foundation's high desert sites and stretch out through several hundred feet of underground caverns. Hillis hopes that a clock "that ticks once a year, bongs once a century, and the cuckoo comes out once a millennium" will help reframe the way people look at the future. Since that first inspiration, the foundation has embarked on several projects to promote long-term thinking.

Long Bets is an online wagering site where anyone can make bets and predictions of social and scientific consequence. All the proceeds plus half the interest go to the charity of the winner's choice; the rest of the interest goes to Long Bets to maintain the service. Since its inception in 02002, bets have covered a diverse set of topics, from when the human population will peak to when solar electricity will become cheaper than fossil fuels.

The Rosetta Project is a compendium of all the world's documented languages micro-etched as readable text onto a three-inch wafer of pure nickel. The disk was designed to last for millennia and act as a key to languages that may become lost or extinct. In 02009, one of the disks was accepted into the Smithsonian's National Anthropological Archive. Just as discovery of the original Rosetta Stone allowed researchers to decipher ancient Egyptian hieroglyphics in the 1800s, this modern version could provide the same service for future civilizations.

All these projects, as well as a monthly seminar series about long-term thinking hosted by Stewart Brand, are attempts to change the conversation. If society only works on problems that can be solved in a four- to eight-year election cycle, then none of the truly large issues can be tackled. Solving problems in education, hunger, health care, macrofinance, population, and the environment all require a diligence and responsibility over decades, if not centuries. If the right time frame is used to solve these issues, what was once intractable can become possible.

Humans are a tenacious species. Chances are that 10,000 years from now, just like 10,000 years ago, there will be people walking on Earth. Just what kind of Earth, and just what kind of life those people may be living, will likely depend on the acorns we sow today that grow into the great oaks of our future.

Alexander Rose
Long Now Foundation
Source: See endnote 14.

carbon fast as a way to demonstrate restraint in consumption and solidarity with people affected by climate change. The call was supported by Ed Milliband, the Minister of Energy and Climate Change in the United Kingdom, and promoted by a development agency, Tearfund, which had enlisted more than 2,000 people for the 2008 fast. Similarly, Muslims in Chicago are being asked to "green Ramadan" by expanding their understanding of the annual ritual fast to include eating locally grown food, reducing their household ecological footprint by 25 percent, switching to cleaner sources of energy, and stepping up the practices of recycling and walking.[16]

Fasting can be conceived more broadly to include a wide range of activities in modern consumer societies. Many possibilities for setting aside consumerist habits already exist. World Carfree Day, for example, established in 2000 to help people experience life without an automobile, is now celebrated in more than 40 countries. Bike to Work day is a similar effort. Earth Hour, which involves turning off lights at a designated time, has become a worldwide phenomenon in the past few years. And TV Turnoff Week encourages families to watch less television and spend more time together.[17]

Meanwhile, in the United States, Buy Nothing Day now stands as a counteroffer to Black Friday, and Take Back Your Time Day offers people the chance to say no to overwork and overscheduling and instead reclaim their time for meaningful activities. Any of these "fasts" could conceivably become ritualized by religious or secular groups to give them deep meaning and impact.[18]

At a personal level, there are many opportunities to ritualize consumption and increase mindfulness about consumption habits. Indigenous practices could be a useful model here, especially the ritual of offering a small act of repentance or gratitude before using a resource. The Tlingit people of Alaska, for example, who use the bark of cedar trees to make clothing and other items, ask permission of the spirits of the tree before harvesting the bark and promise to use only as much as they need. Imagine saying a silent prayer of thanks and a vow not to waste before every act of modern consumption. Such a private ritual would likely bring mindfulness to a person's use of resources.[19]

One example of a more mindful approach to personal consumption comes from Peter Sawtell, a minister in Colorado who explores the link between spirituality and environmentalism. He has proposed that long-distance travel, especially flying, become a ritualized experience, with the Muslim ritual of the Hajj—the once-in-a-lifetime pilgrimage to Mecca—being the gold-standard model. Acknowledging that travel is enlightening, broadening, and even life-changing, Sawtell nevertheless suggests that because of the high environmental impact of trips by air, travel may need to be intentional and sacred now. And while a once-in-a-lifetime trip may be too strict a standard for most people, Sawtell suggests that once a decade or "once a life-stage" (adolescence, adulthood, retirement) might be helpful in thinking about long-distance travel. In the process, he suggests, people may find that less is more: they might appreciate travel and use it more meaningfully than when it was cheap and the environmental impact was ignored. Moreover, intentional travel could easily be ritualized, says Sawtell. "Imagine what it would be like in our churches if we celebrated the value of exceptional trips with special blessings for those who are embarking on this sort of once-in-a-lifetime pilgrimage."[20]

In sum, ritual and taboo figure into many aspects of any human life and help to transmit and shape cultural values. While resistant to cynical manipulation, these ancient human practices will likely find a place in development of new cultures of sustainability. In this epoch that cries out for rapid and comprehensive cultural transformation, human societies need to use every tool in the cultural toolbox.

Environmentally Sustainable Childbearing

Robert Engelman

Although the idea seems pessimistic and is little discussed, it is possible that world population—at 6.8 billion people today and growing by 216,000 a day—has already surpassed sustainable levels, even if everyone on Earth achieved merely modest European rather than lavish North American consumption levels.[1]

Estimates of what could be an environmentally "optimal" population are speculative and contentious. It could even be risky to venture a number, since some people might take it as a target worth aiming at by any means necessary, voluntary or not. Nonetheless, it is clear that with its current range of behavior patterns, humanity is hazardously raising the heat-trapping capacity of the atmosphere, decimating the planet's biological diversity, and risking future food scarcity by depleting freshwater supplies and degrading soils.

What if today's widely varying per capita consumption rates worldwide met in some narrow and modest range—but climate change and environmental deterioration continued anyway? Might it then be time, or is it time already, to evolve cultures that actively promote an average number of children born to each woman so low that world population shrinks in the near future? And if so, how could that be accomplished in ethical and acceptable ways?

The influence of modern culture on childbearing varies widely. The range of modern human fertility suggests this diversity, with women in Bosnia and Herzegovina and in the Republic of Korea having barely more than one child each on average while women in Afghanistan and Uganda average more than six. Women around the world also vary greatly, however, in their access to family planning, which can help them decide whether any given sex act should or should not be open to conception and pregnancy.[2]

So it is not clear which is the larger determinant of fertility: culture and women's (and men's) response to its influence or simply the accumulation of chance pregnancies that result from sexual activity not effectively protected against the risk of pregnancy. Yet with the notable exception of China, where shortages of natural resources are sometimes invoked to justify the government's one-child policy, it would be hard to identify a significant culture in which very small families are promoted to assure environmental sustainability.

Robert Engelman is vice president for programs at Worldwatch Institute and the author of *More: Population, Nature, and What Women Want.*

Paradoxically, according to United Nations surveys, many developing-country governments believe population growth is too rapid in their countries. And out of 41 National Adaptation Programmes of Action submitted by developing countries to the secretariat of the United Nations Framework Convention on Climate Change in recent years, 37 mentioned population density or pressure as hindering the success of adaptation to the impacts of climate change. Outside of China, Viet Nam, and some individual states in India, however, such governmental concerns do not translate into actual pressure on individuals to limit their procreation.[3]

For the people who work most closely with population and reproduction—especially health care providers who help women and their partners prevent pregnancy or enjoy it in good health when they want a child—this is as it should be. In fact, if there is any dominant global cultural paradigm around childbearing, it centers on reproductive health and rights—a social recognition that it is women and their partners, and no one else, who should choose when to bear a child and should do so in good health.

The closest thing to consensus on the perpetually divisive topic of human population is a principle first put in writing at a U.N. conference on human rights in Tehran in 1968 that "parents have a basic human right to determine freely and responsibly the number and spacing of their children." The adverb "responsibly" has sparked some debate, though not much in recent years. It could, nonetheless, become the basis for discussion of what the word might mean in a world where environmental sustainability is challenged by human activities.[4]

Twenty-six years after the Tehran conference, in 1994, another U.N. gathering expanded on reproductive rights when representatives of almost all the world's nations agreed that encouraging healthy and effective reproductive decisionmaking by women and their partners was the sole legitimate basis for governments to try to influence fertility levels and family size within their borders.[5]

USAID

Afghan girls get a meal along with their education.

Abuses of reproductive rights have been more the exception than the rule in six decades or so of global family planning experience. But those abuses—from incentive payments for sterilization to forced abortion documented in India and China and a handful of other countries—have soured policymakers and health care providers on population policies, programs, or media messages aimed at convincing women and couples to have fewer children than they would otherwise choose to have. Absent momentous changes in culture and politics around the world, it is difficult to imagine substantial professional or public support evolving for aggressive promotion of families of just one child or at most two children. The scope for new cultural efforts aimed at convincing couples to forego a wanted second, third, or fourth child for the sake of the environment seems small.[6]

© 2006 Helen Hawkings, courtesy Photoshare

A young family visits a mobile health clinic offering family planning services and basic health care to members of marginalized rural communities in the Dominican Republic.

Does this mean that no conceivable cultural transformation could help shrink the world's population through lower birth rates? Not at all. (And, given the misunderstanding that accompanies this topic, it's worth stating the obvious: population shrinkage based on higher death rates is not something to hope for.) There is much in today's culture that promotes pregnancies that individual women do not seek or want, and these cultural aspects are an easy immediate target for elimination or reversal. Similarly, there is scope for cultural change that might lead couples to change their views about family size, though this route to lower fertility requires vigilance so that the ultimate childbearing choice remains with women and their partners, not with other family members, the government, or the broader society.

Surprisingly, it is likely that global fertility levels would fall low enough to shrink world population if unintended pregnancies could be eliminated, although the reversal of growth would take some time to occur. By the best available estimate, nearly two out of five pregnancies worldwide are not planned or sought by the women who become pregnant. The figures are generally somewhat higher in low-fertility industrial countries than in high-fertility developing ones.[7]

Current average human fertility (2.5 children per woman) is only slightly above the fertility that would yield a stable human population size. (This is currently just above 2.3 children; stubbornly high death rates among the young in many developing countries push the global average above the usually cited figure of 2.1.) Moreover, all countries that offer women and their partners a range of choices of contraception, backed up by access to safe abortion, have fertility rates low enough to end or reverse population growth in the absence of net immigration. A world of fully intentional childbearing might begin to lose population within two or three decades, perhaps sooner.[8]

Moreover, demographic research over several decades makes clear a strong correlation between levels of education and fertility. The number of children women have in fact falls roughly in proportion to their advancement through school. According to calculations by demographers at the International Institute for Applied Systems Analysis, women with no schooling worldwide have on average 4.5 children each. Those with some primary school average 3 children, while those who complete at least one year of secondary school average 1.9 children. And after just one or two years of college, fertility drops to 1.7 children per woman—a rate well below population-maintaining "replacement" fertility.[9]

Given the force with which access to contraception and education for girls reduces fertility, it seems obvious that any cultural constraints on these should be given first priority in any move for reform. Unfortunately, such constraints are deeply rooted in human unease with both sexuality and the idea of gender equality. Cultural transformation must tackle these and advance the principle that all women should have control over their own bodies and fertility and that all should have opportunities equal to those of men—through education, media messages, and the work of policymakers at all levels. Limitations on access to contraception, such as requirements for parental permission or physician prescriptions for routinely safe options, are open to public pressure for legislative or regulatory change.

The use of sex and women's bodies for advertising or easy laughs in television situation comedies fortifies the lower status of women and makes it even more likely that unintended pregnancies will boost population growth rates—not to mention complicate the lives and undermine the aspirations of young people. One study found that the level of exposure to sexual content on television strongly predicted subsequent teen pregnancy, with the 10 percent of teenagers most exposed to television sex more than twice as likely to become pregnant within three years of the exposure as the 10 percent with the lowest exposure.[10]

Such findings illustrate the power of culture—and of media culture in particular—to boost fertility or at least accelerate sexual initiation and subsequent childbearing. Combating such cultural influences thus can play an important role in lowering fertility and contributing to slower population growth. Moreover, there is evidence that media such as television and radio may contribute to lower fertility just as easily as to higher.[11]

Where soap operas designed to model contraceptive use and small family norms are introduced, perceptions on ideal family sizes can fall. For example, after the radio soap opera *Apwe Plezi* (derived from a Creole saying, "after the pleasure comes the pain") was aired in St. Lucia, of the 35 percent of the surveyed population who had heard it, listeners were more likely to trust family planning workers, view extramarital sex as less acceptable, and favor families that averaged 2.5 children as opposed to 2.9 children for those who had not heard the show. While of course other factors also contributed to this shifting norm—such as parallel increases in access to family planning resources—it is clear that the media can play an important role in shaping family size norms.[12]

Another area ripe for cultural transformation is the dominant political view that any jurisdiction in which population stops growing is headed, in the words of a recent *Washington Post* news story, for "slow-motion demographic disaster." A national election in late 2008 in Japan, for example, seemed to revolve in large part on a proposed payment of $276 per month to parents for each child younger than high school age. In Russia, politicians have urged citizens to skip work to have sex and have offered prizes—from refrigerators to a Jeep—to women who have a baby on Russia Day, June 12th. Both countries have declining populations.[13]

There is some evidence that incentives like these can modestly boost a country's fertility, with a greater effect among women with lower incomes. Tax benefits targeted at parents on a per child basis, such as those in the United States, may have a similar impact—and, in fact, U.S. fertility has risen modestly in recent years, as has that in other wealthy countries. (In the case of the United States, fertility has recently risen to roughly the replacement value, which for that country is 2.1 children per woman.)[14]

Politicians justifiably worry that extremely low birth rates will ultimately make it more challenging to support aging populations. But these and similar risks are manageable social challenges that pale in comparison to those the world faces in addressing human-caused climate change, the depletion of renewable freshwater supplies, and the loss of the planet's biological diversity. Anyone who takes these environmental problems seriously has good reason to oppose the efforts of politicians, economists, and the media to promote higher birth rates—as well as those of religious leaders, members of extended families, and others who urge pregnancy on women who have not chosen it for themselves.

Finally, there is the constructive role that education and open discussion about the changing environment and the relation of population to its sustainability can play in shaping reproductive decisionmaking. Studying a lobster-fishing village in Quintana Roo, Mexico, geographer David Carr of the University of California, Santa Barbara, found that cultural attitudes about childbearing had changed as the lobster resource declined. The use of contraception was universal, and the community's birth rates were comparable to those of such low-fertility countries as Italy, Estonia, and Russia. The villagers Carr interviewed explicitly tied their modest family size intentions, so different from those of their parents and grandparents, to the importance of preserving the fishing resource for their children.[15]

Perhaps significantly, many villagers also mentioned the influence on their own reproductive ambitions of television soap operas depicting small North American families. While satellite television may not be considered by many as a positive agent of cultural transformation, in this case it may play a constructive role by spreading an idea—a small family norm—that contributes to environmental sustainability more powerfully than the messages about wealth and consumption might undermine it.

The sharp fall of fertility around the world in recent decades is proof that culturally influenced reproductive behaviors can change surprisingly fast. A family with roughly two children is already a cultural ideal in most industrial countries, albeit no doubt mostly for reasons unrelated to environmental sustainability. If nations soon reach a point where greenhouse gas emissions are actually capped and food and energy prices are high due to a rising mismatch of supply and demand, there is no telling how cultural norms about childbearing and family size might evolve. It is nonetheless hard to imagine that environmentally concerned citizens seeking curbs in human population growth will ever gain much public support for limiting reproductive rights. But the potential for cultural change that would slow and eventually reverse population growth—supporting or at least not undermining individual reproductive choice—is significant and worth pursuing.

Elders: A Cultural Resource for Promoting Sustainable Development

Judi Aubel

There is considerable discussion in western industrialized societies of the need to reexamine the predominant global cultural paradigm of consumerism, which is clearly unsustainable. In efforts to address current challenges to survival, the focus has been on halting environmental degradation and promoting the economic survival of communities around the globe. Unfortunately, the degradation of the social environment and the breakdown in social connectedness have received much less attention.[1]

Another less frequently considered issue is the relevance of the global cultural model of consumerism for other societies that face not only environmental and economic challenges but also problems specific to their history and cultural worldviews. Non-westernized and unindustrialized societies in Africa, Asia, Latin America, and the Pacific are threatened by less tangible forces that are undermining cultural identities and decreasing social cohesion.

One negative consequence of globalization is that western individualistic, consumer-oriented, youth-focused values—communicated through multiple international and national media and institutional channels—are undermining positive traditions and values of more collectivist sociocultural systems. In many cases, these traditions and values provide the basis for the society's sustainable use and development of both natural and human resources.

Respecting the Wisdom of Elders

A community elder in southern Senegal recently lamented the fact that development programs rarely pay attention to local cultural values: "There have been so many programs carried out in our community: to build more school classrooms; to construct a health center; to teach us how to grow more vegetables, how to prevent disease, about the importance of sending girls to school, and of planting trees." His testimony reflects the trend toward carefully targeted development programs that aim to produce "tangible and quantifiable results" corresponding to donor and government priorities but that may fail to address other less tangible cultural parameters that may be equally important for the survival of the communities the programs aim to support. In spite of rhetoric about the need for "culturally adapted" approaches, development policies

Judi Aubel, a specialist in community development and health in developing countries, is executive director of the Grandmother Project.

Judi Aubel

Elder and infant in a village in Rajasthan, India.

and programs often unknowingly convey a set of western values that may be counterproductive to the long-term social development and survival of non-western societies.[2]

One specific and decisive facet of non-western cultures that is rarely even dealt with in discussions on culture and development is the central role played by elders in socializing younger generations, passing on indigenous knowledge and cultural values, and ensuring the stability and survival of their societies. The late Andreas Fuglesang, a well-known leader in development communication, referred to the essential role played by elder community members in more traditional societies as the "information processing unit" of a community. As Malian philosopher Amadou Hampâté Bâ notes, "When an elder person dies in Africa, it is as though a whole library had burned down."[3]

There is clearly incongruity between the centrality of elders in non-western societies and the centrality of young people in development programs—a problem that has gone largely unnoticed. There is a growing clash of cultures between younger members of society, who embrace more global values, and older community members who are holding on to more traditional ones. The tension between the two cultural orientations is seen in the decreased communications and learning between young people and elders. In the past, for example, throughout Africa members of different generations would sit under a large tree in the community to discuss the past, the present, and the future. In French, the designated tree was referred to as "l'arbre à palabres." Today in many communities, while elders still sit and discuss under such trees, young people are more likely to gather around a radio or television to look at images and hear stories of other places.

Yet continuing respect for the wisdom of elders is reflected in a proverb heard widely across Africa, "What an elder sees sitting on the ground, a younger person cannot see even if he/she is up in a tree." In a study in Senegal, community respondents of various ages stated that knowledge is related to age and, consequently, elders are viewed as "knowledge providers" in key domains such as agriculture and health. And in India, Narender Chadha of the University of Delhi finds that, in spite of vast economic and social changes, elders continue to command high respect as "they are considered as the storehouses of knowledge and wisdom within the family and community contexts." This respect for traditional wisdom is similarly found in other collectivist, non-western societies in the Pacific and Latin America.[4]

Respect for the wisdom of elders is also evident in a new effort at the international level to help find solutions to global problems that was initiated in 2007 by Nelson Mandela. He brought together a small number of distinguished world leaders and established a group called The Elders. Mandela's idea was inspired by the role of elders in traditional societies: to bring people together, to encourage dialogue, to provide guidance based on their experience. The Elders are currently working on helping to solve several complex and conflict-ridden problems, including the Israeli-Palestinian situation.[5]

In western individualist societies, however, attitudes toward elders are generally tainted by negative images of aging. With the globalization of culture, increasingly ageist attitudes are being disseminated and slowly permeating non-western cultures as well. And it has been observed that older women suffer from ageist biases even more than men do: they are said to be a bad influence on children and families, illiterate and therefore unintelligent, or too old to learn and to change.[6]

Threats to Intergenerational Relationships

Globalization involves a virtually one-way dissemination of western cultural images and values toward non-western societies. Only recently has there been some concern at the international level about globalization's role in spreading consumerist cultural images and values and the resulting breakdown in intergenerational relationships in non-western societies.

The 2005 *World Youth Report* from the United Nations cautioned, "Young people are increasingly incorporating aspects of other cultures from around the world into their own identities. This trend...is likely to widen the cultural gap between the younger and older generations." Similarly, an analysis of the impact of globalization by the Youth Commission on Globalisation calls attention to an alarming situation: "The youth of the developing world are attracted, lured or forced into non-traditional ways of being by a great many factors...and alienated from their traditional communities. Such cultural disintegration is the primary cause of problems such as the loss of linguistic, historical and spiritual traditions, the break down of family support structures and the loss of a locally organised political voice."[7]

Similar concern about the negative effects of globalization on young people in particular are expressed by Akopovire Oduaran of the University of Botswana, who laments the loss of "the rich African tradition of intergenerational relationships...daily being weakened by the increasing change in our value systems as our communities are opened up to cultural globalization." He argues that with consumerism has come the loss of cultural traditions and weakened bonds and cooperation between family and community members—all disturbing signs of diminished social cohesion.[8]

Yet there is some evidence that young people perceive the dangers of globalization. Members of a Ghanaian youth club noted that "globalisation has brought us a life surrounded by mass-production and mass-consumption.... We see our own cultures giving way to a consumerist monoculture. There is an urgent need to revisit, appreciate and participate in the evolution of our own cultures, which are community-oriented, non-materialistic, eco-friendly and holistic in their worldview." Mamadou, a 20-year-old Senegalese man, stated: "I am part of a whole generation of young people who are lost. We play soccer and watch television but we don't really belong to the western world. Our parents sent us to school but there we didn't learn about our culture and our parents didn't teach us where we came from either. We are lost between two worlds."[9]

How are consumerist values communicated to society at large and specifically to young people in developing countries? Three major institutions are responsible: the mass media and advertising, development organizations and programs, and formal schools.

Mass media and advertising are the major vehicles for diffusion of western values into non-western societies. While there is increased national production of television programs, and even greater local radio programming that integrates local opinions and values, the predominant force remains the global media beamed into the tiniest of villages. The Youth Commission on Globalisation report notes the media's prevalent role in spreading individualist and consumerist values, stimulated by transnational corporations: "Youth are bombarded by advertisements, programming and other media that invite them to seek happiness through the accumulation of wealth and commodities."[10]

Development programs aim to make a positive contribution to communities. But program planners are not always aware of the underlying western values that such programs are inadvertently conveying. A older Malian woman and leader in her community described what happens: "Before the development agents get out of their four-wheel-drive vehicles, we know who they want to talk to, those who have gone to school and who know how to write, i.e. the youth. They almost never ask to see us." While working to improve hygiene or schooling, the attitudes of local development agents like these are inadvertently communicating culturally foreign values regarding who is valued (young people) and who is not (elders). Maternal and child health programs, for instance, invariably focus on women of reproductive age and rarely involve their culturally designated advisors: the senior women (or grandmothers).[11]

Schools are also key institutions in passing on cultural values in society. In a World Bank report, Deepa Srikantaiah maintains that in many countries school curricula do not reflect the cultural values and knowledge of local communities. In Botswana, for example, Pat Pridmore of the University of London analyzed the child-to-child approach used in many developing countries, in which schoolchildren are expected to learn and then teach their parents about "modern" health and hygiene practices. This notion is diametrically opposed to the attitude of hierarchical and collectivist non-western cultures, in which young people are expected to learn from their elders, and it undermines their culturally designated role.[12]

Programs Involving Elders Promote Intergenerational Learning

Numerous intergenerational preschool programs across the United States and Canada involve older adult community members who share their knowledge and provide social support to young children on a volunteer basis. The results include increased self-confidence on the part of children and an increased sense of self-worth on the part of older adults, many of whom are retired but who have extensive knowledge and compassion to share.[13]

In British Columbia, the Elders in Residence program at the Lelum'uy'lh Child Development Centre has helped integrate cultural values and traditions of the Cowichan Tribes into the curriculum with the support of elders through activities such as storytelling, language teaching, and basket-weaving. The program has contributed to greater appreciation of Cowichan culture and to respect for elders' knowledge of Cowichan traditions.[14]

But in Africa, Asia, Latin America, and the Pacific, few organizations or programs explicitly involve elders and promote intergenerational communication. Some that do take this approach are described here.

In Ghana, in a program supported by the United Nations Population Fund entitled "Time with Grandma," grandparents serve as resource persons in educational activities with adolescents dealing with HIV/AIDS prevention and teenage pregnancy. Both young people and elders find these intergenerational activities beneficial as they build on the traditional role of elders as teachers and promote positive cultural values, including abstaining from sex before marriage and respecting elders.[15]

In Malawi, the Ekwendeni Hospital trains grandparents to promote improved family practices related to prenatal care for women and care of newborns. A project review showed that the elder-inclusive strategy has contributed to improved family health–related practices while at the same time improving communication between younger and older community members. This is the first program in which elders have been actively involved, and they say that it has restored their place in society as "teachers of the younger generations."[16]

Judi Aubel

An elder of the village Olo Ologa, Mauritania, shares a story.

In Australian Aboriginal communities, building on the traditional teaching role of elders, senior women leaders in the Yolngu tribe work with alcoholic and drug-addicted teens to increase their pride in their cultural identity by teaching them about Yolngu history and practices, such as hunting and weaving.[17]

Over the past 10 years, grandmother-inclusive and intergenerational approaches have been developed by the Grandmother Project (GMP), a small U.S. nonprofit, and implemented in various countries, including Laos, Uzbekistan, Djibouti, Senegal, Mali, and Mauritania. The programs have dealt with various aspects of women's and children's health and development that older women, or grandmothers, are heavily involved in, including nutrition, newborn care, home care for sick children, early childhood development, and female genital mutilation (FGM). GMP has developed an approach in which multigenerational groups analyze community problems and identify collective actions that can lead to positive and sustainable changes within their own cultural systems.[18]

In Mali (with Helen Keller International) and Senegal (with the Christian Children's Fund), GMP guided development of grandmother-inclusive non-formal health education activities. In both cases these led to improvements in the advice older women gave to pregnant women regarding diet and rest during pregnancy and infant feeding practices. In Mauritania, in both rural and peri-urban areas, GMP in collaboration with World Vision has trained informal grandmother leaders to promote positive nutrition and health practices in their communities.[19]

In Senegal, in a program with World Vision to discourage FGM, participatory educational activities with grandmothers and intergenerational dialogue are key elements of an approach to promote holistic development of young girls. Most programs aimed at decreasing female genital mutilation focus on young people and do not involve grandmothers, who are usually those who do the cutting. In GMP's approach, grandmothers are key actors in promoting abandonment of this practice, while recognizing their positive role within the family as guardians of tradition and a stabilizing factor within the community. One leader in Senegal noted at the end of a two-day workshop that "we never practiced cutting maliciously but rather to educate the girls. Now we understand that as grandmothers we have a responsibility to put an end to this practice."[20]

Even in rural villages in Senegal, western values related to consumerism and sexuality are felt through western-produced television, films, and the Internet. GMP activities have encouraged the use of traditional communication media, such as story-telling, music, and dance in schools and communities in activities that bring young people and elders together. Recognition of grandmothers' story-telling skills has greatly increased this after-dinner activity, and it is reported that children's knowledge of traditional stories has increased while their television watching has declined. Broadcasts of grandmothers' telling stories on the local radio station have also increased the women's self-esteem and young people's interest in traditional knowledge. A young girl named Fatoumata said, "We are happy because now we are learning the traditional stories. If we don't spend time with our grandmas, when we become adults we will be empty inside."[21]

As the urgency to deal with global challenges increases, UNESCO has called for giving more attention to existing cultural realities and resources: "When development recognizes culture it produces change rooted in a community's own values, knowledge and lifestyle and thus tends to be more successful. When development imposes external cultural values it damages the operating system by devaluing indigenous knowledge, and local capacity on which the community is built....the challenge is to find ways of unlocking the cultural resources and assets of the community, to connect with people's own ways of being and enable them to use these creative capacities as a route out of poverty, exclusion and dependency."[22]

Programs that explicitly involve elders and that promote intergenerational learning capitalize on two valuable assets of non-western societies. As the few efforts in developing countries just described have shown, programs that have built on these cultural resources have contributed to positive and sustainable changes in nutrition, health, and education practices while at the same time curbing the spread of consumerism and strengthening the cultural identities and social cohesion of families and communities.

From Agriculture to Permaculture

Albert Bates and Toby Hemenway

Above the door lintels of the cultural museum of Tlaxcala, Mexico's oldest state capital, are murals depicting the rise of civilization. First there appear the hunters, clad in furs, with bows and spears. A woman discovers a small grassy plant and begins to cultivate it. After a time, everyone is planting it, and the newly domesticated plants grow as tall as a person. Special tools appear to prepare the ground, plant, harvest, and process the grain. In the wall panels that follow, civilization arrives, in all its complexity.

Something similar to this story is told in most, if not all, cultures. In the Fertile Crescent of the upper Tigris and Euphrates Rivers there are ancient coins bearing images of a plow drawn by oxen. Images of planters and plows appear on pottery from Egypt and Anatolia and on rice paper from Japan and China, some of it more than 14,000 years old.[1]

As the ice retreated and the climate warmed 20,000 years ago, the area of fertile soil and suitable growing seasons expanded, even as wild game retreated and mammoths and other large animals went extinct. About 8,000 years ago, animal husbandry began to be augmented by the domestication of emmer wheat, einkorn, barley, flax, chick pea, pea, lentil, and bitter vetch. Humans had begun to alter their landscapes in profound ways, clearing forests for fields, building larger villages and cities, and redirecting rivers for irrigation and flood control. By 7,000 years ago, many, if not most, people in the world were farmers.[2]

This might have continued until humanity entered the next Ice Age—a world of cold deserts, land bridges, and massive mountains of ice. But civilization changed that trajectory by harnessing the coal, gas, and oil that fueled the Industrial Revolution. Once more, people altered the planet's rhythms in ways they could not fully grasp.

In the span of a single century—the present one—Earth's climate may warm more rapidly and to a greater degree than in the previous 20,000 years. Agricultural systems will be profoundly challenged, beset by a perfect storm of diminishing fuel supply for tractors, fertilizer, and transportation; by crop-destroying heat waves, expanding pestilences, and declining water supplies for irrigation; by growing and migrating populations clamoring for food,

---

Albert Bates is the director of the Global Village Institute for Appropriate Technology and the Ecovillage Training Center at The Farm. **Toby Hemenway** is Scholar-in-Residence at Pacific University and a biologist consultant for the Biomimicry Guild.

especially for meat and processed foods (see Box 5); and by the financial instability borne of exceeding Earth's limits and having to retrench to an earlier stage of industrial development.[3]

Before the mid-twentieth century, most crops were produced largely without the use of chemicals. Insect pests and weeds were controlled by crop rotations, destruction of crop refuse, timing of planting to avoid high pest population periods, mechanical weed control, and other time-tested and regionally specific farming practices. While these are still in use, changes in technology, prices, cultural norms, and government policies have led to today's industrially intensive agriculture. The dominant system of agriculture now practiced throughout the world, referred to as "conventional agriculture," is characterized by mechanization, monocultures, the use of synthetic chemical fertilizers and pesticides, and an emphasis on maximizing productivity and profitability.

This type of agriculture is unsustainable because it destroys the resources it depends on. Soil fertility is declining due to erosion, compaction, and destruction of organic matter; water supplies are being depleted and polluted; finite fossil energy supplies are being exhausted; and the economies of rural communities are left in shambles as agricultural outputs are shipped to distant markets. The shortage of productive cropland, decreasing soil fertility, and the enormous waste and imprecise management associated with industrial-scale food economics are responsible for the world's recurrent and accelerating food and water shortages, malnutrition, mass starvation, and loss of biodiversity. In addition, agriculture accounts for about 14 percent of greenhouse gas emissions, and from 1990 to 2005 global agricultural emissions increased by 14 percent.[4]

Humanity now confronts a critical challenge: to develop methods of agriculture that sequester carbon, enhance soil fertility, preserve ecosystem services, use less water, and hold more water in the landscape—all while productively using a steadily compounding supply of human labor. In short, a sustainable agriculture.

Defining Sustainable Agriculture

Fortunately, for the past half-century some pioneers have been preparing the agriculture of the future, and their ideas are now moving to center stage. Organic no-till, permaculture, agroforestry, perennial polycultures, aquaponics, and biointensive and biodynamic farming—long considered fringe ideas—are now converging as serious components of a sustainable agriculture.[5]

One of the foundation stones was laid early in the twentieth century, when Franklin Hiram King journeyed to China, Korea, and Japan to learn how farms there had been worked for thousands of years without destroying fertility or applying artificial fertilizer. In 1911 King published *Farmers of Forty Centuries: or Permanent Agriculture in China, Korea and Japan*, which described composting, crop rotation, green manuring, intertillage, irrigation, drought-resistant crops, aquaculture and wetlands farming, and the transport of human manure from cities to rural farms.[6]

King's work was inspiration for many, including Sir Albert Howard. In 1943 he published *An Agricultural Testament*, which described building compost piles, recycling waste materials, and creating soil humus as a "living bridge" between soil life, such as mycorrhizae and bacteria, and healthy crops, livestock, and people. At the heart of Howard's work was the idea that soils, nutritious crops, and organisms in general are not just arrays of minerals but are parts of a complex ecology of cycling organic matter, and that these life-supporting cycles are critical for a self-regenerative agriculture.[7]

Howard became embroiled in a mid-twentieth century conflict. On one side were disciples of chemists such as Carl Sprengel and

Box 5. Dietary Norms That Heal People and the Planet

While many different combinations of foods will meet a person's dietary needs, dietary norms are for the most part shaped by the individual's culture, typically very early in life. Traditionally, these preferences were in large part shaped by the foods that were available to people in their bioregion.

In today's globalized world, however, more people can choose from a wide array of foods. While increased choice is theoretically a good thing—giving people variety and the opportunity to choose diets that are healthy and have little ecological impact—dietary norms have been reshaped in an increasingly unhealthy and unsustainable manner. Easy access to high-fat, high-sugar foods combined with billions of dollars spent annually on marketing have dramatically shifted what is considered a "normal" diet—from the number of calories per meal to the amount of meat, sugar, and refined flour consumed. All of these in turn have contributed to rising obesity levels and have had significant ecological impacts.

Today 1.6 billion people are either overweight or obese, and 18 percent of greenhouse gases are produced by livestock that are raised to feed humanity's growing demand for meat. In 2007, people ate 275 million tons of meat, about 42 kilograms per person worldwide and 82 kilograms in industrial countries (2.7 servings every day).

By cultivating new dietary norms, food can contribute to good health and possibly even help heal the planet. A study of several of the longest-lived peoples in the world found that they ate just 1,800–1,900 calories a day, no processed foods, and minimal amounts of animal products. By comparison, the average American consumes 3,830 calories a day.

Food writer Michael Pollan explains succinctly what a healthy, restorative diet could look like: "Eat food, not too much, mostly plants." By food, Pollan means that people should avoid food-like products with so many additives, preservatives, flavors, and fillers that their nutritional value may be compromised.

And by eating fewer calories (while ensuring those calories are high in nutrients), overall health and longevity can be increased—a finding that has been borne out in several different animal species, including humans. Moreover, eating fewer calories means having a smaller ecological impact. For example, if a person starts adhering to an 1,800-calorie-a-day diet at age 30, he could live to the age of 81 before consuming the same amount of calories as a person who follows the typically recommended 2,600-calorie diet would by the age of 65.

Eating "mostly plants"—not necessarily completely vegetarian but, as in many cultures throughout history, eating meat infrequently or perhaps even just ritually—will have significant ecological benefits. According to agricultural researcher David Pimentel, a vegetarian diet needs one third fewer fossil fuels than a meat-based diet. Another study found that producing just 1 kilogram of beef involves as much carbon dioxide emissions as the average European car being driven 250 kilometers.

Unfortunately, today the dietary norm that is spreading around much of the world—driven by the media, government subsidies, advertising, and even by parents—is the consumer diet of high quantities of meat, processed foods, refined flours, and sugar.

What is needed is the intentional cultivation of sustainable dietary norms—an effort that is getting started, thanks to books like *In Defense of Food*, documentaries like *Food Inc.*, government programs that promote healthier eating, social enterprises selling healthier food, and movements like "Slow Food" that enourage people to consider carefully what they eat.

—*Erik Assadourian and Eddie Kasner*

Source: See endnote 3.

Justus von Liebig, who advocated fertilizing principally with nitrogen, phosphorus, and potassium minerals and promoted a mechanical approach, arguing that plant growth is boosted by adding the scarcest, or limiting, mineral. This soon became a widely accepted agronomic principle and the basis for the Green Revolution. On the other side were the organic advocates, adhering to Howard's view that crop health depends on maintaining soil ecology by returning to the soil not just the minerals lost in farming but also the organic matter that supports the nutrient cycles of soil life. Howard's position was, in the words of biologist Janine Benyus, that it is life that best creates the conditions conducive to life.[8]

Courtesy Maya Mountain Reasearch Farm

The face of agroforestry at the Maya Mountain Research Farm, Belize.

Howard lost that battle but may yet have won the war, as it becomes apparent that many aspects of industrial agriculture are unsustainable, from the topsoil loss that approaches 75 billion tons annually to the looming depletion of the critical fertilizer phosphorus and the negative returns typified by crops that use 10 calories of fuel energy to produce one calorie of food energy.[9]

Twentieth-century agriculture has badly degraded nearly every ecosystem it has encountered while consuming roughly 20 percent of world energy production. The style called "conventional" depends for nearly all of its workings on a dwindling and increasingly expensive supply of fossil fuels.[10]

Sustainable agriculture, in contrast, can be pursued indefinitely because it does not degrade or deplete the resources that it needs to continue. Since most of Earth's arable land is already under cultivation and human populations are continuing to expand, an even better goal would be to actually improve the capacity of the land to produce.

Some net gain approaches are coming into view, but they are not magic elixirs. While optimized farming practices can increase the capacity of the land to produce over the long term, they cannot be considered in isolation; a robust solution to humanity's continued existence on this planet must include adopting sustainable lifestyles and maintaining human population at sustainable numbers.

Organic Agriculture: An Overview

Key features of organic agriculture are the use of biologically produced fertilizers such as carbon-enhanced manures instead of manufactured inorganic nitrates and phosphates, infrequent use of biologically derived pesticides rather than routine application of synthetic and systemically toxic compounds, and—most critically—maintenance of soil ecology and organic matter through cover crops, green manures, crop rotation, and composting.[11]

A long-term comparison done by the Rodale Institute from 1981 to 2002 found that organic systems provided crop yields equivalent to those of conventional methods. The trials showed that when rainfall was 30 percent less than normal—typical drought levels—organic methods yielded 24–34 percent more than standard methods. The researchers attribute the increased yields to better water retention due to higher soil carbon levels.[12]

Data gathered from the trial have revealed

that soil under organic agriculture management can accumulate about 1,000 pounds of carbon per acre-foot each year. This is equal to about 3,667 pounds of carbon dioxide per acre (4,118 kilograms per hectare per year) taken from the air and sequestered into soil organic matter. Also, organic methods used 28–32 percent less energy and were more profitable than industrial methods. These results suggest that organic methods offer great promise for reducing fossil fuel use and greenhouse gas emissions. The study suggested that converting the 64 million hectares of U.S. cropland currently planted in corn and soybeans to organic methods would sequester 264 million tons of carbon dioxide; this is the equivalent of shutting down 207 (225-megawatt) coal-fired power plants, about 14 percent of the installed coal electric capacity in either the United States or China.[13]

Perennial Polycultures

Wes Jackson and his colleagues at The Land Institute in Salina, Kansas, have been developing new perennial crops to replace annual grains that must be replanted every year. These grains are grown in polycultures, mixed with other perennial species that fix nitrogen for fertility, and produce seed oil for food, fuel, and lubricants. These polycultures mimic the plant communities that make up wild prairie.[14]

"Here's where we have to be thinking deeply," Jackson says. "Agriculture had its beginning 10,000 years ago. What were the ecosystems like 10,000 years ago, after the retreat of the ice? Those ecosystems featured material recycling and they ran on contemporary sunlight. Humans have yet to build societies like that. Is it possible that embedded in nature's economy are suggestions for a human economy in which conservation is a consequence of production?" Ecological wealth, Jackson argues, is a more reliable sponsor of human food systems than fossil fuels, bank loans, or government subsidies are.[15]

Land Institute research shows that compared with annuals, perennial food plants provide more protection against soil erosion, manage water and nutrients more effectively, sequester more carbon, are more resilient to pests and stresses, and require less energy, labor, and fertilizer. Yields are currently low compared with annual crops, but they are rising. Studies performed in Africa suggest that many grains, fruits, and vegetables now farmed in annual monocultures will produce similar results when farmed in perennial polycultures.[16]

Agroforestry

Agroforestry combines trees and shrubs with annual crops and livestock in ways that amplify and integrate the yields and benefits beyond what each component offers separately. Like other methods of sustainable agriculture, it is based on observing productive natural ecosystems and mimicking the processes and relationships that make them more resilient and regenerative.

In one form of agroforestry, called alley cropping, grains or other non-woody crops are planted in strips between rows of nut, fruit, timber, or fodder trees. Cattle, poultry, or other livestock can be pastured in the alleys or fed from the crop yields.

Near the town of San Pedro Columbia, in Southern Belize, Christopher Nesbitt has been growing food crops in this traditional forest style at his Mayan Mountain Research Farm for the past 20 years. He mixes some fast-growing native tree species, some annual crops, and some intermediate and long-term tree crops to build soil and produce continuous harvests. Some of his trees are leguminous and hold nitrogen by the microbial attraction of their roots. Some are pollinator-friendly and attract bees and hummingbirds to transfer the fertile pollen of important food plants. Understory trees like coffee, cacao, cassava, allspice, noni, ginger, and papaya benefit from intercropping with high canopy trees like breadfruit, açai and coconut palm, cashew, and mango. Fast-yielding crops such as avocado, citrus, banana, bamboo, yams, vanilla, and climbing squashes provide an income for the farm while waiting for the slower harvest of samwood, cedar, teak, chestnut, and mahogany to mature.[17]

The World Agroforestry Centre reports that methods like these can double or triple crop yields while reducing the need for commercial fertilizers. A U.N. Environment Programme report estimates that if best management practices were widely used, by 2030 up to 6 gigatons of CO_2 equivalent could be sequestered each year using agroforestry, which equals the current emissions from agriculture as a whole.[18]

No-till and Low-till

Some of the nutrient-accumulating and -conserving features that allow natural ecosystems to build and sustain soil fertility include minimum soil disturbance, the presence of a protective layer of plant residues covering the soil surface with no large bare areas for any length of time, and a constant covering of living plants to take up and store any nutrients that become available through decomposition. These nutrient-building and -conserving features can be incorporated into cropping systems by converting to no-till or low-till methods, such as reducing the period of bare fallow, planting cover crops, reincorporating stubble and plant residues, keyline plowing, and reducing aeration of the soil.[19]

On his 2,000-hectare farm near Wellington, New South Wales, in Australia, Angus Maurice is convinced that permanent pasture and what he calls "no kill" cropping systems will be the future of grain production. "We have seen significant recruitment of perennial grasses in the past five years, which is encouraging," he says, "but we realize to reach the system's full potential we would have to eliminate the use of herbicides altogether, which is something we can achieve through fine-tuning and successful recruitment of the right grasses."[20]

Long-term research studies reveal average losses of 328 pounds of organic matter per acre per year with plowing, whereas no-till studies report an average increase of 956 pounds of organic matter per acre per year. Erosion from a conventionally tilled watershed has been found to be 700 times greater than that from a no-till watershed. No-till systems that use high-residue cover crops build soil organic matter content and slow the movement of water over the soil surface, allowing more of it to penetrate. In New South Wales, Maurice reports his most interesting finding: soil carbon levels were significantly higher in areas of perennial grass in the remnant vegetation—about 4 percent, compared with 1.5 percent in paddocks coming out of the old continuous-cropping system.[21]

Permaculture

The term permaculture, a contraction of "permanent agriculture," was coined by Australians Bill Mollison and David Holmgren and refers to a systems approach for designing human ecologies, from farms to houses to cities, that mimics the relationships found in natural biomes. It integrates concepts from organic farming, sustainable forestry, no-till management, and the village-design techniques of indigenous peoples. It applies ecological theory to understand the characteristics of and potential relationships among different design elements.[22]

The discipline uses a set of principles adopted from ecosystems science. One principle is to use the cradle-to-cradle model of recycling all resources and producing no waste. Another is to promote interactions between components so that needs and yields are integrated within the design. For example, a chicken needs food, water, safe habitat, and other chickens, and it produces eggs, feathers, meat, and manure, as well as services such as weed-eating and insect control. A design that integrates chickens would meet their needs from on-farm resources and allow the chickens' outputs to meet the needs of other elements in the design, such as crops or an aquaculture system.

Once set in motion, permaculture designs evolve naturally, capture synergies, and produce a high density of food and other products with diminishing labor and energy inputs over time. One example of a permacultural strategy is the combining of crops in synergistic alliances called guilds, such as the traditional blending of corn, beans, and squash. Researchers have found that these combinations can increase total yields two- to threefold over monocultures of single crops.[23]

One of the better known examples of successful permaculture is found in one of the least hospitable places on Earth for farming. In the Kafrin area of the Jordan valley, 10 kilometers from the Dead Sea, the nearly flat desert receives only two or three light rainfalls in winter. The fine-grained silt is salty, and even the wells in the area are too saline to be used for irrigation.

It was there that Geoff Lawton and his team of permaculturists set up a small, 5-hectare farm and in 2001 began digging swales—2-meter-wide mounds and shallow trenches that crossed the farm in wavy lines on contour. They planted leguminous forest trees in the mounds to fix nitrogen and make leaf fodder. Each tree was given a drip-node from an irrigation line coming from a water dam built to capture road runoff; the lake formed by the dam was stocked with tilapia and geese, which contributed organic fertilizers for the trees.[24]

In the moist trenches, they planted olive, fig, guava, date palm, pomegranate, grape, citrus, carob, mulberry, cactus, and a wide range of vegetables. Barley and alfalfa were planted as legumes and forages for farm animals between the swales. Tree and vegetable plantings were mulched with old newspapers and cotton rags, and animal manure was added before and after planting. Animals raised on the farm included chickens, pigeons, turkey, geese, ducks, rabbits, sheep, and a dairy cow. They were fed from the farm once there were enough trees and plants growing to harvest regularly without overtaxing the system.

Within the first year the soil and well-water began showing a marked decline in salinity, and the garden areas had significant increases in growth. Pests were minor and largely controlled by the farm animals. The combining of plants and animals brought about the integration of farm inputs and outputs into a managed ecosystem of continuous production, water conservation, and soil improvement. In less than a decade a permacultural balance had been achieved, with lessening inputs and improving outputs.

Transitional Agriculture

The early corn-growers depicted in the murals in Tlaxcala would not have imagined they were transforming humans' relationship with Earth's ecology. Although it might be inspiring to have a grand mission like restoring the balance of nature, most farmers who venture into sustainable agriculture are simply interested in improving crop yields or saving labor or money. While tradable credits for sequestering carbon could soon provide another farm revenue stream, many farmers will likely go into sustainable agriculture simply because gas-and-oil-dependant agriculture is becoming more expensive.[25]

As Angus Maurice's family farm in Australia demonstrates, sustainable agriculture is not an either/or proposition, and there will necessarily be a period of transition from the current system to a more sustainable one. Even if most farmers do not go all-organic or apply permaculture principles, they can still improve their fortunes—and that of the planet—by adopting bits and pieces, a little at a time.

Education's New Assignment: Sustainability

For a shift away from consumerism to occur, every aspect of education—from lunchtime and recess to class work and even the walk home—will need to be oriented on sustainability. Habits, values, preferences—all are shaped to a large degree in childhood. And throughout life, education can have a transformative effect on learners. Thus, harnessing this powerful institution will be essential in redirecting humanity toward cultures of sustainability.

No educational system is value-neutral, but all teach and are shaped by a certain set of ideas, values, and behaviors, whether that be consumerism, communism, religious beliefs, or sustainability. As UNESCO states, "Education is not an end in itself. It is a key instrument for bringing about the changes in the knowledge, values, behaviours and lifestyles required to achieve sustainability and stability within and among countries, democracy, human security and peace. Hence it must be a high priority to reorient educational systems and curricula towards these needs. Education at all levels and in all its forms constitutes a vital tool for addressing virtually all global problems relevant for sustainable development."[1]

The more sustainability can be integrated into existing school systems—whether at a Catholic school, a private university, or a public elementary school or through less-formal educational institutions such as museums, zoos, and libraries—the more people will internalize teachings of sustainability from an early age, and these ideas, values, and habits will become "natural." If education can be harnessed, it will be a powerful tool in bringing about sustainable human societies.

This section investigates a sampling of what is happening around the world as educators work to shift from a cultural pattern of consumerism to one of sustainability. Ingrid Pramling Samuelsson of Gothenburg University and Yoshie Kaga of UNESCO describe the formative role that early education can play in teaching children to live sustainably when effectively incorporating key environmental lessons into curricula. Susan Linn of the Campaign for a Commercial-Free Childhood focuses on how important it will be to reclaim childhood from marketers and provide children with unstructured and creative playtime that does not stimulate consumerist values or desires.

Kevin Morgan and Roberta Sonnino of the University of Cardiff explain that school meals are a particularly important part of the school day that could be better used to teach environmental awareness, while helping establish dietary norms that are healthy and sustainable. And David Orr of Oberlin College considers the two important roles that universities play in reorienting learning on sustainability: teaching environmental thinking to students and modeling sustainability both for students and surrounding communities.

Included within these articles are several shorter discussions of other important developments: the benefits of getting children and adults back into nature, toy libraries that have started up in dozens of countries, the effort of one museum to become a center of sustainability education, the role of professional schools in cultivating a sustainability ethic, and the proposed Millennium Assessment of Human Behavior, which could be used to mobilize the academic community to investigate how best to shift human cultures.

Incorporating sustainability education into teacher training and school curricula and providing lifelong opportunities to learn about sustainability will be essential in cultivating societies that will thrive long into the future. The key now will be to expand programs like the ones described here and embed them deeply into leading educational institutions. This will help transform education's role from one that too often reinforces unsustainable consumer behaviors to one that helps to cultivate the knowledge essential to living sustainable lives.

—*Erik Assadourian*

Early Childhood Education to Transform Cultures for Sustainability

Ingrid Pramling Samuelsson and Yoshie Kaga

In view of the unprecedented challenges presented by continuing population growth, environmental destruction, and ever-shrinking resource availability, education at all levels should be reviewed to give a stronger focus on its role of promoting values, attitudes, practices, habits, and lifestyles that promote sustainability. As part of this effort, the education of children in their youngest years deserves special attention.

Research shows that the human brain and biological pathways develop rapidly and that children's experiences before they start primary school shape their attitudes, values, behaviors, habits, skills, and identity throughout life. Thus the first years of life provide a window of opportunity for nurturing children's love of nature and the habits, practices, and lifestyles that favor sustainability. (See Box 6.) Basic life skills such as communication, cooperation, autonomy, creativity, problem-solving, and persistence are acquired in these early years, and the motivation to learn is put in place.[1]

This is an ideal time to look at how to connect early childhood education programs to a sustainability agenda because these programs have increased dramatically in recent years, in part due to changing family structures and the increased number of women in the workforce. About a third of young children in western industrial countries are now being looked after outside the home from the age of one or younger, and most children are in early childhood programs for at least two years before they start primary schooling. Between 1999 and 2006, the global pre-primary percentage of children aged one to five who were enrolled in a kindergarten or the equivalent grew from 33 to 40 percent. The share of children in such educational settings varies widely around the world, however. By 2006 the figures were 14 percent in sub-Saharan Africa, 18 percent in the Arab states, 45 percent in East Asia and the Pacific, 65 percent in Latin America and the Caribbean, and 81 percent in North America and Western Europe.[2]

Early Childhood Education Can Help Make the Shift

Early childhood education can help build a culture of sustainability if it is framed in terms of sustainable development, if curriculum and pedagogical guidelines are oriented toward

Ingrid Pramling Samuelsson is a professor in early childhood education at Gothenburg University in Sweden. **Yoshie Kaga** is a program specialist in early childhood care and education at UNESCO.

Box 6. Sustainability and the Human-Nature Relationship

Humans depend on the natural world to meet all their basic needs, including air, water, food, energy, and shelter. Studies suggest that contact with natural environments, living creatures, and ecological systems is also critical for healthy human development, particularly the development of a healthy self-concept. Psychologists have observed that children as well as adults benefit from "ecological development" in which they create or move toward an understanding of themselves in relation to the non-human world.

Yet many people are increasingly isolated from nature. According to the *2008 Outdoor Recreation Participation Report*, participation in outdoor activities among U.S. youngsters aged 6 to 17 declined 11.6 percent between 2006 and 2007, with the sharpest drops among 6- to 12-year-olds. The time that young people spend indoors is associated with increasing levels of computer, video, and technology use and decreasing levels of physical activity. The negative health effects of this trend, from depression to obesity and diabetes, are well documented.

Research indicates that repeated, regular, and sustained positive experiences in natural environments are influential for attaining sustainable behaviors and lifestyles. Journalist Richard Louv, in his 2005 book *Last Child in the Woods*, points to the psychological and physical benefits of greater interaction with nature. Children in particular can benefit from opportunities for unstructured play in semi-natural environments close to home. Such informal outdoor experiences may be more powerful than the formal, classroom-based environmental education that has gained ground in many countries in the last 30 or so years.

In the United States, Louv's book inspired the drafting in 2007 of the No Child Left Inside Act, designed to guarantee every American child (in particular, impoverished inner-city youth) effective and educationally significant access to outdoor nature. Although this legislation has not yet been adopted, it signals rising concern in the world's largest consumer culture about the next generation's experience in and with natural environments—the outdoors.

Outdoor and environmental education have a long-standing tradition in countries such as Germany, Norway, the United Kingdom, Australia, and New Zealand. Their formal educational efforts are often complemented by a strong outdoor recreation and wilderness tradition. Examples include the *Wandervogel* and youth hostelling movement in Germany, *friluftsliv* in Norway (open air life/life in nature), and the scouting and outdoor education tradition in the United Kingdom, Australia, and New Zealand. Summer camps in the United States and Canada, along with Canada's cottage culture, promote active interaction with natural environments. What unites all these activities is the intent to develop a relationship in which the effects of human behavior on nature and "self" become felt, experienced, and valued.

For many people, especially those in the westernized world, the most direct relationship they have with nature (apart from the air they breathe) is through the food and water they consume. Efforts to live more sustainably though food choices are thus a critical and integral element in the systemic shift to a culture of sustainability. Trends such as the bioregional movement, the rise in community and market gardens, increasing interest in local and organic food, and the embrace of vegetarianism all suggest an attempt to restore a more direct, immediate, and enriching human-nature relationship.

—Almut Beringer
Director of Sustainability,
University of Wisconsin

education for sustainability, if staff training in this field is reinforced, and if parents and communities are involved in the process.

In May 2007 an international workshop on the role of early childhood education in a sustainable society brought early childhood professionals and experts from 16 countries to Gothenburg, Sweden. Participants recognized that there was a great deal in the traditions of early childhood pedagogies that aligns with education for sustainability, such as the interdisciplinary approach, the use of the outdoors for learning, learning through concrete experiences and real-life projects, and the involvement of parents and communities. A subsequent conference in Gothenburg in November 2008 recommended that early childhood education should be conceived as a first step in learning to live sustainably, should be given more priority in policy development, should receive more resources, and should involve cross-sectoral support and collaboration.[3]

It is important that the goals and content of early childhood curriculum be aligned with education for sustainability. In this exercise, environmental education is not the only component. In addition to fostering love for and respect toward nature and promoting an awareness of problems due to unsustainable lifestyles, early childhood education must encourage the outlook and basic skills that enable children to take informed actions responsibly. Instead of the 3Rs of reading, writing, and arithmetic, early childhood education can follow the 7Rs—reduce, reuse, recycle, respect, reflect, repair, and responsibility:

- *Reduce* is about reducing the consumption of food, materials, and resources, which may involve working with parents on the problem of children's exposure to advertisements promoting endless consumption.
- *Reuse* is about showing children that materials can be used many times for different purposes in preschool and at home.
- *Recycle* can be encouraged by asking children to bring recyclable materials to school and integrating them into a range of activities.
- *Respect* is about nurturing understanding of and respect for nature and natural processes and reducing the extent to which they are violated.
- *Reflect* is a habit and skill everybody will benefit from in working for sustainability.
- *Repair* involves taking care of broken toys and other objects and repairing them.
- *Responsibility* is about trusting children to take care of something or do something they can feel proud about.[4]

There is much in the world that is unknown to children. Working toward making the unknown visible to them means creating opportunities to discover the unknown in what they do and work with. This puts demands on early childhood teachers to be aware of what a child's learning should be directed toward.[5]

At the same time, there are also unknown phenomena for the teachers, particularly concerning the future. From a pedagogical perspective, this is a difficult challenge. One way to deal with this might be to try to identify what all children may benefit from having in the future. Eva Johansson suggests that courage, integrity, critical thinking, and responsibility are necessary personal attributes in order to be prepared for an unknown future. Also, it is important to nurture the ability to recognize injustice, as well as to be skilled and creative in solving complex questions. If children are given ample opportunities to be challenged, to make mistakes, and to enjoy seeking possible answers, they will be better equipped to confront the complex questions raised by sustainable development.[6]

At the heart of teaching and working with young children should be the notion of the rich and competent child and active citizen, being in an equal position as his or her teacher, constructing understanding and meanings with others. The "project approach" is a teaching strategy that addresses children's intellectual dispositions, allowing children to examine the basis of their own opinions, ideas, and assumptions. This strategy will help them examine the behaviors of their own cultures and others in terms of implications for sustainable development.[7]

Courtesy Earth Sangha

Young students plant a vegetable garden at their elementary school in Washington, D.C.

It is not necessary to invent entirely new pedagogies in order to "do" education for sustainability in the early years; it is possible to build on pedagogical traditions instead. Arjen Wals points to the qualities in the pedagogical tradition of early childhood education that are particularly useful for education for sustainability—qualities that other levels of education may lack: "So let us return to kindergarten and explore why kindergartens offer more for moving towards a more sustainable world than many of our universities. Kindergarten ideally is or can be places where young children live and learn, explore boundaries, in a safe and transparent world without hidden agendas....There are no dumb questions in kindergarten and there's always time for questions and questioning."[8]

Research shows that the traditional subject-based teaching of knowledge that is common in schools does not give the best results in learning about issues related to sustainable development, which are interdisciplinary in nature. Furthermore, modeling behavior is found to be more effective than direct teaching or preaching in helping young children internalize values and develop desirable attitudes and leanings. Children should have role models who can make these values and characteristics visible and "lived" in daily settings, including early childhood centers, schools, and families, as well as through various public media.[9]

Families, indeed, are the child's first educators. They have the greatest influence in shaping young children's attitudes, values, behaviors, habits, and skills. So they have a central role to play in educating their children for sustainable development. And grandparents often have age-old wisdom about ways of life that favor living together, the preservation of nature throughout generations, and cohabitation with different species—wisdom that should be tapped. Thus where formal early education programs are not available, non-formal education can be set up—as an integral component of community programs or otherwise—to provide parents and grandparents with opportunities to discuss what could be done differently in daily life in order to encourage or enable sustainable development. Where an early childhood education program does exist, the participation of parents can strengthen the link between what takes place in the education setting and at home.[10]

Case Studies on Young Children and Sustainability

The May 2007 workshop in Sweden highlighted numerous examples of how to get young children involved with questions about sustainable development. In one case in Australia, for example, children have numerous opportunities to act as agents of change for sustainability. They work on such mini-projects as litter-less lunches, responsible cleaning, reusing and recycling things, a vegetable garden, a register of native plants, environmental aesthetics, efficient use of natural resources, and construction of a frog pond. They also worked on lifestyle questions such as waste management as well as the "eco-friendliness" of their outdoor environment. The teacher skillfully designs the activities based on the children's interests. They work collaboratively and ensure that informed, reflective practice infuses interactions and deliberations.[11]

Another example is a case study from Japan, where the project approach was practiced in a preschool in relation to the cycle of the silkworm, a fascinating insect. Silk and silkworms have a long use and cultural meaning in traditional clothes in Japan, yet the mulberry trees—which provide the natural food of silkworms—are disappearing in the school's neighborhood. Children learned the whole ecological cycle surrounding silkworms by experiencing, hands on, the growth of cocoons into caterpillars in less than 25 days, observing how caterpillars eat and when silk fibers are produced. While the project was mainly focusing on nature, culture and economy were included as well when the teachers discussed silk clothing and the silk industry in Japanese society.[12]

The last example is from Sweden. The Swedish national curriculum for early childhood education and care clearly spells out that teachers are responsible for promoting respect for the intrinsic values of each person as well as for the shared environment. It also very specifically focuses on children acquiring a caring attitude to nature and the environment as well as an understanding that they are part of nature's regeneration process. The curriculum asks teachers to address ethical dilemmas, and it regards gender equality as a precondition for a sustainable society.[13]

Current Challenges in Early Childhood Education

Although an individual's capacity to learn is most receptive during the first years of life, these are the years that traditionally receive the least support in the education world. Policymakers must pay more attention to this area, given the crucial importance of quality early childhood education, staffed by competent educators, for nurturing active and responsible members of society.[14]

Other areas and levels of education can learn a great deal from the pedagogical strengths of early childhood education, such as the hands-on approach, use of the outdoors as a teaching tool, interdisciplinarity, the whole-project approach, encouraging children's initiatives and interests, and connecting with parents and communities.

With the growing concern about producing a competitive workforce in a globalized knowledge economy, early childhood institutions are increasingly pressed to place school readiness and the acquisition of formal skills at the heart of their goal. But these schools and other preschool bodies need to resist pressures to become packed with hurried and scheduled curricula with predefined goals that are implemented through second-hand learning. These years are the ideal time for children to develop a love of the environment and to learn the basic 7Rs of caring for it.[15]

Commercialism in Children's Lives

Susan Linn

Marketing is linked to a host of public health and social problems facing children today. The World Health Organization and other public health institutions identify marketing to children as a significant factor in the worldwide epidemic of childhood obesity. In addition, advertising and marketing have been associated with eating disorders, sexualization, youth violence, family stress, and underage alcohol and tobacco use.[1]

Among the most troubling ramifications of allowing marketers unfettered access to children is the erosion of creative play, which is central to healthy development. The commercial forces that are preventing the development of children's natural capacity for play are daunting. But there is a burgeoning movement to reclaim childhood from corporate marketers and a resurgence of interest in protecting and promoting hands-on, unstructured, child-driven "make-believe."[2]

Why Play Matters

Play is both culturally universal and fundamental to children's well-being—factors that led the United Nations to list it as a guaranteed right in its 1989 Convention on the Rights of the Child. Play is critical to healthy development, and ensuring children's right to play is an essential building block toward a sustainable world. Yet in the twenty-first century, hands-on creative play is an endangered species. Perhaps the most insidious and powerful threat to what is every child's birthright is the escalation of commercialism in young people's lives.[3]

The ability to play creatively is central to the human capacity to experiment, to act rather than react, and to differentiate oneself from the environment. It is how children wrestle with life and make it meaningful. Spirituality and advances in science and art are all rooted in play. Play promotes attributes essential to a democratic populace, such as curiosity, reasoning, empathy, sharing, cooperation, and a sense of competence—a belief that the individual can make a difference in the world. Constructive problem-solving, divergent thinking, and the capacity for self-regulation are all developed through creative play.[4]

Children at play may enthusiastically conjure cookies out of thin air or talk with creatures no one else can see, yet they still remain grounded in the "real" world. Once children develop the capacity for simultaneously recognizing

Susan Linn is with the Campaign for a Commercial-Free Childhood and Harvard Medical School.

an object for what it is and what it could be, they are able to alter the world around them to further their dreams and hopes and to conquer their fears. When children are given the time and opportunity, they turn spontaneously to "pretend play" to make sense of their experience, to cope with adversity, and to try out and rehearse new roles. They also develop the capacity to use pretend play as a tool for healing, self-knowledge, and growth.

It is traditionally assumed that when children have leisure time, they are engaged in some kind of self-directed, or "free," play, the motivation for which generates from within, rather than from external forces. But for the first time in history, that is not the case. Between 1997 and 2002, in just five years, the amount of time that six- to eight-year-old children in the United States spent in pretend play—such as dress up or play based on imaginative transformations—diminished by about a third. More than half of parents in Japan and France characterize shopping as a play activity. An international survey of 16 countries found that only 27 percent of children engaged in imaginative play, and only 15 percent of mothers believed that play was essential to children's health.[5]

Babies are born with an innate capacity to play. When commercial interests dominate a culture, however, nurturing creative play can become countercultural: it is a threat to corporate profits. Children who play creatively are not as dependent on consumer goods for having fun. Their playfulness, as well as their capacity for joy and engagement, rests mainly within themselves and what they bring to the world rather than what the world brings to them. They are active rather than reactive, and they do not need to be constantly entertained.

Children who engage readily in make-believe are masters of transformation. They can conjure something out of nothing and readily turn a mere stick into, for instance, a wand, a sword, the mast of a boat, or a tool for drawing in the sand. Their enjoyment does not depend on the novelty of acquisition but rather on what they can make of their environment. They are thus more likely to have the internal resources to resist messages that push them toward excessive consumption.

© 2006 Javahar, courtesy Photoshare

On a beach in India, a hole in a soccer net makes a game.

There have been no longitudinal studies exploring the long-term ramifications of children deprived of creative play. But a survey of 400 major employers across the United States found that many of their new young employees, whose childhoods have been shaped by intensifying commercialization, lacked critical thinking and basic problem-solving skills, as well as creativity and innovation, all of which are nurtured in creative play.[6]

The Rise of Commercialism

The fervor for government deregulation that began in the United States in the 1980s, in combination with the digital revolution, has resulted in an unprecedented escalation of commercialism in the lives of children. In 1983,

U.S. marketers spent some $100 million targeting children, a paltry sum compared with the $17 billion they are spending today. While much of the impetus for marketing to children originates in the United States, the trend is promulgated worldwide by multinational corporations. (See Table 7.) Food companies alone spend about $1.9 billion annually marketing directly to children around the world.[7]

Commercial entertainment generated in the United States has long been one of the country's most profitable exports. Mickey Mouse was recognizable around the world long before the escalation of advertising and marketing to children in the 1980s. But the combination of globalization, sophisticated media technology, and U.S. anti-regulatory policies has made the world's children more of a target than ever before. Technological advances such as video and DVDs, as well as cable and satellite television stations, increase marketers' access to children. With the Internet and video games now accessible on MP3 players and cell phones, the pathways to children are increasing.

The mere introduction of electronic screen media into a culture can profoundly influence societal norms such as standards of beauty, diet, and interpersonal interactions. A classic study showed the rise of eating disorders among women in Fiji after television was introduced to the island in 1995. The introduction of specific programming also has an effect. In 1994, just after World Wrestling Entertainment television programming came to Israel, social scientists documented what they described as an epidemic of schoolyard injuries caused by children imitating wrestling moves.[8]

The two companies that dominate the world toy industry, Hasbro and Mattel, create films and television programs to promote their products worldwide. In 2009, Hasbro announced plans to form its own U.S. children's cable television station in partnership with the Discovery Channel, featuring popular brands such as Tonka and My Little Pony. In a recent international study of children's leisure activities, researchers expressed surprise at how little differentiation there now is in how children around the world spend their leisure time.[9]

Critics of globalization characterize the commercialization of childhood as a powerful vehicle for inculcating capitalist values in very young children. The underlying message of nearly all marketing, regardless of the product being advertised, is that the things people buy will make them happy. Aside from the fact that research on happiness shows this to be false, immersing children in a message that material goods are essential to self-fulfillment

Table 7. Childhood Marketing Efforts from Around the World

Disney English Language Program	In China, parents pay $1,000 per semester to send children to Disney-themed language programs. Some children reportedly learn as few as four words, yet their efforts are rewarded with Disney trinkets and access to government-banned Disney films.
McDonald's Happy Meals	As McDonald's expands its presence in India, increasing numbers of children are sampling toys from films such as *Ice Age Three* and *Madagascar* with their chicken burgers and French fries.
SpongeBob SquarePants	A "live" version of the most popular animated character on Viacom's Nickelodeon channel recently visited schools in Namibia. The show is aired in 171 markets around the world in 25 languages.

Source: See endnote 7.

promotes the acquisition of materialistic values, which have been linked to depression and low self-esteem. Research shows that children with more materialistic values are also less likely to engage in environmentally sustainable behaviors such as recycling or conserving water.[10]

The Impact of Commercialism on Play

Children's favorite leisure activity these days, in both industrial and developing countries, is watching television. In the United States, children spend more time in front of television screens than in any other activity besides sleeping: about 40 hours a week outside of school. Nineteen percent of U.S. babies under the age of one have a television in their bedroom. In Viet Nam, 91 percent of mothers report that their children watch television often, as do more than 80 percent of mothers in Argentina, Brazil, India, and Indonesia.[11]

Research indicates that the more young children engage with screens, the less time they spend in creative play. Unlike other media such as reading and the radio, which require people to imagine sounds or visual images, screen media does all of that work. While there is some evidence that certain screen media can encourage children to play creatively and enhance specific kinds of learning, when screens dominate children's lives—regardless of content—they are a threat, not an enhancement, to creativity, play, and make-believe.[12]

The ability to view programs on DVDs, MP3 players, and cell phones, as well as on TIVO and other home recording devices that provide programming "on demand," makes multiple viewings of the same program a new fact of children's lives. Across platforms, electronic screens are the primary means for marketers to target children. Loveble media characters, cutting-edge technology, brightly colored packaging, and well-funded marketing strategies combine in coordinated campaigns to capture the hearts, minds, and imaginations of children—teaching them to value that which can be bought over their own make-believe creations.

Today, more than ever, children need the time, space, tools, and silence essential for developing their capacities for curiosity, creativity, self-reflection, and meaningful engagement in the world. But when consumerism and materialistic values dominate society, creative play is no longer valued. The toys that nurture imagination—blocks, art supplies, dolls, and stuffed animals free of computer chips and links to media—can be used repeatedly and in a variety of ways, diminishing the need to spend money on new toys. Toy libraries are another way to reduce spending on yet another new item. (See Box 7.)[13]

Leonid Mamchenkov

Watching TV with teddy, in Cyprus.

The electronic wizardry characterizing today's best-selling toys makes for great advertising campaigns. They look like fun. But they

Box 7. Toy Libraries

A clever way that many parents are reducing consumerism in childhood is through toy libraries. These are like book libraries—except children check out toys and games instead.

Located in the heart of a community, toy libraries bring families together to share collective goods. One estimate found that 4,500 toy libraries are scattered across 31 countries. In New Zealand, for example, 217 toy libraries serve over 23,000 children.

By providing toys and games, the libraries help parents save money. Based on local community values, toy librarians can also screen out toys that lack educational value or reinforce negative consumer values, like Barbie dolls and toy cars and guns.

The libraries also resolve an important dilemma facing parents: how do you fulfill children's basic right to play with varied and stimulating goods and still avoid excessive consumption and waste? In addition, the toy library helps parents decrease the influence of the marketplace on their children. Parents often find that shopping and buying for children at toy stores is fraught with stress and conflict. Borrowing at the toy library offers children an abundance of goods from which to choose and a wealth of challenging toys.

Sharing collective goods also teaches children many valuable lessons, such as generosity, empathy, and environmental values. These positive sharing experiences appear to be viral, and parents expand into other such experiences such as donating toys, engaging in children's clothing swaps, giving second-hand goods as gifts, joining book cooperatives, sharing cars, and joining time banks.

—*Lucie Ozanne, Marketing Professor, University of Canterbury, New Zealand*
—*Julie Ozanne, Marketing Professor, Virginia Tech University*
Source: See endnote 13.

are created with a kind of planned obsolescence. They are not typically designed with the goal of engaging children for years, or even months. They are designed to sell. If interest wanes, so much the better—another version will soon be on the market. Toys that talk and chirp and do back flips all on their own take much of the creativity, and therefore the value, away from play activities.

Brand-licensed toys are an especially large business, bringing in an estimated $6.2 billion just in the United States in 2007.[14] Toys that represent familiar media characters whose voices, actions, and personalities are already set rob children of opportunities to exercise their own creativity—especially if kids are familiar with the program on which the character is based. Unless some way is found to prevent marketers from targeting children, their play activities will foster imitation, reactivity, and dependence on screens rather than creativity, self-initiation, and active exploration.

Nurturing Play in a Commercialized World

Protecting children's right to play is inextricably linked with their right to grow and develop without being undermined by commercial interests. Laws protecting children from corporate marketers vary widely, with many countries relying primarily on industry self-regulation. The most stringent laws are in the Canadian province of Quebec, which prohibits television advertising to children under 13, and in Norway and Sweden, which prohibit such advertising to children under 12. In Greece, toy ads cannot be aired before ten o'clock at night, and ads for war toys are prohibited entirely. France has banned programs on broadcast television aimed at children under the age of three.[15]

Because of the Internet and satellite broadcasting, however, marketers are increasingly able to target children in any country, making adequate regulation a complex but even more essential task. Changes in regulatory policy take time and are often met with strong and well-funded resistance from commercial stakeholders. As a result, the task of "saving" play in a commercialized world rests on the efforts of nongovernmental organizations (NGOs) and professional groups that are working to influence policy, set limits on marketers' access to children, and help parents and schools encourage creative play. Public institutions, such as libraries and museums, can offer alternative creative educational opportunities. (See Box 8.)[16]

Organized efforts to stop the commercial exploitation of children are in their infancy, but they continue to grow. Pressure from NGOs has led the U.K. government to regulate the marketing of certain foods on television. In Brazil, thanks to efforts by the national advocacy group Criança e Consumo, the state television station in São Paulo no longer markets to children, and a bill prohibiting marketing to children is being considered in the national legislature.[17]

In the United States, which regulates marketing to children less than most industrial democracies, pressure from groups such as the Campaign for a Commercial-Free Childhood has forced companies like Disney and McDonald's to alter some of their marketing practices. The Federal Communications Commission recently launched a review of its rules for children's television with the goal of meeting the new demands of digital technology. And professional organizations such as the American Academy of Pediatrics and American Psychological Association have issued recommendations that include no screen time for children under the age of two, limited screen time for older children, and restricted advertising and marketing to children under eight.[18]

Ad hoc groups of health care professionals and educators have come together to issue strong statements about the importance of play and the need to limit commercial access to children. In the United Kingdom, diverse luminaries such as the Archbishop of Canterbury, children's book author Phillip Pullman, and members of Parliament have joined with educators and health care professionals to deplore the state of childhood in the country, urging limited commercial access to children and advocating for increased opportunities for creative play.[19]

Efforts to limit children's exposure to commercialism and promote creative play are aided by a growing recognition of the need for children to connect with nature. Studies indicate that children play more creatively in green space. As a result of grassroots efforts from NGOs like the Children & Nature Network, the U.S. Congress is currently considering the No Child Left Inside Act, which provides funding for teachers to use schoolyards and local green spaces for lessons. In the Netherlands, conservation and environmental activists—in cooperation with the Minister of Agriculture, Nature and Food Safety—are urging Parliament to support major efforts to help children connect with nature. In Germany, Waldkindergärtens—preschools where young children spend their school time out in nature—are flourishing.[20]

Previous generations took it for granted that children used their leisure time for play. But that is no longer true. Play is an endangered species, and there needs to be a conscious, concerted effort to save make-believe for future generations. The consequence of millions of children growing up deprived of play is a world bereft of joy, creativity, critical thinking, individuality, and meaning—so much of what makes it worthwhile to be human. We need to let children play.

Box 8. Transformation of the California Academy of Sciences

Periodic reinvention is important for all institutions, but particularly for natural history museums, which often seem to be more concerned with the past than the future—more "cabinets of curiosities" showcasing life's historical forms than institutions grappling with the most challenging problems of today and tomorrow.

Helping people of all ages learn about nature and the science of life is an obvious role for natural history museums. Public engagement should not be their secondary mission, but one that is primary. Considering this and financial realities—museums have expenses and depend on paying visitors—exhibits have to be scientifically accurate as well as engaging for a wide array of people.

One institution that has tackled this issue is the California Academy of Sciences in San Francisco. No function has been untouched. The challenge was to be green and sustainable—intellectually, financially, educationally, and operationally—while remaining faithful to the Academy's core mission: to be the most engaging natural history museum in the world, to inspire visitors of all ages to be curious about the natural world, expand their knowledge of it, and feel a responsibility to preserve it; to encourage young visitors to pursue careers in science; to improve science education at all levels; to carry out the highest-quality research on questions of major importance; and to be successful financially.

The Academy started its reinvention with a new building completed in 2008—a necessity after an earthquake damaged the old one in 1989. This building earned the highest possible rating in the Leadership in Energy and Environmental Design rating system: Platinum. Actually, by exploiting a variety of green building technologies and strategies, including recycled building materials, natural ventilation, solar energy generation, and a living roof, it exceeded the threshold for Platinum certification. Today the new Academy uses about 30–35 percent less energy than typical for a building of its type, generates 213,000 kilowatt-hours of solar electricity, and prevents 3.6 million gallons of run-off with its living roof, which is also a popular exhibit for visitors.

Along with a new physical structure, the Academy has made some innovative new program additions in order to engage broader audiences. A few highlights include:

- Free admission one day each month, and always free for visiting classes.
- A glass-walled "project laboratory" where visitors can view scientists' work and learn about the details on connected video screens.
- A robust Web site providing lesson planning material, scientists' blogs, and a live video feed to Farallon Islands, a nature reserve otherwise closed to visitors.
- A Teacher Institute on Science and Sustainability that engages elementary school teachers each year.
- A program called NightLife to attract the age group least represented as visitors—21- to 40-year-olds. Every Thursday evening, visitors 21 and older can enjoy the exhibits, scientific presentations, drinks, and lively DJs—all of which make the Academy what has been rated the "steamiest" date spot in San Francisco.

Whether for NightLife or for class visits, the Academy is challenging more people to consider two essential questions of our time: How did life arise and evolve, and how can it be sustained?

Gregory C. Farrington
Executive Director,
California Academy of Sciences
Source: See endnote 16.

Rethinking School Food: The Power of the Public Plate

Kevin Morgan and Roberta Sonnino

For the vast majority of children in industrial countries, school food is something that has to be endured rather than enjoyed—a rite of passage to an adult world where healthy eating is the exception, rather than the norm, as evidenced by the burgeoning problems of diet-related diseases. Millions of children in developing countries have to endure something far worse, of course, because school food is still conspicuous by its absence in many cases.

In parts of Europe, North America, and Africa, things are changing today. People have moved beyond debates on whether public bodies are capable of delivering a healthier school food service. The jury is in: it is indeed possible—because public bodies are already doing it. When properly deployed, public procurement—the power of purchase—can fashion a sustainable school food service that delivers social, economic, and environmental dividends while also promoting a culture of sustainability. Healthy school food is also generally associated with behavioral improvements, especially in terms of children's concentration levels and learning capacity.[1]

Although the power of purchase has been deployed to great effect to meet strategic priorities—most notably, to create military technologies in the United States or nuclear energy in France—it is rarely used for such prosaic things as fresh food for schools, hospitals, and extended care facilities. Fortunately, more and more people are beginning to realize that healthy eating must in itself be a strategic priority in order to truly value human health, social justice, and environmental integrity—the key principles of sustainable development.

The school food service is a litmus test of a society's political commitment to sustainable development because it caters to young and vulnerable people whose physical tastes and habits of thought are still being formed. But delivering a sustainable school food service is more challenging than it appears. Indeed, despite the stereotype of being a simple service, school food is part of a quite complex ecology in which many variables have to be synchronized. To be effective, school food reform requires changes throughout the system, given the interdependencies involved in the process that brings food from farm to fork.

Kevin Morgan is professor of governance and development and **Roberta Sonnino** is a lecturer in environmental policy and planning in the School of City and Regional Planning at Cardiff University.

Creating New Generations of Knowledgeable Consumers

Being part and parcel of their communities, schools cannot solve societal problems on their own, especially when it comes to something as complex as people's dietary habits. In virtually every society where it has been broadcast, the "healthy eating" message has faced two formidable obstacles: it has been overwhelmed by the "junk food" message, which dwarfs it in terms of advertising spending, and the public health community has naively assumed that getting the right information to the public would be sufficient to induce cultural change.

A disposition for healthy eating is a socially acquired facility, the result of learning with family and friends at home and at school. A "whole-school" approach—one that embeds the healthy eating message into a wider educational package that stresses the positive links between food, fitness, health, and both physical and mental well-being—can have a positive influence on what children eat in and outside of school, and to that extent it plays a key role in fostering the demand for healthier food in schools.

Crucially, though, the healthy eating ethos has to inform every aspect of the school environment—the classroom, the dining room, the vending machine, even the school grounds—to ensure that the landscape and the mind-set of the school are compatible and mutually reinforcing. Where it is fun, stimulating, and enabling, the whole-school approach can deliver handsome dividends even in the most challenging social environments, creating the single most important ingredient of a sustainable school food service: knowledgeable consumers who care about the origin of their food.

Fashioning Sustainable Food Chains through School Food Reform

Whereas the role of school meals in forging new generations of informed consumers is immediately evident, people do not necessarily think of schools as markets for quality food producers. Yet many countries are using school food reform as a tool to develop new supply chains that set a high premium on the use of "quality" food, which is generally equated with fresh, locally produced food.[2]

In the United States, securing food from local suppliers is one of the hallmarks of the Farm-to-School movement, which has been helping schools to reconnect with local food producers. So far more than 1,000 schools in 38 states are buying fresh products from local farms. "Home-grown school feeding" has also become a priority in many developing countries, where the World Food Programme of the United Nations has been trying to replace food imports (on which conventional school feeding programs were based) with locally grown foods. The chief aim of this revolutionary initiative, which has been especially successful in Brazil

Dylan Oliphant

Room for improvement: a high school cafeteria lunch in the U.S.

and Ghana, is to create markets for local producers in the process of promoting the health and education of the children involved.[3]

Sustainable food systems are not wholly synonymous with local food systems. Although there is no reason to assume that locally produced food is inherently better than imports, there is no doubt that the demand for healthier school food creates important opportunities for economic development if local suppliers have the appropriate produce and the infrastructure to distribute it. Thus school food reform has an important role to play in creating new opportunities for small producers who have too often been marginalized, if not displaced, by globalization of the food system.[4]

Tapping the Power of Purchase

Public procurement is the most powerful instrument for creating a sustainable school food service, but its potential has been stymied in some countries by narrow interpretations of what constitutes "value for money." In cost-based contracting cultures, like those of the United Kingdom and the United States, the biggest barrier to sustainable procurement has been a systemic tendency for low cost to masquerade as best value—a tendency that procurement officers and catering managers often justify by referring to the wider regulatory context of their work. In Italy, in contrast, as described later, best value embodies cultural as well as financial attributes, allowing local authorities to take account of the qualitative features of the service when awarding contracts.

In the United Kingdom, European public procurement regulations have often been seen as a barrier to school food reform. But when the U.K. approach is compared with that traditionally adopted in Italy, which is subject to the same European Union (EU) regulations, it is clear that the problem is one of interpretation. Where the United Kingdom was conservative, Italy was bold; where the United Kingdom stressed value for money in the narrow economic sense, Italy sought values in the broadest sense of the term. The explanation for these divergent interpretations is to be found in the interplay of cultural values and political willpower, which in Italy's case sets a high premium on the creative procurement of produce that is strongly associated with seasonality and territoriality. In short, EU procurement rules are not barriers if public bodies have the competence and confidence to deploy the power of purchase within these rules.[5]

In the United States, too, procurement rules have been interpreted as a barrier, preventing school districts from purchasing locally produced food in the school lunch program. The U.S. Department of Agriculture interprets the rules very conservatively, claiming that school districts are not allowed to specify local geographic preferences when they issue their tenders—an interpretation that is fiercely contested by other legal experts. Nothing will do more to promote the cause of local school food procurement in the United States than a clarification of the regulations so that local sourcing is positively and explicitly encouraged by federal and state legislation.[6]

Pioneers of the School Food Revolution

Each of the reforms just described—the whole-school approach, the creation of sustainable food chains, and creative public procurement—is a major challenge in itself. But the biggest challenge of all is to synchronize the reforms so that they have a mutually reinforcing, synergistic effect. This is what the pioneers of school food reform have in common: they all recognize the ecological and interdependent character of the school food service.

Even though all over the world people are becoming increasingly aware of the role of school food in promoting the objectives of sustainable development, two countries can be considered pioneers of the school food revolution: Scotland and Italy. Indeed, in these countries all three fundamental aspects of the school reform process have been taken into account, reflecting a new vision of the service that is beginning to transform cultural values at all stages of the school food chain—among children and their parents, school staff, procurement officers, suppliers, and policymakers.

Scotland pioneered the British school food revolution long before *Jamie's School Dinners*, a popular TV series that in 2006 widely exposed the general public to the problems of the British school meal service. By then Scotland had just ended the first stage of its school food reform, which included an investment of £63.5 million (some $104 million) to redesign the school meal service. This process started in 2002 with the publication of *Hungry for Success*, a report commissioned by the Scottish government that explicitly promoted the whole-school approach. In addition to emphasizing the need to echo the message of the classroom in the dining room, this seminal report introduced new nutrient-based standards to improve the quality of food served in schools and suggested that the school meal service was closer to a health rather than a commercial service.[7]

The rural county of East Ayrshire, in central Scotland, has gone farthest in implementing the government's recommendations. Making the most of the power of purchase gained through *Hungry for Success*, in 2004 East Ayrshire introduced a pilot scheme in one of its primary schools based on the use of fresh, organic, and local food. The initiative was so successful among children, parents, and the catering staff that one year later the Council decided to extend the reform to another 10 primary schools. Today, all primary schools in the county are involved in the program.[8]

Central to the process was the adoption of a creative procurement approach that aimed to help organic and small suppliers become involved in the school meal system. For example, some of the "straightness" guidelines for Class 1 vegetables were made more flexible to attract organic suppliers; the contract was divided into smaller lots to help smaller suppliers cope with the scale of the order; and award criteria were equally based on price and quality. At the same time, the Council actively worked to create a shared commitment to the ideals of the reform all across the food chain. Specifically, training sessions on nutrition and healthy eating were organized for catering managers and cooks. Farmers were invited into the classroom to explain where and how they produce food. Parents were also taken on board through a series of "healthy cooking tips demonstrations."[9]

In East Ayrshire, school food reform has delivered important outcomes from a sustainable development perspective. As a result of the Council's sourcing approach, food miles have been reduced by 70 percent and packaging waste has decreased. Small local suppliers have been provided with new market opportunities, while users' satisfaction with the service has increased significantly. A recent survey found that 67 percent of children think that school meals taste better, 88 percent of them like fresh food, and 77 percent of the parents believe that the scheme is a good use of the local council's money. Even more important perhaps, the school food revolution in this deprived rural county has created a new shared vision of sustainable development that is cutting across the realms of consumption, production, and procurement, challenging widespread misconceptions about the potential for procuring quality food.[10]

In Italy, the whole-school approach is traditionally embedded in the school meal service, which is considered an integral part of citizens' right to education and health. As a result, as noted earlier, best value there is not at all synonymous with low cost; in fact, the qualitative characteristics of the service and its compatibility with the curriculum (specifically, local traditions) are always taken into account in the tendering process. Not surprisingly, then, Italian schools have been sourcing locally for decades, often complementing their emphasis on local products with a wide range of educational initiatives for children and their parents that emphasize the values of seasonality and territoriality. Unlike what happens in most other countries, these strategies are supported by the national government, which enacted a law in 1999 that explicitly promotes "the use of organic, typical and traditional products" in school and hospital canteens.[11]

When this law was passed, the city of Rome was governed by a Green Party administration that, like many others in Italy, was interested in the potential of organic catering in schools. What made the situation in Rome different from other cities was the size. Some 150,000 children who eat at school in Rome consume approximately 150 tons of food per day. To avoid the shock that such massive demand would have created on the organic food market, the city chose a progressive procurement approach. In the beginning, catering companies were required to supply only organic fruit and vegetables, but an incentive system was created for them to increase the range of organic products for schools. At the same time, award criteria were designed to stimulate bidders to improve the socio-environmental quality of the products and services offered—including, for example, criteria that reward initiatives to improve the eating environment for children or to provide products certified as Fair Trade (which are used as a tool to teach children the value of solidarity with developing countries).[12]

Like East Ayrshire, Rome understood the importance of creating a new collective culture of sustainability around school food. Contracted suppliers have been ensured a constant dialogue with city authorities through the creation of a permanent round table, which aims to foster "a shared willingness of going in a certain direction," as the director of one catering company explained. At the same time, they have been asked to introduce food education initiatives among service users, who have been given the opportunity to participate in the reform through Canteen Commissions. These consist of two parents who can inspect the school premises and provide feedback on children's reaction to the changes being introduced.[13]

Peiling Tan

Doing a better job? A high school lunch in Grenoble, France.

After years of efforts and continuous improvement, Rome is in the vanguard of the school food revolution. Today, 67.5 percent of the food served in the city's schools is organic, 44 percent comes from "bio-dedicated" food chains that focus exclusively on organic products, 26 percent is local, 14 percent is certified as Fair Trade, and 2 percent comes from social cooperatives that employ former prisoners or that work land confiscated from the Mafia. As the reform process continues to unfold, a new type of quality-based food system is beginning to emerge—and with it new cultural values that are educating civil society to the values and meanings of sustainability.[14]

From School Food to Community Food

The examples of Scotland and Italy demonstrate that properly designed and delivered school food reform can play a crucial role in creating new forms of "ecological citizenship" that lead people to think more critically about their interactions with the environment, engage practically with collective problems, and assume responsibility for their conduct. In simple terms, school food reform is creating new generations of knowledgeable consumer-citizens.[15]

Much more could be achieved if the power of purchase were to be harnessed across the entire spectrum of the public sector—in hospitals, nursing homes, colleges, universities, prisons, government offices, and the like. In the context of climate change and food security, extending the benefits of school food reform to larger, more significant social and spatial scales is more and more an imperative, not just an option.

Many cities around the world are beginning to move in this direction through the development of a range of food strategies that are designed to ensure access to healthy food for all citizens. As planners and policymakers begin to redesign the urban foodscape of cities like New York, London, Belo Horizonte, and Dar es Salaam, among others, new challenges continue to arise in the realms of infrastructural development, transport, land use, and citizens' education, to name just a few.[16]

In this context, one fundamental lesson can be learned from school food reform. If sufficient political will could be mustered for a new "ethic of care" that has a global as well as a local reach, as has happened in Rome and East Ayrshire, community food planning could play an invaluable role in promoting human health, social justice, and environmental integration—the hallmarks of sustainable development.

What Is Higher Education for Now?

David W. Orr

Education does not occur in a vacuum. It begins with different and often unstated "pre-analytic" cultural assumptions about how, why, and what people learn and the kind of aptitudes and skills necessary to support and pass on a particular kind of society—whether theocratic, democratic, industrial, or what is now being called sustainable. The specific goals of education and the art and science of instruction further depend a great deal on whether those being educated are presumed to be empty vessels to be filled with knowledge or to have inborn qualities that can be drawn out and disciplined. In general, pre-collegiate and collegiate education in the United States was modeled on the former belief: that people are born ignorant and so must be improved in order to increase public virtue, support democracy, provide the skills necessary for economic growth, and more recently serve the information economy and the development of high and ever higher technology. That model has become dominant virtually everywhere.

It is now generally accepted, however, that the modern project of economic growth and domination of nature has gone badly awry. The excesses built into the industrial system threaten the living systems of the planet, moving toward massive biotic impoverishment and potentially catastrophic climate change. It is reasonable to assume that the disordering of ecological systems and Earth's biogeochemical cycles reflects a prior disorder in thinking about humanity's role in ecological systems. If so, ecological problems originate in how people think and so are first and foremost problems of education having to do with the substance and process of formal schooling and higher education. That recognition, in turn, requires comprehension of the problem of education, not just problems in education. The ideas on which modern higher education worldwide is founded reflect a world that disappeared long ago.

When Locke and Rousseau developed their influential views on education in the seventeenth and eighteenth centuries, world population was perhaps 800 million. It is now approaching 7 billion. When Thomas Jefferson designed his "academical village," the fastest mode of transport was a good horse or a frigate in a strong wind. When John Dewey published his treatise on democracy and education in 1916, the first aircraft were bi-wing planes

David Orr is the Paul Sears Distinguished Professor of Environmental Studies and Politics at Oberlin College in Ohio.

capable of speeds of 125 miles per hour.

But rapid technological change is now reshaping the social, cultural, and ecological landscape everywhere. In short order, humans are creating a different planet, arguably a different human nature, and a global culture that is evolving faster than people can comprehend and adapt to. The challenges of conceiving and building a durable civilization, in other words, are sweeping. But the dialogue about sustainability has been almost exclusively focused on how to arrest environmental deterioration—as if the evolution of machines and prosthetic devices is unrelated and unproblematic.

Under these circumstances, it is appropriate to ask, What is education for? What kind of education will enable the rising generation to deal with increasingly complex and portentous global issues? What do they need to know and how should they learn it? And what is the role of professional educators and institutions of higher learning in equipping the young to live full and productive lives relevant to the larger topography of their time? Whatever the specifics, the answer must be the kind of education that enables students to live sustainably, competently, and decently in recognition of their dependence on the web of life. It would be a kind of education that extends their sense of obligation and possibility to a farther time horizon. This will require fundamental changes in the curriculum, changes in the design and construction of schools and campuses, and a more expansive view of the role of educational institutions.

The Development of Environmental Education

The idea that education ought to be harnessed to advance the related causes of environmental sustainability and justice has gathered considerable momentum in recent years. In the Tbilisi Declaration of 1977, organized by UNESCO and the U.N. Environment Programme, representatives from 66 countries called for the inclusion of environmental education in national educational programs. Among their recommendations were 12 guiding principles to make education interdisciplinary and a lifelong process that integrated environmental science and issues across the entire curriculum.[1]

The principles from Tbilisi and similar documents since then have been clearly stated, plausible, and well intentioned, but they have not led to change commensurate with the scale of the problems they addressed. Virtually everything about the modern educational enterprise—from teacher training programs to the stranglehold of disciplines and the procedures for attaining tenure in the modern academy—conspired to undermine changes or render them marginal. The goals did not fit the organizational and professional structures built up over many decades. Plus, the underlying assumptions of education in general included the unstated belief that the environment was both too vast to be significantly affected by human actions and otherwise useful mostly as a resource to be exploited for economic growth.

Still, against considerable resistance, significant progress has been made in the past three decades. But the purposes of environmental education remain deeply controversial, reflecting much of the ambiguity inherent in attempts to define sustainability and chart a plausible course to a more durable, decent, and just future. Many unresolved questions remain on overall purposes and specific issues (see Box 9), but there is no legitimate question that the human presence in nature is increasingly precarious and that the biosphere is uncomfortably close to the threshold of irreversible changes in Earth systems. Even so, there will be no early consensus on the meaning of loaded and complicated words like "sustainability" or agreement about what schools, colleges, and universities should do

Box 9. Unresolved Questions in Environmental Education

- Is it necessary to "love" nature or do people just have to have a basic ecological competency in order to live in harmony with it?
- Will the end of the era of cheap fossil fuels significantly threaten the systems that provide food, energy, and materials and so require skills necessary to a greater degree of local self-reliance? If so, how should practical skills be included in the modern curriculum?
- To what extent does an adequate response to environmental deterioration require a cultural "paradigm shift"? Or can humans be "rich, numerous, and in control of the forces of nature," as Herman Kahn once put it, and also be sustainable? If so, the curriculum would be mostly more of the same with greater emphasis on science and technology.
- To what extent is nature still "natural" and not an artifact of human manipulation? Is there something inherently wrong with "plastic trees," which is to say an increasingly contrived nature, and if so, exactly what? What is natural and what is not? And what difference, if any, does that difference make?
- What is the purpose of environmental education of any sort when nature is being radically altered by the twin forces of rapid climate change and the loss of biological and landscape diversity?
- Is there any place left for role models like Aldo Leopold, Wangari Maathai, and Rachel Carson? Or might natural capitalists, carbon traders, and entrepreneurs making the big deals and the big money suffice to create a sustainable future? If so, environmental education ought to emphasize the management of carbon.

Source: See endnote 2.

about it, however it is defined.[2]

A wide diversity of environmental education programs are found in U.S. colleges and universities, with some stressing environmental science and others the social sciences and humanities. (See Box 10.) Many institutions now offer a major in environmental studies; others, only a minor. Some, like College of the Atlantic and Arizona State University, are integrating environmental issues and systems thinking throughout the entire institution. Institutions such as Carnegie-Mellon University have developed imaginative cross-disciplinary programs in engineering and architecture. Almost everywhere, institutions are engaging environmental issues on two levels: curriculum and campus design and operations.[3]

Curriculum and Education

In the United States, the belief that the environment ought to be given special priority in the curriculum of higher education came of age in the 1960s and 1970s with the creation of environmental studies programs at Williams College, Middlebury, and Brown University. In the late 1980s Tufts University created the first university-wide program encouraging faculty to include environmental issues in courses across the curriculum.

In October 1990, Tufts University president Jean Mayer convened a meeting of 22 university presidents and chancellors at Talloires, France, that culminated in the Talloires Declaration. The document included 10 goals, including leadership to increase awareness of environmental challenges, fostering environmental literacy throughout the campus, and changing operations to reduce environmental impacts. By 2008, some 360 presidents in 40 countries had signed the Declaration.[4]

Even with such promising beginnings, few observers could have imagined the growth of environmental education on college and university campuses worldwide in the following decades. Today environmental studies pro-

Box 10. Maximizing the Value of Professional Schools

To spread cultures of sustainability, the transformation of higher education cannot stop at the undergraduate level but will need to permeate professional schools as well. The good news is that this is starting to happen. More law schools are offering environmental law programs, agricultural schools are teaching sustainable agricultural techniques, and medical schools are greening their labs—all indicators that sustainability is being incorporated into a vast array of professional programs. Business schools may be the ones that are most actively adopting ideas of sustainability.

Many business schools have started to rethink what makes a good business manager. A few have devoted themselves fully to "sustainable management," such as the Presidio School of Management and Bainbridge Graduate Institute. Many others have started to incorporate sustainability into their curricula more broadly. A bi-annual survey by the Aspen Institute tracks over a hundred business schools around the world to measure their commitment to environmental education and research. In 2007 the survey found that 63 percent of business schools required students to take a course on business and society, up from 34 percent in 2001. And since 2005, elective courses on social and environmental issues have increased by 20 percent.

Business students are taking on social and environmental issues outside of the classroom as well. In 2009, more than half of the graduating class at the Harvard Business School took the equivalent of a doctor's Hippocratic Oath. Students vowed to act with the "utmost integrity," not to make choices that "advance my own narrow ambitions but harm the enterprise and the societies it serves," and to "strive to create sustainable economic, social, and environmental prosperity worldwide." Within a few months, the oath's organizers at Harvard received inquiries from 25 schools from around the world, and students from some 115 countries had taken the oath.

But this is just the newest twist on a trend that is nearly a decade old. One group, Net Impact, has been organizing business students "to create positive social and environmental change through business" since 2002. It has more than 200 chapters on six continents and 15,000 members who are business (as well as other) students, business professionals, and academics. And Net Impact has been working toward its goal in several innovative ways. Along with standard efforts to teach members how to green their campuses, the organization provides members with tools and guidance on how to encourage school faculty to add sustainability and social responsibility courses to their schools' curricula. Net Impact also helps members use their business training to make community organizations more effective—a valuable benefit, as these groups tend to lack business-trained staff members.

Between student-led and institutional efforts, business schools may help develop a whole new meaning of the role of business, as well as a new generation of sustainable business managers.

—*Erik Assadourian*

Source: See endnote 3.

grams exist in one form or another on perhaps half of U.S. campuses and are increasingly prominent in universities worldwide. Chalmers University (in Gothenburg, Sweden) has created a partnership with the Massachusetts Institute of Technology, the Swiss Federal Institute of Technology, and the University of Tokyo that annually brings hundreds of scientists together to discuss environmental issues. Individual campuses such as the Technical University of Catalonia (Spain), TERI University (India), and Kyoto University have developed imaginative and diverse environmental curriculum. UNESCO sponsors chairs in sustainable development at 45 universities in 27 countries as well as conferences on "Higher Education for Sustainability." The success of the *International Journal of Sustainability in Higher Education* reflects a growing maturity and self-reflection in the field.[5]

But one study shows that there is no "common path" toward change. Rather, education for sustainability is flourishing because of many factors, including committed faculty, imaginative leadership, student activism, response to specific opportunities, and larger societal changes.[6]

Despite great progress in environmental education, there is good evidence that it is clearly an inadequate counterweight to the conventional curriculum and an inadequate response to the mounting environmental crisis. The National Wildlife Federation, for example, concluded in its *Campus Environment Report: 2008* that between 2001 and 2008 "the amount of sustainability-related education [in the U.S.] did not increase and may even have declined." That conclusion is supported by global poll data that consistently show a majority of the public—including college graduates—to be uninformed, sometimes misinformed, and otherwise confused about the fundamentals of ecology and science in general.[7]

Campus Design and Operations

Alongside efforts to increase ecological awareness and literacy are others aimed to change the "design" of campuses by improving energy efficiency, lowering carbon emissions, reducing waste, recycling, and building the high-performance buildings that have become mainstream virtually everywhere. The beginnings of such efforts are found in April Smith's Masters thesis at UCLA in 1988, "In Our Backyard," and the Meadowcreek Project's early study of campus food systems at Hendrix, Carleton, and St. Olaf colleges in 1988–89.[8]

By the mid-1990s the first studies of campus ecology had grown into larger studies of campus resource flows of food, energy, materials, water, and waste in which the campus became a laboratory for education and also the foundation for better campus management. The National Wildlife Federation's campus ecology program, ably led by Julian Keniry, brought increased awareness of environmental issues to campuses and developed materials useful for improving efficiency and integrating campus management with curriculum. Walter Simpson created and directed the first successful university-wide energy efficiency program at the State University of New York–Buffalo. Others, like Will Toor at the University of Colorado, created effective campus-wide recycling and low-impact transportation programs. The emergence of organizations such as the North American Association for Environmental Education and the American Association for Sustainability in Higher Education (AASHE) amplified and coordinated otherwise disparate campus ecology efforts.[9]

In the late 1990s, two factors significantly focused attention on what Keniry had called the campus ecology movement and the design of the campus. The first was the rapid growth of the green building movement in the United States, the United Kingdom, Europe, and Asia. The result has been an effort to reduce the environmental impacts of new construction on college and university campuses. Dramatic improvements in energy and materials technology and the practice of the integrated design necessary to build low-impact, high-performance buildings created large opportunities to incorporate environmental goals into campus buildings while lowering costs for operations and maintenance. The first substantially green building on a U.S. college campus was the Adam Joseph Lewis Center at Oberlin College, constructed in the late 1990s, which is still the only entirely solar-powered, zero-discharge building on a U.S. college campus. Other and larger, more complex buildings, including science facilities, followed on hundreds of other campuses, so that green building criteria have become standard for new academic construction worldwide.[10]

The second driver in the green campus movement has been increasing concern about rapid climate change. The four assessment reports of the Intergovernmental Panel on Climate Change (in 1991, 1995, 2001, and 2007) and a large and growing body of scientific evidence has established beyond legitimate dispute that climate is changing and that humans are the culprits. It is now clear that the speed, scale, and duration of climatic change are at or beyond the worst-case scenarios of even a few years ago.[11]

The first call for carbon-neutral campuses appeared in the *Chronicle of Higher Education* in 2000. But the effort to organize both professional organizations and academic leadership began in earnest with the efforts of 12 college and university presidents, in collaboration with Second Nature, AASHE, and ecoAmerica, to get other presidents and professional academic societies to publicly commit to move their institutions toward carbon neutrality. More than 600 college and university presidents to date have signed the pledge. The results could be both the reduction of a significant fraction of U.S. carbon emissions and a sterling example of leadership for other sectors. Architect Edward Mazria calculates, however, that the addition of only four new medium-sized coal plants anywhere in the world would eliminate the gains even if all U.S. institutions of higher education were to eliminate their carbon dioxide emissions entirely.[12]

Future Initiatives

Despite considerable progress since the Tbilisi Conference in 1977, there is a great deal more to be done to create the permanent institutional and cultural wherewithal to educate people around the world about systems and ecology and equip them with the capacity to think across the lines of professional and disciplinary specializations. But promising efforts are under way. (See Box 11.)[13]

Beyond institutions of higher education, many diverse organizations—from Schumacher College in Devon in the United Kingdom to the Center for Eco-Literacy in Berkeley, California—offer teacher training, expertise in curricular reform, and forums for rethinking core assumptions underlying education and the broader culture. Formerly U.S.-centric organizations like the Bioneers are becoming important nodes in the global conversation about the intersection of ecology, education, and justice. The authors and organizers of the Earth Charter are similarly creating a transnational dialogue about education rooted in international law, philosophy, and ecology.

There are signs of a larger shift in the role of institutions of higher education in the transition to sustainable economies as well. The Universitat Autònoma de Barcelona has developed collaboration with the Barcelona City Council to enhance the sustainability of public events. On a larger scale, Judith Rodin, as president of the University of Pennsylvania from 1994 to 2004, led an imaginative and sweeping transformation of West Philadelphia using institutional investment to leverage several billions of dollars of outside funds. The efforts reversed urban decline in dozens of blocks surrounding the university and offer a brilliant example of not only urban renewal but leadership in higher education. Joined to the campus ecology movement, colleges and universities everywhere might become catalysts for prosperous post-fossil-fuel

Box 11. A New Focus for Scientists: How Cultures Change

It seems clear that unequivocal knowledge of civilization's biophysical peril alone is insufficient to spark the changes required to avoid its collapse. We also need a greater understanding of how cultures change, which underlines the desperate need for global society to focus its attention on the need for a cultural revolution. Providing that focus is the goal of a Millennium Assessment of Human Behavior (MAHB) that is in the early stages of development.

In light of the success of the Intergovernmental Panel on Climate Change, a small group of natural and social scientists and humanists is working on starting the MAHB. It is currently being organized by the Global Sustainability Alliance, with member groups in the United States, Norway, Sweden, Ghana, and China. It will ideally be launched with a global conference involving scholars, politicians, and a broad spectrum of stakeholders. This would be followed by workshops, regional conferences, worldwide policy debates, and research activities.

Major roles of the MAHB will include generating public discussion on the causes of self-destructive behavior such as climate change and biodiversity loss, debating its ethical dimensions, and investigating how cultural evolution can be steered toward creating a sustainable world society. That is the direction almost all human beings presumably desire—a chance for their children and grandchildren to lead lives as rewarding as or better than their own.

The organizers' basic goal is to find ways to reframe people's definitions of and solutions to sustainability problems and to promote a global discussion about what human goals should be. The MAHB will invite people from literature and the arts to develop narratives and visual materials as signposts to guide civilization toward sustainability. People need visions of futures that do not include perpetual growth of consumption or human numbers, the idea that having gadgets is the ultimate goal of human life, or the notion that gross domestic product is the best measure of human well-being.

One of the MAHB's early tasks will be to secure governmental buy-in and enlist the support of key decisionmakers in industry, academia, the media, religious communities,

regional economies while equipping students with the analytical skills, knowledge, and inspiration to design and build a decent, fair, and sustainable world.[14]

The government of Bhutan offers perhaps the most far-reaching example of national leadership in education. Having replaced the yardstick of "gross national product" with one that measures "gross national happiness" (GNH) in 1972, the government is now sponsoring an effort to educate its citizenry for happiness, sustainability, justice, and peace. Led by Prime Minister Lyonchen Jigmi, the first goal is to integrate GNH principles such as the interdependence of humans and nature into the curriculum at all levels. The second goal involves creating a model of GNH in central Bhutan where civil servants and teachers alike can "take long and short courses that renew their commitment to environmental protection, sustainable economic development, and responsible and accountable leadership." The goal is to create a self-perpetuating system that joins individual psychology with larger ecological and cultural systems.[15]

Institutions of higher education—indeed, all schools—must aim to create an ecologically literate and ecologically competent citizenry, one that knows how Earth works as a physical system and why that knowledge is vitally

Box 11. *continued*

foundations, and so on. It must mobilize appropriate stakeholders to participate in the discussion and help accelerate needed changes in cultural practices and institutional structures. Indeed, the task of assembling such support is at the core of the overall challenge and will determine whether the infant MAHB (see mahb.stanford.edu) survives to tackle its global task.

The MAHB envisions establishing an "observatory" on humankind's collective behavior. It would gather evidence on dimensions of cultural change from existing documents and databases as well as from a variety of global stakeholders. The observatory would explore the role of values in well-being to determine what institutional and cultural barriers stand between declared values and actual practices. It would examine the factors that drive human happiness and fulfillment across cultures and their implications for ecological sustainability. It will use modern communications systems to assess how diverse societies measure success and happiness, to depict the links between global environmental risks and lifestyle choices, to explore cultural differences in attitudes toward the environment and sustainability, and to embed the human narrative in a deeper understanding of humankind's relationship to nature. The behavioral observatory would include an interactive portal sharing up-to-date information about particular environmental problems, human factors relating to these problems, and frameworks to deal with them.

Once established, the MAHB could be a powerful new tool to mobilize people who have devoted their careers to studying behavioral change to help solve the largest threat humanity has ever faced: unsustainable practices undermining the very systems people depend on. Natural scientists have already shown the way toward a sustainable future by elucidating the problems and outlining many solutions. Now it is time to figure out how to frame these in ways that will motivate people to respond—a job well-suited to the MAHB, whose public outreach and debate functions could play a major role in generating the changes needed.

—*Paul R. Ehrlich and Anne H. Ehrlich*

Source: See endnote 13.

important to them personally and to the larger human prospect. There are many challenges to actually making this a reality, not the least of which is the very real possibility of growing despair and nihilism among young people in the face of what will likely be a time of increasingly dire news and seemingly unsolvable social and economic problems.

The scientific evidence suggests that the years ahead will test coming generations in extraordinary ways. Educators are obliged to tell the truth about such things but then to convert the anxiety that often accompanies increased awareness of danger to positive energy that can generate constructive changes. Environmental education must be an exercise in applied hope that equips young people with the skills, aptitudes, analytic wherewithal, creativity, and stamina to dream, act, and lead heroically. To be effective on a significant scale, however, the creative energies of the rising generation must be joined with strong and bold institutional leadership to catalyze a future better than the one in prospect.

Business and Economy: Management Priorities

Business is not just a central component of the global economy, it is a leading driver of societies, cultures, and even the human imagination. And while today business is primarily shaping a cultural vision centered on consumerism, this vision could as readily be centered on sustainability—given new management priorities.

Priority number one will be to gain a better understanding of what the economy is for and whether perpetual growth is possible or even desirable. As environmentalist and entrepreneur Paul Hawken explains, "At present we are stealing the future, selling it in the present, and calling it gross domestic product. We can just as easily have an economy that is based on healing the future instead of stealing it."[1]

In this section, Robert Costanza, Joshua Farley, and Ida Kubiszewski of the Gund Institute for Ecological Economics first describe how redirecting the global economy is possible through a variety of means such as creating new sustainable economic metrics, expanding the commons sector, and mobilizing leading economic and governmental institutions.

Another key economic shift will be the better distribution of work and working hours among the global workforce, as Juliet Schor of Boston College describes. Right now, many people work excessive hours earning more money and converting that income into increased consumption—even as others search for work. Dividing work hours in a better way will not only address unemployment and provide more people with the means for a basic standard of living, it will free up time to enjoy life outside of the workplace. And it will reduce the amount of discretionary income people have, which at the moment encourages them to consume more than necessary.

Another priority will be to reassess the role of corporations. Consider their vast power and reach: in 2006, the largest 100 transnational corporations employed 15.4 million people and had sales of $7 trillion—the equivalent of 15 percent of the gross world product. A sustainable economic system will depend on convincing corporations, through an array of strategies, that conducting business sustainably is their primary fiduciary responsibility.[2]

Ray Anderson of Interface, Inc., Mona Amodeo of idgroup, and Jim Hartzfeld of InterfaceRAISE note that some corporations have already figured out the importance of a thriving Earth to their business and are working to put sustainability at the heart of their corporate cultures. Understanding how to shift business cultures and finding the resolve to do so will be an essential step in creating a sustainable economic model.

Beyond the corporate system, there are opportunities to completely reinvent the purpose and design of business, also a key priority. Johanna Mair and Kate Ganly of IESE Business School describe social enterprises that are turning the mission of business upside down. Business does not have to be only or even primarily about profit, but profit can provide a means to finance a broader social mission. Social enterprises worldwide are addressing pressing social problems, from poverty to ecological decline, and are doing so profitably.

Local businesses are also starting to crop up, like pioneer species in disturbed ecosystems. As most corporations fail to respond to increasing concern for social and environmental injustices, people are creating local alternatives—from grocery stores and restaurants to farms and renewable energy utilities. Michael Shuman of the Business Alliance for Living Local Economies notes that these local enterprises can have improved environmental performance, treat workers better, provide healthier and more diverse products, and—in worst-case scenarios—provide a layer of resilience to global disruptions by being rooted locally. Moreover, the rise of social enterprises and local businesses should provide additional pressure to stimulate change within corporate cultures.

Throughout the section, Boxes describe other sustainable business innovations, such as redesigning manufacturing to be "cradle to cradle," a new corporate charter that integrates social responsibility directly into the legal code, and a carbon index for the financial market. There is also a Box that examines the absurdity of the concept of infinite economic growth.

Business is a powerful institution that will play a central role in our future—whether that future is an era of sustainability or an age of reacting to accelerating ecological decline. With a combination of reform of current interests and the growth of new socially oriented business models, the global economy can help avert catastrophe and instead usher in a sustainable golden age.

—Erik Assadourian

Adapting Institutions for Life in a Full World

Robert Costanza, Joshua Farley, and Ida Kubiszewski

Today's dominant worldviews and institutions emerged during the early Industrial Revolution, when the world was still relatively empty of humans and their built infrastructure. Natural resources were abundant, social settlements were more sparse, and the main limit on improving human well-being was inadequate access to infrastructure and consumer goods.[1]

Current ideas about what is desirable and what is possible were forged in this empty-world context. "Cheap" fossil fuels have provided the abundant energy necessary for economic growth and helped societies overcome numerous resource constraints. Fertilizers, pesticides, and mechanized agriculture have allowed humanity to stave off Thomas Malthus's predictions of population collapse. As a result, the world has changed dramatically over the past two centuries. It is now a "full" world, where increasingly complex technologies and institutions, mounting resource constraints, and a decreasing energy return on investment have made human society more brittle—and hence more susceptible to collapse.[2]

Laws and policies that incorporate the empty-world vision are legion. The 1872 Mining Act in the United States, for example, was designed to promote minerals mining and economic growth. It did this by essentially giving away the right to mine on public lands while collecting no royalties and requiring no environmental protection. The act is still in force, even though conditions have changed dramatically. The consequence has been massive environmental destruction and a giveaway of public wealth to private interests.[3]

Today's prevailing worldviews, institutions, and technologies are failing to meet humanity's needs in a rapidly changing world. Climate change, declining oil supplies, biodiversity loss, rising food prices, disease pandemics, ozone depletion, pollution, and the loss of life-sustaining ecosystem services all pose serious threats to humanity. Yet most of these threats were not even imagined when today's worldviews, institutions, and laws were being formed.

All these crises can be traced back to one overarching problem: we have failed to adapt our current socioecological regime from an empty world to a full world.

Robert Costanza and **Ida Kubiszewski** are with the Gund Institute for Ecological Economics and the Rubenstein School of Environment and Natural Resources at the University of Vermont. **Joshua Farley** is also with the Gund Institute and with the Department of Community Development and Applied Economics at the University of Vermont.

Brian Burger

Landscape consumption in British Columbia, Canada: logging roads, clearcuts, and slash piles.

Under Stress in an Increasingly Full World

There are three fundamental reasons why the current regime no longer serves humanity in a full-world context. The first is that unlimited increases in resource and energy use are physically impossible on a finite planet. (See Box 12.) All economic production requires the transformation of raw materials and energy, making these inputs less available to serve as the structural building blocks of the ecosystems that provide life-support services for all species. The global climate crisis is just one example of an ecosystem service—climate regulation—that is being consumed at an unsustainable rate.[4]

The use of fossil fuels not only depletes a nonrenewable resource, it also creates waste emissions that further degrade ecosystem function. But even advances in energy technology cannot create energy out of nothing. While the development of renewable energy sources is a priority, no currently feasible energy alternative can sustain today's rate of resource-intensive global economic growth.

The second reason why the current regime no longer serves humanity in a full-world context is that unlimited increases in resource and energy use do not continue to increase well-being. Unlimited conventional economic growth (that is, growth in the gross domestic product (GDP)) is not only impossible, it is undesirable. GDP measures marketed income, not welfare. What is really needed is to provide satisfying lives with less economic activity, raw materials, energy, and work required. When GDP rises faster than life satisfaction, this efficiency declines.

The genuine progress indicator (GPI) is one alternative measure of welfare designed to adjust for the inadequacies of GDP, subtracting factors such as the costs of crime and pollution, and adding factors such as the value of household and volunteer work. In the United States, GPI neared its per capita peak in 1975, at a time when per capita GDP was about half what it is today. (See Figure 3.)[5]

Subjective measures of well-being, such as the share of people who consider themselves "very happy," have also not increased since 1975. Empirical evidence suggests that a return to 1970s per capita consumption levels would not make people worse off but would instead lower resource depletion, energy use, and ecological impacts by half. People would actually be better off because they would have more time and resources to invest in public, non-consumption goods produced by natural and social capital.[6]

The final reason why the current regime no longer serves humanity in a full-world context is that today's institutions are designed to maximize energy and resource use and are poorly adapted to the needs of a full world. Market institutions, for example, enhance economic growth, but they deal well only with pri-

Box 12. The Folly of Infinite Growth on a Finite Planet

Although the climate challenge is receiving a lot of attention these days, the global temperature increase is but a symptom. The planet has a fever, and it is essential to identify the disease in order to prescribe the right medication. Could the real disease be expanding levels of consumption, growing national economies, and ballooning populations?

Nearly 40 years ago, Jay Forrester warned of the challenge of exponential growth and its implications for a finite planet. This challenge can be illustrated by a biological experiment: If the conditions are right, bacteria will double in number every day, filling the surface of a container by the fiftieth day. But the surface will only be half covered on the forty-ninth day. Humanity may already be on its forty-ninth day and—like a bacteria colony—may completely consume its home if it does not somehow change course.

The ecological capacity of Earth is not expanding, while humanity's footprint is. Global ecological capacity was used up more than 20 years ago. Thus industrial economies, to free up resources for Earth to function and allow developing countries to meet their populations' needs, need to contract significantly.

Many economists believe the opposite, however: that the world economy must continue to grow and that a simple, low-consumption life is a threat to the prevailing economic model. Yet John Stuart Mill, the founding father of modern capitalism, would not support that view. He realized that industrial society, by its very nature, could not last for long and that the stable society that must replace it would be a far better place. "I cannot regard," wrote Mill in 1857, "the stationary state of capital and wealth with the unaffected aversion so generally manifested towards it by politicians of the old school."

Economist Kenneth Boulding went even one step further by claiming that gross national product (GNP) be considered a measure of gross national cost and that people should devote themselves to its minimization. And it has become increasingly clear that GNP does not couple well with actual well-being, as can be seen in measures like the Genuine Progress Indicator and others. The need for a fundamental rethinking of modern economics is perhaps most eloquently put by Paul Hawken, Amory Lovins, and Hunter Lovins in their book *Natural Capitalism*.

Yet instead of becoming outmoded, the perpetual growth model is now spreading worldwide. From 1958 to 2008 the number of cars increased from 86 million to 620 million. Air passengers skyrocketed from 68 million in 1955 to 2 billion in 2005. The ecological effects of these trends are catastrophic.

The challenge in terms of our fixation on growth is how to get started on a new course. Obviously nobody can expect the Chinese or the Indians to take the initiative on non-growth thinking. At the moment, it looks rather unlikely that any major industrial country will lead the way. But maybe a rich, well-educated country could—a country like Norway or Sweden. With a small population and ample resources, perhaps Scandinavia could lead the way and demonstrate the feasibility of a vision of what the good life in a steady state economy would look like: less hours worked, less stuff, less stress, more time with family and friends, more time for civic engagement, more leisure.

It will not be easy, but it is necessary. It will require a new consumption culture, a new technology culture, and a new intellectual culture—all based on ecological intelligence. In fact, it will demand a fundamental reordering of global priorities.

—*Øystein Dahle*
Chairman, Worldwatch Norden
Source: See endnote 4.

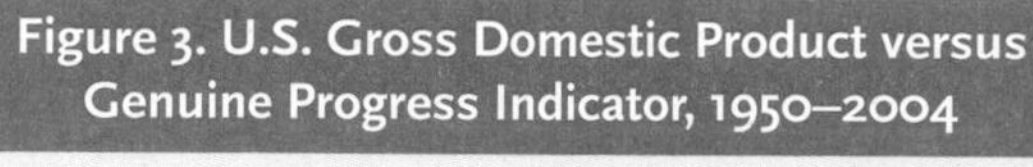
Figure 3. U.S. Gross Domestic Product versus Genuine Progress Indicator, 1950–2004

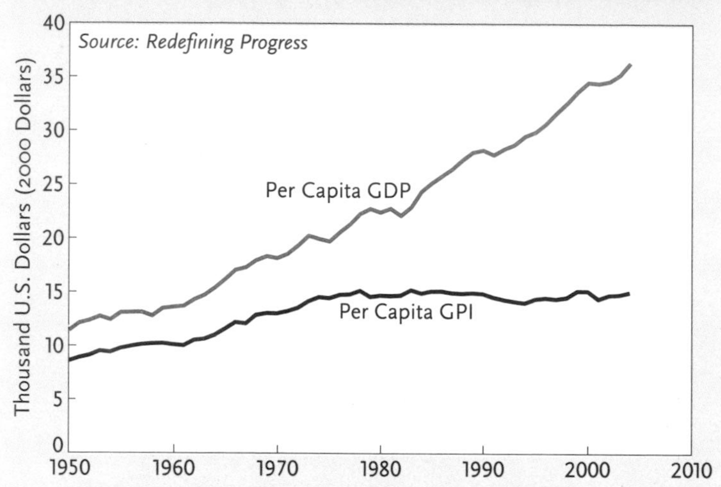

privatize it. Continuous material economic growth in wealthier countries is a major cause of this biophysical crisis.[8]

Global climatic stability and ecological resilience are global public goods that require cooperative global solutions, whereas fossil fuels are market goods that promote competition and resource struggles. The transition to sustainability demands new energy sources that are "non-rival," such as energy from the sun and wind. (For example, U.S. development of cheap and efficient solar power will not limit China's use of this resource; moreover, China would likely improve the technology, thus conferring benefits to other users.) Unfortunately, international trade institutions such as the World Trade Organization give priority to private, market goods and services at the expense of public goods. Countries that cannot afford renewable energy technologies will continue to burn coal, preventing the new technologies from helping to address climate change. Open access to information about renewable energy technologies is needed to solve this problem.

vate goods and services. They often provide these at the expense of public goods and services—such as education, infrastructure, public health, and ecosystem services—that would most significantly improve quality of life in today's full-world context. A 1997 study valued worldwide ecosystem services at approximately $33 trillion, more than the value of the gross world product at that time.[7]

Many governments have long-standing policies that promote growth in market goods at the expense of non-market, public goods that are generated by healthy ecosystems. These policies include the more than $2 trillion in annual subsidies for market activities and externalities that degrade the environment; the privatization or reduced protection of common (shared) resources, such as forests and fisheries; and inadequate regulation and enforcement of existing regulations against environmental externalities. Perhaps the most serious environmental externality facing the world today is climate change. To solve this "mother of all market failures," the world needs to deal with the atmosphere as a global common asset, not

Toward a New Sustainable and Desirable Regime

Regime shifts can be driven by collapse or through conscious and integrated changes in worldviews, institutions, and technologies. New goals, rules, and tools can be developed. These changes provide the opportunity to move away from unsustainable practices and to avoid social, economic, and ecological collapse. This section looks at five ideas to stimulate and seed this transition.

Redefine well-being metrics. In today's

full-world context, the goal of an economy should be to sustainably improve human well-being and quality of life. Material consumption and GDP are merely means to that end, not ends in themselves. Both ancient wisdom and new psychological research confirm that material consumption beyond real need can actually reduce overall well-being by creating an unending and unsatisfying drive for more stuff.

Such a reorientation leads to specific tasks. For a start, efforts should be made to identify what actually contributes to human well-being and include the substantial contributions of natural and social capital, both of which are under increasing stress. It is important to distinguish between real poverty (in terms of low quality of life) and merely low monetary income. Ultimately, it is necessary to identify what the economy actually is and what it is for, and to establish a new model of development that acknowledges today's full-world context. Many efforts are under way to develop better well-being measures, including the GPI, but a global effort is needed to build consensus that will allow these alternative measures to gain broad acceptance and credibility.[9]

Ensure the well-being of populations during the transition. It will be important that any reductions in economic output and consumption that accompany the shift to a new regime fall on those who will be hurt the least—that is, the wealthy. Presently, the U.S. tax code taxes the third wealthiest man in the world, Warren Buffett, at 17.7 percent, while his receptionist is taxed at the average rate of 30 percent. Appropriate monetary policies can enhance employment, moderate the gap in income, restore the natural environment, and invest more in public goods while overall consumption decreases. For example, ecological tax reform could be implemented that would change consumption patterns and tax the wealthy more because they pollute more, while reducing taxes on social security or other benefits, which will benefit those who rely more fully on these payments.[10]

Reduce complexity and increase resilience. History offers lessons about the collapse of societies as well as examples of successful adaptation. While environmental factors often contributed to societal declines, it was cultural and institutional resiliency and adaptability that most influenced a society's chances of survival. Resilience depends on cultural values as well as the ability of political, economic, and social institutions to respond.[11]

Many societies have collapsed due to insufficient resources to sustain their complex structures. The Western Roman Empire, for example, was a thriving, highly complex system as long as increasing resources were available through conquests. But when the limits of conquest were reached, the empire began to tax farmers heavily in an attempt to retain the resource influx, eroding the system's ability to absorb shocks and making it vulnerable to barbarian invasions and other pressures. Maintaining resilience in a full world means shifting the emphasis away from growth and expansion and toward sufficiency and sustainable prosperity.[12]

Expand the "commons sector." During the transition to a new regime, it is important to greatly expand the "commons sector" of the economy, the sector responsible for managing existing common assets and creating new ones. Some assets, such as resources created by nature or by society as a whole, should be held in common because this is more just. Other assets, such as information or ecosystem structures (for example, forests), should be held in common because this is more efficient. Still other assets, such as essential common-pool resources and public goods, should be held in common because this is more sustainable.

One option for expanding and managing the commons sector is to create "common asset trusts" at various scales. Trusts, such as the Alaska Permanent Fund and regional land trusts, can propertize the commons without privatizing them. At a larger scale, a proposed Earth Atmospheric Trust could help to massively reduce global carbon emissions while also reducing poverty. This system would comprise a global cap-and-trade system for all greenhouse gas emissions (preferable to a tax, because it would set the quantity and allow price to vary); the auctioning of all emission permits before allowing trading among permit holders (to send the right price signals to emitters); and a reduction of the cap over time to stabilize atmospheric greenhouse gas concentrations at a level equivalent to 350 parts per million of carbon dioxide.[13]

The revenues resulting from these efforts would be deposited into the Earth Atmospheric Trust, administered transparently by trustees who serve long terms and have a clear mandate to protect Earth's climate system and atmosphere for the benefit of current and future generations. A designated fraction of the revenues derived from auctioning the permits could then be returned to people throughout the world in the form of a per capita payment. The remainder of the revenues could be used to enhance and restore the atmosphere, invest in social and technological innovations, assist developing countries, and administer the Trust.

Use the Internet to remove communication barriers and improve democracy. Unlike with television and other broadcast media, very low technological and financial barriers exist to establishing a presence on the Internet. This has the effect of decentralizing the production and distribution of information by returning control to the audience, providing a venue for dialogue instead of monologue. Opinions and services that were previously controlled by small groups or corporations are now shaped by the entire population. Television news networks, sitcoms, and Hollywood productions are being replaced by e-mail, Wikipedia, YouTube, and millions of blogs and forums—all created by the same millions of people who are the audience for the content.

The 2008 U.S. presidential election marked the first election year where more than half of the nation's adult population became involved in the political process by using the Internet as a source of news and information. Rather than simply receiving uni-directional news, approximately one fifth of the people using the Internet used Web sites, blogs, social networking sites, and other forums to discuss, comment, and question issues related to the election.[14]

Conclusion

Changes in worldviews, institutions, and technologies will be necessary to achieve lifestyles that are better adapted to today's full-world context. To a certain extent, people can design the future they want by creating a new vision and new goals. If societal goals shift from maximizing growth of the market economy to maximizing sustainable human well-being, different institutions will better serve these goals. It is important to recognize, however, that a transition will occur in any case and that it will almost certainly be driven by crises. Whether these crises lead to decline or collapse followed by ultimate rebuilding or to a relatively smooth transition to a sustainable and desirable future depends on people's ability to anticipate the required changes and to develop new cultures and new institutions.

Sustainable Work Schedules for All

Juliet Schor

Discussions of ecological sustainability typically focus on greenhouse gas emissions, biodiversity, and other measurements of the natural world. They may include economic and social trends in production or population. But they rarely feature time use. Yet patterns of human time use are key drivers of ecological outcomes. People combine time, money, and natural resources to carry out their daily lives and activities. Firms combine time, physical capital, and natural capital to create production. To a great extent, time and natural resources are substitutes for each other: doing things faster usually takes a greater toll on Earth. So time-stressed households and societies tend to have heavier ecological footprints and greater per capita energy use.

In the transition to sustainable cultures and economies, people are going to have to adapt to new schedules and temporal rhythms. The culture of long working hours and excessive busy-ness that characterizes a number of wealthy countries will need to be replaced by more sustainable patterns of time use. While there will be adjustment costs, a slower and more humane pace of life brings social benefits to family, community, and individual well-being.

The Connection Between Productivity, Hours, and Ecological Footprint

Productivity growth is at the core of contemporary market economies. When productivity increases, it is possible to produce a larger quantity of goods and services, or output, with a given level of resources. Productivity can be measured in terms of natural resources such as land—how much crop yield is possible from a given acreage—as well as labor—how many automobiles or garments or computers a worker can produce in any given unit of time. When those measures rise (after taking due account of changes in natural "capital" or natural resource stocks), productivity has grown.

Growth in labor productivity creates a tremendous benefit. It becomes possible to produce a given level of goods and services in a shorter period of time, thereby giving workers more free time away from the job, or to produce more goods and services by keeping working hours constant. How a society manages that "choice," which all economies with productivity growth have, is crucial to achieving sustainability. If "too much" productivity

Juliet Schor is a professor of sociology at Boston College and the author of *Plenitude: The New Economics of True Wealth.*

growth goes into additional production, the eco-impact is too high. What constitutes "too much" varies over time, however, and partly depends on trends in technological impact and population. From the standpoint of climate change, for example, it is clear that the world has gone beyond what the planet can tolerate.

In the United States, it looks like "too much" productivity growth has been channeled into additional production. Since the early 1970s, labor productivity has roughly doubled. At that time, Americans worked on average about 1,700 hours a year. (That works out to a 32-hour workweek, as it includes part-timers and full-timers; full-time schedules were closer to the 40-hour norm.) Had Americans opted to put all the bounty of productivity growth into shorter hours, the average work year today would only be 850 hours, or just over 20 hours per week. Instead, the hours worked actually rose, and by 2006 the average schedule topped 1,880 hours a year. In addition, more people are in paid employment, as the United States is increasingly work and market-centered. In 1970, just 57.4 percent of the population was employed. In 2007, before the recession, the figure had risen to 63 percent.[1]

This experience is in stark contrast to earlier U.S. history. In the nineteenth century, hours were grueling, and it is estimated that people worked about 3,000 hours per year—a 60-hour workweek. Beginning in 1870, total hours began to fall, and they continued to fall for decades as a significant portion of productivity growth was used to create leisure time. By 1929, before the Great Depression, work hours had been reduced by more than 600, to 2,342. By the 1970s, at least another 400 hours had been taken off. That 1,000-hour total is the equivalent of half a job, assuming a 40-hour workweek and a 50-week work year. But for a number of reasons—having partly to do with the cost structures facing firms as well as the absence of union pressure to reduce hours—the trend of reduced work hours stalled in the United States in the 1970s.[2]

In contrast, West Europeans have commonly chosen to use productivity growth to reduce hours of work, with the result that average annual hours of work are much lower. Short schedules do not entail austerity: these are wealthy societies with plenty of material comforts. In case these differences seem deeply cultural or unbridgeable, it is worth remembering that 50 years ago the United States had much shorter working hours than Europe. Today many Europeans get six-week vacations, additional holidays, and daily work schedules that give them plenty of time for family life, leisure activities, and community participation. (See Figure 4.) Shorter hours are also more common in other parts of the world.[3]

This lifestyle is far easier on the planet. Studies of the relationship between working hours

Figure 4. Annual Hours of Work in Selected Countries, 2007

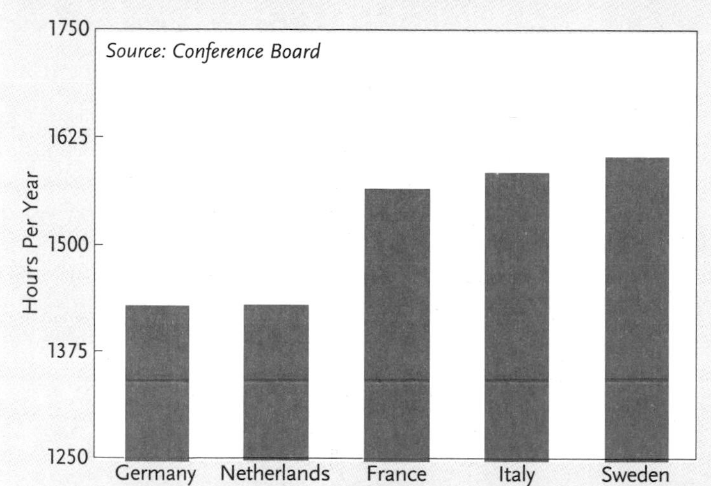

and ecological footprint find that as hours rise, so does the environmental impact. This relationship has shown up at the household level, where people who downshift their hours are found to have lower ecological footprints. It is also true across nations. Countries with shorter average working hours have smaller footprints, even controlling for income and other factors.[4]

This is true for several reasons. Most important, long hours typically occur when productivity growth is being channeled into production and consumption, which means more environmental degradation. A second effect is the energy usage associated with commuting. A third is that people who are "time-poor" (that is, they work long hours) tend to make lifestyle choices that are more resource-intensive. Their travel is more carbon-intensive. They eat out more often. In one study, they were found to have larger houses, which in turn used more energy. Time stress also limits engagement in low-impact, time-consuming activities, such as vegetable gardening or Do-It-Yourself projects. A study by the Center for Economic and Policy Research found that if the United States were to shift to West European patterns of time use, energy use there could decline by 20 percent even without changes in technology.[5]

There are also human benefits to working less. Long hours of work are stressful, undermine family functioning and social connections, and cause physical and emotional illnesses. Overworked employees are more likely to be depressed, more likely to experience stress, and less likely to take care of themselves. Excessive work hours also reduce sleep, which in turn erodes health. People who work too much are unable to engage in other activities, primarily social ones, that improve their well-being. And finally, the additional money earned by working more hours yields less benefit than people expect. A growing mountain of research shows that more income has a huge beneficial effect on people in poverty. But once a middle-class income is attained, the additional well-being available from increased income is surprisingly limited.[6]

Library of Congress

A lantern slide of the original Ford Motor Company assembly line in Detroit, Michigan.

The View from Business

Reductions in working hours may be better for people and the planet, but can businesses thrive in an environment of schedule shortening? The historical record suggests yes. The United States and Western Europe have both gone through long periods when hours of work were in decline and economic performance and profits were robust. Starting in 1870, a good portion of productivity growth went to giving people more leisure time, as the grueling schedules of the Industrial Revolution gave way to pressures from the 10- and 8-hour workday movements, the establishment of Sunday and then Saturday as a day of rest, and the emergence of the modern vacation. Far from undermining economic performance, shorter hours were an integral part of creating strong and profitable economies with healthy middle classes.[7]

Robert Scoble

A Seagate hard drive factory in Wuxi, China.

A second vantage point is competitiveness, and here the issue is not how many hours each individual person spends on the job but how productively those hours are worked and how they are compensated. If shorter hours come courtesy of productivity growth, that is a trade-off of income for time, and it can be cost-neutral. Across nations, similarly competitive countries have significant divergences in hours of work. Shorter hours can enhance productivity as work intensity rises. Better schedules reduce employee stress and improve retention and morale. Shorter hours can also reduce joblessness, which is now at crisis levels and rising in many places.[8]

In the United States, the major obstacle to hours reduction has been that health insurance is paid per employee, which means it costs employers much less to hire fewer people and work them longer. If there were a single-payer health care system, or even if businesses pro-rated medical and other benefits and government helped finance the remainder, shorter hours would be much more cost-effective.

The Road to "Time Affluence"

So if reducing work time is better for the planet, and better for people, shouldn't society be moving in that direction? Millions of people have already come to that conclusion. For more than a decade, a significant fraction of the American population has been making voluntary lifestyle changes that give them more time off the job. They are shifting to part-time, opting out of paid employment altogether, or changing to positions with less demanding schedules. This "downshifting" trend has helped to ease the extreme stress that characterized U.S. culture in the 1990s and is part of the reason that the escalation of annual hours slowed after its rapid increase in the 1980s and early 1990s. A subset within the downshifting group has taken the lifestyle change farther—embracing voluntary simplicity, a way of living that requires little income and is therefore usually associated with short hours of paid work.[9]

Downshifters report high levels of satisfaction with their new lifestyles, even those who have absorbed significant income reductions. A 2004 national survey by the Center for a New American Dream found that 85 percent of people who reported making lifestyle changes that reduced their incomes were happy about the change.[10]

Change is also happening at a more systemic level. Employers in some of the most demanding professions have made it possible to maintain successful careers even working fewer hours than the norm. Flexible arrangements have become more common in law, medicine, and academia, although there are still career penalties, and short hours are less common at the pinnacle of those fields. The changes have been the most far-reaching in accountancy. Since the 1990s all the large multinational firms instituted major family-friendly schedules, including fewer days per week, in a bid to retain high-productivity female talent.[11]

In the aftermath of the financial collapse of 2008, reductions in hours of work have spread throughout the private, public, and nonprofit sectors. Employers have attempted to avoid layoffs by instituting company-wide cutbacks in schedules, furloughs, and other work reduction measures. This ethic of sharing work has not been widely seen in the United States since the 1930s. Since the recession began, average weekly hours in the private economy have fallen by nearly an hour.[12]

Surveys of large employers show that reducing workweeks and mandating furloughs or unpaid work time have become widespread. A Hewitt Associates study of 518 large companies found that 20 percent cut hours. A Towers Perrin study recorded even higher numbers: 40 percent reported they had instituted a furlough and 32 percent, a shorter workweek. High-tech employers in the Pacific Northwest, such as Hewlett-Packard, Siltronic, and Tektronic, have reduced hours and pay (but usually not benefits).[13]

State and local governments have also been changing schedules in order to cut costs. The best-known case is the state of Utah, which switched 17,000 employees to a four-day, 10-hour schedule. Although not technically a cut in hours of work, it has allowed employees to reduce their commuting time. The change allowed the state to close offices on Fridays, and resulted in a 13-percent reduction in the state's energy costs and a decline in greenhouse gas emissions. Absenteeism and overtime also fell. Employees have been overwhelmingly positive about the change, as 82 percent reportedly want to maintain the compressed workweek even when the recession is over.[14]

Other states and cities have instituted furloughs and unpaid leave programs. The city of Atlanta has closed many of its services on Fridays; California has mandated unpaid days off. At the University of California, furloughs of 11–26 days have been introduced. If past recessions are a guide, many workers—particularly those who get a three-day weekend—will adjust to the lower incomes and decide not to resume a five-day schedule.[15]

Looking forward, it is increasingly clear that work-time reduction should be high on the sustainability agenda. This will require some policy changes in the United States, especially with health care, to alter the incentive structure facing businesses. It will require some cultural flexibility, to make sure busy-ness and long hours of work are not a status symbol. And consumption-driven competitions will need to be dampened. But if these challenges can be met, the result will be a slower, saner pace of life that is good for people and the planet.

Changing Business Cultures from Within

Ray Anderson, Mona Amodeo, and Jim Hartzfeld

The current Industrial Age was born out of the Enlightenment and the unfolding understanding of humanity's ability to tap the power and expansiveness of nature. The mindset that was developed early in the Age was well adapted to its time, when there were relatively few people and nature seemed limitless. Unfortunately, this mindset is poorly adapted to the current reality of nearly 7 billion people and badly stressed ecosystems. A new, better-adapted worldview and global economy are being born today from a greater understanding of how to thrive within the frail limits of nature.

Vital to the transition of the economy is the very institution that serves as its primary engine: business and industry. To lead this shift, business must delve much deeper than just the array of eco or clean technologies that are in vogue, to the core beliefs that drive actions. While a few visionary companies have been founded on the principles of sustainability, most businesses will require radical change. In the coming decades, business models and mindsets must be fundamentally transformed to sustain companies' value to their customers, shareholders, and other stakeholders.

More and more organizations are turning to sustainability as a source of competitive advantage. Yet many companies are trapped and frustrated by their limited understanding of this challenge; many see it only as a set of technical problems to solve or a clever marketing campaign to organize. Perhaps the greatest danger is that these superficial approaches give companies a false sense of progress, which in the long run will very likely lead to their demise.

On the other hand, businesses that are willing to address change at the deeper cultural level have the opportunity to embrace a new paradigm built on the values of sustainability. Those willing to lead the way will reap the "first-mover" benefits, while supporting and accelerating the fundamental societal shifts that are becoming increasingly apparent. Every company's sustainability journey will be unique, but a basic road map, using what has been learned from pioneering companies and researchers, can help those that are interested in the journey to travel at a faster pace.[1]

Ray Anderson is founder and chairman of Interface, Inc. **Mona Amodeo** is president of idgroup, a consulting and creative firm on branding, organization change, and sustainability. **Jim Hartzfeld** is founder and managing director of InterfaceRAISE.

The Need for Transformational Change

At the societal, business, and personal levels, the understanding and adoption of sustainability practices is limited less by technical innovation than by people's inability to challenge outdated mindsets and change cultural norms. Paraphrasing Edwin Land, physicist Amory Lovins has observed that "invention is the sudden cessation of stupidity...[that is,] that people who seem to have had a new idea often have just stopped having an old idea."[2]

A company's rate of adoption of new ideas, and therefore business opportunities, can be increased significantly by understanding the stages of change and the strategic decisions needed to support the evolving belief systems necessary for culture change. Personal change of this magnitude rarely occurs overnight, and changing an organization is often an even longer process.

Much can be learned from businesses that have moved beyond surface-level change to fully embrace sustainability and in doing so have created deep changes within their organizational culture. Experience suggests that sustainability derives its greatest power and effect in organizations when it is deeply embraced as a set of core values that genuinely integrate economic prosperity, environmental stewardship, and social responsibility: profit, planet, and people.[3]

To achieve this degree of change, leaders must put forth bold visions—so bold that they take the breath away—and they must engage their organizations in different, deeper conversations about the purpose and responsibility of business to provide true value to both customers and society. Moreover, the whole enterprise must be proactively engaged in such a systemwide way that mental models become explicit, multiple stakeholder perspectives are incorporated into the process, and collective interaction yields new knowledge, structures, processes, practices, and stories that can drive the organization forward.

When organizations embrace sustainability in this way, it is fully woven into every facet of the enterprise. Sustainability becomes definitional, revealing itself in every decision—a strategic and emotional journey that enhances the entire enterprise. After all, can anyone really make "green" products in a "brown" company?[4]

A Framework for Culture Change

The U.S.-based global carpet manufacturer Interface, Inc. offers a valuable case study of a company that has embraced and achieved transformational change toward sustainability. Interface reports being only about 60 percent of the way toward achieving its Mission Zero 2020 goals, but the company has come far in its 15-year journey to sustainability. It has reduced net greenhouse gas emissions by 71 percent, water intensity by 74 percent, landfill waste by 67 percent, and total energy intensity by 44 percent. It has diverted 175 million pounds of old carpet from landfills, invented new carpet recycling technology, and sold 83 square kilometers of third-party certified, climate-neutral carpet. In the process, Interface has generated substantial business value in its brand and reputation, cost savings of $405 million, attraction and alignment of talent, and industry-leading product innovation.[5]

Interface's sustainability leadership has been recognized internationally in multiple Globescan surveys of "global sustainability experts," receiving the number one ranking in 2009. But the company's transition was not choreographed in advance. During the first decade of the journey, Interface went through five developmental phases of change, driven by key levers that propelled its progress. (See Figure 5.) Deep changes in the identity, values, and assumptions about "how we do things here" moved the company to a new view of purpose, performance, and profitability within the larger con-

Figure 5. Culture Change Model

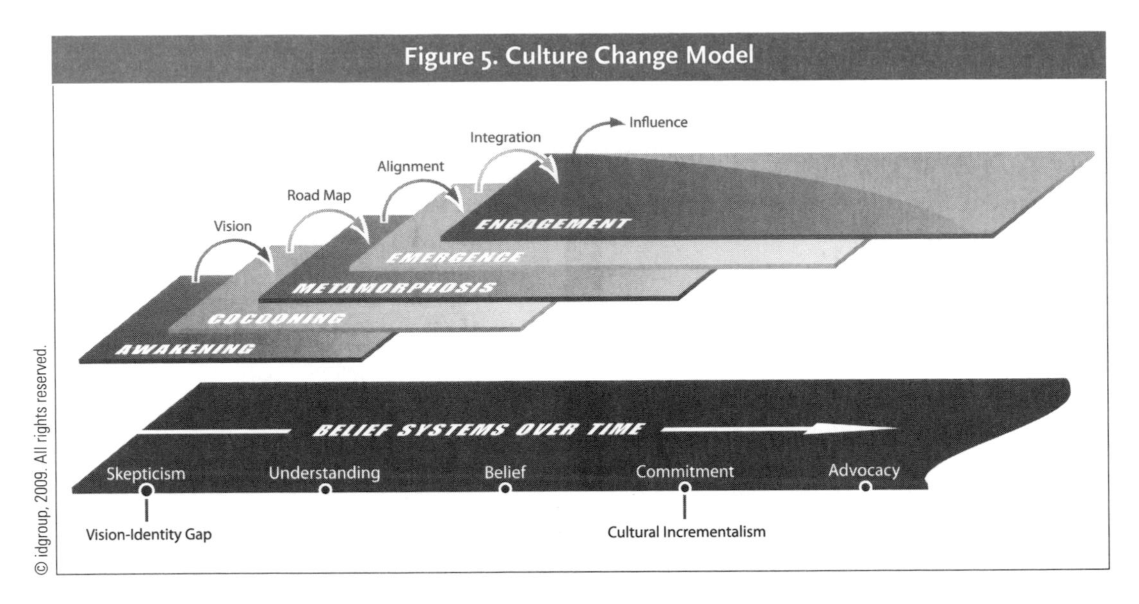

text of environmental and social responsibility.[6]

The Interface model of cultural change represents a journey of the head and heart, facilitated by strategic decisions and deepening connections to the values of sustainability. These interacting factors closed the initial gap between the vision—a future sustainable company—and the actual unsustainable existing company, by incrementally transforming the culture through successive phases along a time continuum. As the company went through the phases of transformative change (awakening, cocooning, metamorphosis, emergence, and engagement), an evolution of belief systems also occurred, moving from skepticism to understanding, belief, commitment, and advocacy. This psychological progression worked in tandem with strategic decisions (vision, road map, alignment, integration, and influence) to create deep culture change.

Over time, the transformation can be envisioned as a dynamic process where new and ongoing connections, relationships, and networks come into being and flourish through an infusion of knowledge, wisdom, and grassroots experience. Early skepticism gives way to understanding as an organization confirms the validity of the values of sustainability, which in time comes through successes actually experienced. As the collective identity of the organization changes, new behaviors associated with these values are reinforced and become more embedded in the culture. Understanding is augmented by belief and commitment.

New ways of thinking, believing, and doing emerge incrementally as strategic decisions are confirmed, and sustainability becomes fully embraced as "the way we do things around here." This shifting paradigm produces innovations in technologies, sustainable business practices, and new leadership capacity, as well as a sense of pride, purpose, and commitment on the part of those associated with the organization. Externally, the organization realizes increasingly strong connections and levels of trust with its marketplace.

The Stages of Change

Awakening: defining the vision. To allow change to occur, a company must first be open to sensing and considering aberrant signals that may suggest or uncover new challenges or opportunities. The source of the signal can be internal or external, subtle or cacophonous. Likewise, a company's awareness of the need to address sustainability can be stimulated in many ways, including inspired leadership, a grassroots internal uprising, a technical or physical challenge, or an unanticipated shock in the cost or availability of key resource inputs. At some point the magnitude of the emerging risks or opportunities become "real" enough to cause the organization to begin to seek more information and direction.

At Interface, the persistent and aggressive voice of a single customer caught the ear of the founder, Ray Anderson. At Walmart, the impetus was inspired leadership stimulated by a barrage of external challenges on multiple fronts. At Nike, it was the outrage sparked by a 1996 *LIFE* magazine article about child labor in Pakistan, which featured a photo of a 12-year-old boy surrounded by Nike-brand soccer balls he had been stitching. Other examples of external stimuli for increased sustainability awareness include Greenpeace's pressure on Electrolux and the Rainforest Action Network's pressure on Mitsubishi.[7]

Once a general direction is suggested, a small group of innovators or "scouts" may explore the magnitude of the problem and what it means to the organization and then propose a potential vision of the future. During this stage, it is important to suspend skepticism and engage the top leadership in a deep and honest exploration of the facets of sustainability—what it means to each person as well as to the organization. Investing the time, energy, and effort in individual and organizational reflection will establish the necessary tension to propel change and determine the level of commitment needed to move forward.

A natural sense of curiosity and the persistent ability to resist the pressures of the dominant paradigms (and existing structures) is important to allow new and unusual signals to penetrate and to overcome the natural response of defending the status quo. At this point, the leadership makes a go/no-go decision. A clear vision is created, and the process of expanding the engagement of others in the organization begins—with the leadership acting as the messenger, evangelist, teacher, and cheerleader.

At Interface, Ray Anderson was inspired to declare his vision of sustainability for the company after reading the groundbreaking 1993 book *The Ecology of Commerce*, which proposed a culture of business in which the natural world is allowed to flourish. Jeff Mezger, CEO of U.S. home construction company KB Home, recently directed his leadership team to explore what goals and commitments they should make toward sustainability, even in the teeth of the industry's historic downturn. In July 2008, he communicated this vision in the company's first sustainability report.[8]

At Walmart, CEO Lee Scott and members of his leadership team took a year to personally explore, challenge, read, and tour settings around the world, from the ecologically crippled state of Montana described in Jared Diamond's book *Collapse* to cotton fields in Turkey and an Interface carpet mill in Georgia. Only after that year of exploration did Scott announce the company's direction in a landmark speech in October 2005, "21st Century Leadership." Even while stating ambitious goals for Walmart—"to be supplied by 100 percent renewable energy, to create zero waste, and to sell products that sustain our resources and environment"—Scott admitted that he was not sure how to achieve them.[9]

Cocooning: creating the road map. With a vision defined, a company must then determine how to translate the vision into action. In addition to deeper planning and early prototypes, the learning of the awakening phase is taken deeper and shared more widely across the organization and beyond. The result is a road map of action that normally includes goals, timelines, resource allocation, and—most important—metrics.

During this stage, the company is engaged

in activities that further "awaken" people in the organization to sustainability—the problems, challenges, and opportunities—with the view that people will typically only defend and support that which they help build and create. Frequently, an organization's "whole" cannot be changed until the collective is assembled to work together to shape a new potential future. It is important at this point to tap into the organization's creative intelligence and its stakeholders through dialogue, collaborative inquiry, community building, and cutting-edge methods of change that support new ways of thinking and transforming.[10]

At Interface, Ray Anderson sought to simultaneously engage a wide range of his internal leadership team, who were already associated with the company's QUEST waste initiative, as well as the most visionary collection of external experts he could find, eventually named the EcoDream Team. Through an intense 18-month process, Interface's Seven Fronts initiative (later renamed Seven Faces of Mt. Sustainability) was identified and published in the company's first sustainability report in November 1996. The document laid out the sustainability challenge and proposed solutions in detail, with supporting metrics that outlined an extensive list of everything the company "takes, makes and wastes."

Nike, following the media storm created by the 1996 *LIFE* article, went rapidly into cocooning with its internal staff and external experts and commissioned various university studies—taking nearly two years to develop a code of conduct for labor and environmental practices. CEO Phil Knight unveiled the code at a widely publicized 1998 speech at the National Press Club in Washington, D.C.[11]

An extreme example of externally engaged cocooning is Walmart's Sustainable Value Networks and quarterly Milestone Meetings. Announced in December 2005, Walmart created 14 teams to address major product categories and cross-cutting issues such as waste, packaging, and transportation. A stunning facet of this period was the extent to which Walmart proactively engaged environmental groups and its suppliers. As evidence of the company's key words for the era, "collaborate and innovate," Walmart convened collective learning opportunities for the entire network. One early meeting included 800 attendees and began with Interface's "Global Village Exercise," where Ray Anderson and Jim Hartzfeld facilitated an interactive session highlighting global environmental and social conditions. In another meeting, Al Gore appeared at the screening of his film "An Inconvenient Truth," and U.S. evangelical leader Jim Ball spoke on the alignment of scripture and concern for the environment.[12]

Metamorphosis: aligning the organization. Once a well-defined road map and early prototypes are established, the hard process of driving widespread change in the company begins. As with a caterpillar's metamorphosis, the process will likely require the creative destruction of entrenched mindsets and processes. Sustainability cannot be a program confined to a specific grouping of staff. Instead, it must be broadly aligned, integrated, and institutionalized into corporate systems, structures, and processes.

This is a period of intense learning and experimentation. During this often messy time, it is important for the leadership to continually and consistently remind the organization of the vision, while at the same time meeting people where they are. Leaders should be prepared to support the push toward new innovations while maintaining a high tolerance for the associated risk of failure. Permission to fail is essential to empowering people to innovate at their best.[13]

Structures and programs that support organizational learning by rewarding and celebrating success will reinforce the organization's commitment and provide the motivation needed to keep people going. Providing the necessary resources, both financial and human, is of paramount importance. And while there can be great power in telling the sustainability story to internal and external audiences, it is also critical that the story be authentic—not to let the "talk" get in front of the "walk." Outspoken commitment serves as a strong reinforcing mechanism for organizational members—a source of pride and pressure. Incorporating the sustainability story into marketing communications programs also creates increased marketplace recognition, trust, and connection.

At Interface, this process extended to all functions and regions of the company, including cascading dialogue with employees about dominant corporate values, as well as incentives and rewards. At Walmart, "metamorphosis" began with the creation in 2007 of the Personal Sustainability Project, designed to eventually engage all 1.8 million employees by connecting the company's sustainability journey to the personal lives of its employees. Taking it one step further, Walmart created a supplier "packaging" score card that gave clear direction to its more than 60,000 suppliers that the company sought to engage everyone it was connected with, and not just the few early innovators, in its sustainability journey.[14]

During this stage, companies often falter after gathering the low-hanging fruit associated with technical changes. But the metamorphosis stage can also reveal the payoff of the "consciousness-raising" work done during the cocooning stage. If individuals in the organization move beyond understanding to belief, the organization will progress beyond minor improvements or adjustments that have little impact on the core of the organization. As a result, new innovations will begin to emerge as members begin to dismantle existing paradigms by asking new questions.

Emergence: ongoing integration. As the metamorphosis reaches critical mass, engaging more people and demonstrating success, the momentum is accelerated by the positive energy of the process. Early successes drive learning, which stirs further innovation. Good metrics inform positive feedback loops of learn, do, measure, recognize—reinforcing the values and belief systems. At some point, the company's identity must be fully invested in sustainability, and the associated beliefs and behaviors must become ingrained into the DNA, or cultural assumptions, of the organization. If this level of cultural integration is not achieved, the organization will never really achieve liftoff.

Engagement: influencing others. Even many years into a company's sustainability journey, engagement is a continuing effort. Each level of success reveals new questions and challenges. This ongoing search for answers spirals to new levels of understanding about what is possible. Relative to the model presented, the stages of the process are continuous and recursive with deeper learning and innovation at every new loop in the spiral.

As an organization becomes more committed to sustainability, educating and influencing others becomes an important part of the change process. This advocacy role is beneficial to both the company and to the larger societal cause. In addition to helping others along in their journey and building the company's image, additional learning and expanding knowledge come through collaboration and teaching others. Interface, for example, formed a consulting subsidiary, InterfaceRAISE, to help other companies move more quickly up the learning curve and through the phases of their journey. The company also developed an extensive speakers' bureau consisting of Interface associates for general public and business education.[15]

Conclusion

Business and society are in a period of crisis as well as potential. Doing the same things a little differently, better, or faster will not bring about the transformational changes needed to address today's challenges or grasp new opportunities. The Industrial Age can be supplanted by a new age of evolving human wisdom and emergent innovations, but only if businesses are willing to challenge existing paradigms and proactively discover new answers through collective inspiration. (See Box 13.)[16]

Business and industry—the most dominant institutions on the planet in both size and influence—can bring about organizational awakening that can catalyze more sweeping societal change. If business models are grounded in the values of sustainability, the people who work in those firms will also likely accept and adopt the behaviors associated with sustainability as the "way things are and should be." This offers business and industry a unique opportunity to accelerate the tipping point needed to correct society's current trajectory. To achieve this shift, companies must explore new worldviews and discard the old flawed views by encouraging personal reflection and new dialogue about the purpose and responsibility of business.

Box 13. Upgrading the Corporate Charter

Many U.S. businesses are redesigning their corporate charters to incorporate the interests of all stakeholders—customers, employees, communities, and the planet—rather than just those of their shareholders. Since 2007 the nonprofit organization B Labs has had a thoroughgoing certification process that identifies and validates precisely these types of businesses as B Corporations (the B stands for "benefit").

By expanding legal responsibility, B Corporation certification allows businesses to alleviate the pressure to pursue nothing but the exclusively profit-centered "bottom line." In addition, the designation helps to distinguish the corporations that are truly committed to socially valuable and environmentally sustainable practices from those just wanting to "greenwash" their operations. A B Corporation can also use the rigorous standard by which it is certified to monitor its own sustainability performance—a useful tool for companies that genuinely want to have a positive impact on society and the environment.

In order to be certified as a B Corporation, a company must submit responses to an extensive survey, which is then reviewed by B Labs. The company is subsequently audited in order to validate compliance with the B Ratings System. A minimum passing score of 80 indicates that the organization is eligible for certification, at which point it is obligated to submit a new corporate charter amended with the B Corps Legal Framework.

The B Corporation brand has already certified more than 190 companies spread across 31 industries with revenues totaling over $1 billion. Although its financial depth is admittedly a drop in the bucket compared with the roughly $14-trillion U.S. economy, this innovative tool could have lasting impact as corporations strive to reach B Corporation standards and, in so doing, acknowledge their increasing responsibility to pursue social and environmental benefits that extend beyond the traditional constraints of the profit motive.

—Kevin Green and Erik Assadourian

Source: See endnote 16.

Social Entrepreneurs: Innovating Toward Sustainability

Johanna Mair and Kate Ganly

In May 2009, U.S. President Barack Obama announced the creation of a $50-million Social Innovation Fund and a new White House Office that will coordinate the fund's efforts "to identify the most promising, results-oriented non-profit programs and expand their reach throughout the country." This commitment to supporting and nurturing a diverse range of decentralized alternative solutions to intractable social problems taps a wave of global popularity and public awareness that has been building around the phenomenon of "social entrepreneurship" for several years. Social entrepreneurs use a variety of organizational forms—from social businesses and cooperatives to leveraged nonprofits, hybrids, and pure charities. But they all have one thing in common: the innovative use and combination of resources to pursue opportunities to catalyze social change.[1]

Social entrepreneurial initiatives (SEIs) are influenced by local conditions both in the opportunities they have to address a social or environmental need and in the regulatory architecture that affects their form. In Europe, a dominant form of social entrepreneurship deals with work integration for marginalized groups such as migrants, youth, and the disabled. This has been encouraged by government support in France, Spain, and Portugal, where such initiatives are addressing the persistence of structural unemployment among particular groups. La Fageda, to cite just one example, is a dairy in Catalunya that has a cooperative of 250 employees, 140 of whom suffer from mental illness. In both Italy and the United Kingdom, specific legislation was introduced in 2005 to recognize and foster "social purpose ventures." For instance, the U.K. "community interest company" is a limited liability company designed to operate for community benefit: it has a cap on dividends and individual profits, which ensures that revenues and assets are retained for community purposes.[2]

A Growing Movement

Social entrepreneurs existed long before they were labeled as such. Since the Grameen Bank and its founder Muhammad Yunus were jointly awarded the Nobel Peace Prize in 2006, however, media coverage of this growing phenomenon and accompanying accolades have made social entrepreneurs highly visible.

Johanna Mair is a professor of strategic management at IESE Business School. **Kate Ganly** is a research affiliate of the IESE Platform for Strategy and Sustainability.

Grameen provided an early model of an SEI when in the late 1970s it started offering credit to the poorest of the poor in rural Bangladesh without the borrowers needing to provide collateral for their loans. The Bank's micro-credit program expanded rapidly, and by mid-2009 nearly 8 million people were receiving loans, 97 percent of whom were women.[3]

While definitions vary, social entrepreneurship can generally be seen as a label for initiatives that proactively address social or environmental issues through delivery of a product or service that directly or indirectly catalyzes social change. To ensure that change is sustainable, a large part of what social entrepreneurs do is challenge or disrupt existing institutions. As used here, the term institutions includes taken-for-granted collective behaviors such as consumption that dominate daily routines. Excessive consumption, environmentally unsustainable practices, and a culture of individual private gain over shared community or public benefit are just some of the institutionalized behaviors that social entrepreneurs seek to change. Often these goals are tied up with other, more specific aims.

Reliable comparative data on SEIs are hard to come by, primarily because countries define and recognize social entrepreneurship differently. Italy first created a legal form for "social cooperatives" in 1991; by 2001 there were approximately 7,000 such organizations employing 200,000 workers and benefiting 1.5 million people. As mentioned, the United Kingdom has also championed SEIs: its 2005 Annual Survey of Small Businesses found that 55,000 social enterprises existed with a combined turnover of £27 billion, contributing £8.4 billion per year to the U.K. economy. The United Kingdom is also one of the few countries to measure social entrepreneurial activity as part of the annual Global Entrepreneurship Monitor. Data from 2006 indicate that 3.3 percent of the U.K. population was involved in creating or running an early-stage SEI, while another 1.5 percent ran an established SEI. This represents a significant chunk of the population compared with the figure for mainstream early-stage entrepreneurship at 5.8 percent. In Japan, where a legal form for nonprofits was introduced in 1999, the number of SEIs grew from 1,176 in that year to over 30,000 in 2008. This sector contributed approximately 10 trillion yen to the economy in 2005, accounting for 1.5 percent of Japan's gross domestic product.[4]

The origin of the phrase "social entrepreneur" can be traced to Bill Drayton, a former business management consultant who in 1980 set up Ashoka, the first foundation to support and fund such individuals. Today Ashoka has over 2,000 "fellows" in more than

Courtesy of Waste Concern

An illustration from a Waste Concern poster promoting rural waste composting technology.

60 countries and continues to expand. Other important global support organizations include the Schwab Foundation, which invites fellows to attend the World Economic Forum in Davos, and the Skoll Foundation, which also holds an annual world forum. The latter was set up by eBay founder Jeff Skoll, one of a group of high-profile "new philanthropists" funding SEIs—a group that includes Bill Gates and George Soros.[5]

Indeed, the number of venture philanthropy organizations and investment vehicles is rapidly increasing. They include everything from the Acumen Fund (launched in 2001 with seed capital from the Rockefeller Foundation and Cisco Systems), which now has hundreds of investor partners—from companies to individuals—to online platforms such as Global Giving, which lets individuals invest in small projects of selected social change organizations located anywhere in the world.[6]

The popularity of social entrepreneurship is also apparent in academia, as evidenced by the growing number of research centers, publications, international conferences, dedicated faculty appointments, and student demand for courses. But perhaps the biggest boost for social entrepreneurship has been endorsement from celebrities, business leaders, and political leaders such as President Obama. This kind of support has stimulated popular interest and generated broad exposure for social entrepreneurship, setting it well on the path to become a defining trend of the twenty-first century.

Challenging What Is Taken for Granted

One of the most powerful ways that social entrepreneurs are able to bring about change is by challenging accepted ways of doing things and demonstrating alternatives. (See Box 14 on recent challenges to design principles.) In Egypt, for example, the SEI Sekem challenged the automatic acceptance that desert land far from the Nile could not be made fertile, and it overturned conventional thinking about the necessity of chemical pesticides. Founded by Ibrahim Abouleish in 1977 with the intention to "heal the land and its people," today Sekem is a multi-business company with more than 2,000 employees; it encompasses seven for-profit companies producing organic food products, cotton, textiles, and medicinal herbs and includes a range of nonprofit entities—from education and health facilities for its staff and their families to a research and development institute and a university.[7]

Similarly, Waste Concern in Bangladesh proved that Dhaka's waste problem could be turned into a resource by taking a radical new approach to waste processing and collection. The founders set up small-scale composting plants that employed waste-pickers to collect and process the compost. Instead of burning or flaring solid waste, they created fertilizer from organic and enriched compost, which reduced pollution while creating jobs.[8]

In Thailand, the Population and Community Development Association (PDA) challenged traditional attitudes to sex and contraception. In addition to training rural women to sell birth control pills and condoms, PDA used humor—such as through the creation of a restaurant chain called "Cabbages and Condoms" and "Miss Condom" competitions in Bangkok's notorious red-light district—to create a proactive awareness to help limit an exploding population and, later, to halt the spread of HIV.[9]

An important contribution of social entrepreneurs that is related to challenging what is taken for granted involves demonstrating "proof of concept"—that is, showing how new approaches and ideas can actually work. SEIs often create new markets, opening up a space for customers and competitors and fostering supply and demand. In this respect, social entrepreneurs are path breakers, paving

Box 14. Cradle to Cradle: Adapting Production to Nature's Model

Many of today's business strategies fall short of a model that truly sustains planetary systems. Instead, most responses to these challenges seek to limit the impact of human activity by minimizing pollution and waste—focusing on being more "eco-efficient" instead of being "eco-effective."

But there is another way. We often say that design is the first signal of human intention, which raises the question: what are our intentions? Do we intend to create things that have only positive effects? Or just fewer negative ones?

Imagine buildings, neighborhoods, transportation systems, factories, and parks all designed to enhance economic, environmental, and social health—that reach beyond sustainability to enrich lives. To help realize this vision, production can be based on three key operating principles of the natural world that allow business to apply the intelligence of natural systems to human artifice.

Waste equals food. In nature, the processes of every organism contribute to the health of the whole. One organism's waste becomes food for another, and nutrients flow perpetually in regenerative, cradle-to-cradle cycles of birth, death, decay, and rebirth. Design modeled on these virtuous cycles eliminates the very concept of waste: products and materials can be designed of components that return either to soil as a nutrient or to industry for remanufacture at the same or even a higher level of quality.

Use current solar income. Nature's cradle-to-cradle cycles are powered by the energy of the sun. Trees and plants manufacture food from sunlight—an elegant, effective system that uses Earth's only perpetual source of energy income. The wind, a thermal flow fueled by sunlight, can be tapped and along with direct solar collection can generate enough power to meet the energy needs of entire cities, regions, and nations. Developing wind and solar power transforms the energy infrastructure, reconnects rural areas to cities through the cooperative exchange of energy and technology, and can one day end the reliance on fossil fuels.

Celebrate diversity. Healthy ecosystems are complex communities of living things, each of which has developed a unique response to its surroundings that works in concert with other organisms to sustain the system. Each organism fits in its place, and in each system the most fitting survive. Abundant diversity is the source of an ecosystem's strength and resilience. Businesses can celebrate the diversity of regional landscapes and cultures and grow ever more effective as they do so.

With these three principles in mind, businesses participate ever more creatively with nature. They harvest the energy of the sun and capture rain. Food and materials grown in the countryside, using implements and technology created in the city, are absorbed by the urban body and returned to their source as a form of waste that can replenish the system. Thus, human settlements and the natural world flourish side by side.

The goal of cradle-to-cradle design is a delightfully diverse, safe, healthy, and just world, with clean air, water, soil, and power—economically, equitably, ecologically, and elegantly enjoyed. In the end, the success of our efforts will be measured against how we have answered what we have found to be the fundamental question: how do we love all the children, of all species, for all time?

—William McDonough and Michael Braungart
McDonough Braungart Design Chemistry

Source: See endnote 7.

the way toward a more sustainable and humane future.

Sekem, for example, pioneered organic agriculture in Egypt and demonstrated that cotton, a major crop, could be successfully grown without pesticides—an innovation later instituted by the Egyptian government, thereby eliminating the spraying of 30,000 tons of chemicals annually. In Bangladesh, Waste Concern developed a method of organic composting that produced a rich fertilizer and applied it to the vast problem of Dhaka's solid waste buildup. Yet in developing a solution to one problem the founders managed to address another: the issue of Bangladesh's soil degradation due to the overuse of chemical fertilizers. Not only did Waste Concern's actions create a market for organic fertilizer, they led the company to become a leader in carbon trading through the Clean Development Mechanism set up under the Kyoto Protocol and a role model for U.N. projects. (See Box 15 for another innovation on carbon regulation.)[10]

While innovations in technology, energy, and industry are important, it is the more difficult and elusive collective changes in behavior and thinking that may have the biggest impact in the transition to sustainability. It is important to understand that this is an interconnected and globalized world, but that real and sustainable behavioral change often happens locally and painstakingly slowly. Social entrepreneurs have an important role to play in initiating such changes by challenging the taken-for-granted assumptions and the institutionalized behaviors that contribute to maintaining the status quo.

SEIs that specifically address the issue of conscious consumption are being seen more often. One example is the fair trade movement. Small handicraft fair trade outfits have existed in the United States and the United Kingdom for more than 50 years, but it has been SEIs such as Transfair USA, founded in 1998, that helped establish fair trade labels for a much wider range of products. And Rugmark, founded in 1994, combines a campaign to end child labor with certification for ethically produced rugs. These groups and the many SEIs promoting and supporting organic coffee producers, poor country artisans, and the like have made social entrepreneurship a global commercial phenomenon. These organizations are helping people question what, why, and how they consume and consider the repercussions of their collective actions.[11]

These and other initiatives that recognize a global need to source products in ways that sustain communities and the environment have often been initiated and driven forward by SEIs in the West, but they are now spreading to the East and global South. While the governments and indeed large companies in many affluent countries have begun to respond to this need, it has remained a gap in the system to be exploited by social entrepreneurs in many other places. In Latin America there is a new wave of initiatives mobilizing consumers to use their purchasing power to influence business practices for more responsible consumption. El Poder de Consumidor in Mexico, Interrupcion in Argentina, and the Akatu Institute for Conscious Consumption in Brazil are just some of these. Poland experienced the consequences of rapidly advancing consumerism after the fall of socialism: massive amounts of waste and terrible pollution were problems that people had no models for dealing with as they were used to such issues being addressed by a central authority. Several Polish social entrepreneurs sprang up to deal with this and other specific problems caused by the transition.[12]

Local Efforts Have Global Impacts

Although most SEIs initially develop in response to quite local issues, today the repercussions of their actions cannot be isolated because they are linked globally. One of the

Box 15. A Carbon Index for the Financial Market

The World Federation of Exchanges reports that in 2008, more than $113 trillion in stocks, futures, and options was traded on its 51 publicly regulated exchanges. The 46,000 or so listed companies had a total market capitalization of more than $33 trillion. Meanwhile, the world derivatives market—including both over-the-counter and exchange-traded derivatives—has been estimated at some $791 trillion, 11 times the size of the world economy.

Most of the world's financial capital is traded with no carbon regulation, causing a "free flow" of carbon dioxide into the global economy. Shares, or units of ownership in a corporation, can propel or mitigate greenhouse gas emissions. Adoption of a Carbon Index for the stock market—and for financial markets as a whole—would broaden the transparency of the global finance system, disclose the carbon footprints of corporations and investors, and create a new platform for decarbonization in financial markets, aligning the financial industry with the low-carbon economy. A complementary DCarb Index could measure the level of decarbonization, shaping standards for low-carbon financial flows.

Positive signs of change are emerging in the exchange markets. The Dow Jones Sustainability Indexes, launched in 1999, track the financial performance of leading sustainability-driven companies worldwide, providing objective benchmarks for managing sustainability portfolios. And in June 2009, NASDAQ OMX Group, Inc. and CRD Analytics introduced a Global Sustainability 50 Index that enables investors to track the top 50 companies in sustainability reporting—disclosing information such as their carbon footprints and workforce diversity.

In March 2009, Standard & Poor's introduced the S&P U.S. Carbon Efficient Index, a subset of companies listed on the S&P 500 that have a relatively low carbon footprint (calculated as annual emissions divided by revenue). According to Standard & Poor's, the average annual carbon footprint of companies listed on the index through 2008 was 48 percent lower than that of the S&P 500.

To provide guidance for low-carbon policy decisions, the U.S. Environmental Protection Agency (EPA) has proposed mandatory reporting of greenhouse gas emissions from large sources in the United States. Suppliers of fossil fuels or of industrial greenhouse gases, manufacturers of vehicles and engines, and facilities that release 25,000 tons or more per year of emissions would need to submit annual reports to the EPA. Compiled, this information would inform investors of both "high" and "low" carbon tendencies by company or sector, orienting large quantities of capital toward sustainability.

Expanded more widely, the use of Carbon Indexes could lead to greater protection of the economy's natural support systems. For example, development of an Amazon STOXX Index, based on the Dow Jones STOXX Index, could help build investment knowledge for profitable eco-oriented businesses to conserve the world's largest tropical forest. Brazil's BM&FBOVESPA, the second largest exchange operator in the Americas by market value, has the opportunity to support these low-carbon businesses—attracting investors and promoting economically, socially, and environmentally integrated profits.

With such initiatives, the "low-carb" market, a symbol of the new eco-economy, can compete with high-carbon initiatives, stimulating greener investments. Because of its clout, the global financial market is one of the strongest and most flexible tools to build a low-carbon, sustainable economy.

—Eduardo Athayde
Worldwatch Institute publisher, Brazil
Source: See endnote 10.

strongest links is financial: the amount of venture philanthropy money available in North America, Europe, and Japan to be invested in poorer parts of the world is large and growing. The World Bank Institute, for example, estimated that private net capital flows to developing countries in 2007 totaled $590 billion.[13]

Social entrepreneurs are setting trends and sparking movements that are spreading across the world. These could have far-reaching effects in different locations and future scenarios. Efforts of SEIs in industrial countries to help people consume less, use energy more efficiently, and limit environmental damage could provide valuable lessons for developing countries with burgeoning consumer classes, massive urbanization, and potentially huge environmental problems. At the same time, innovative and low-cost responses to the lack of resources at the grassroots in developing countries are providing appropriate technology solutions (such as solar lighting for villages that have never been electrified or biogas plants using cow or pig manure) that may be valuable in industrial countries still battling consumption-related problems.

What is most important about social entrepreneurs, wherever they operate, is that they challenge existing rules and institutions and create innovative vehicles to achieve their social goals. These may end up directly provoking markets through competition or providing alternatives, or they may indirectly put pressure on industries by creating awareness and stimulating behavioral and attitude change. Achieving this kind of change is a long and bumpy road, but one of the most distinctive characteristics of social entrepreneurs is persistence. The challenge remains extending the adoption of these ideas across both the public and the private sector and throughout society so that they do not become isolated efforts but penetrate all economic, social, and political domains.

Courtesy of Dawn Starin

Art created with out-of-date condoms and birth control pills raises awareness at the Bangkok restaurant Cabbages and Condoms.

Recent events have highlighted the need to create a balance between economic growth—which is irrevocably tied to enrichment and consumption but also to a better quality of life and human development—and an approach to markets and governance that is based on ethical needs and that recognizes global interlinkages and inequalities. The good news is that the momentum for social entrepreneurship has never been greater and the timing never better to shock the world into collective cultural change.

Relocalizing Business

Michael H. Shuman

To see what a "culture of sustainability" might really look like, pay a visit to Bellingham, Washington, recently named by the Natural Resources Defense Council as the #1 "Smarter" small city in the United States. This coastal town two hours north of Seattle has pioneered an economic development strategy that is radically different from the traditional preoccupation with attracting and retaining global businesses. Thanks to the leadership of a nonprofit called Sustainable Connections, Bellingham has focused on nurturing its local businesses and organizing them into a powerful collaborative network to rebuild the community economy from the ground up.[1]

Here is some of what Sustainable Connections has accomplished in less than a decade. Its Local First campaign—now widely copied around the United States and Canada—uses festivals, store signs, posters, advertisements, and coupon books to motivate residents to buy local. An independent survey by Applied Research Northwest found that 69 percent of Bellingham consumers are now paying attention to the local character of businesses, 58 percent have begun localizing their purchasing habits, and business proprietors regard Local First as one of the most compelling reasons they are thriving. Sustainable Connection's energy program has mobilized 1 in 10 residents to buy local "green power"—the second highest percentage in the United States. The number of farmers in surrounding Whatcom County marketing directly to consumers increased 44 percent between 2002 and 2007, twice the state-wide rate. The value of direct sales—a key strategy for boosting farmers' income—has increased 125 percent over the same period, quintuple the state rate.[2]

Bellingham is among a growing number of communities worldwide that see their future sustainability and prosperity grounded in local businesses. The Business Alliance for Local Living Economies (BALLE) has more than 70 member communities in North America. Another 50 or so communities are affiliated with the American Independent Business Alliance. Internationally, more than a thousand communities are beginning to undertake similar work through organizations like Transition Towns and Post-Carbon Futures.[3]

As these organizations see it, local business has two meanings. One is ownership. In a locally owned business, more than half the

Michael Shuman is director of research and public policy for the Business Alliance for Local Living Economies.

owners live where the firm operates. By this definition, local ownership actually characterizes the vast majority of sole proprietorships, partnerships, nonprofits, cooperatives, and public-private partnerships operating in the world. Even most privately held corporations are local. Really, the only kind of business clearly not local is a publicly traded company. The other meaning of local is the proximity of its stakeholders, like suppliers and consumers. Because locally owned businesses tend to give priority to using local labor, land, and capital and producing goods and services for local markets, these two concepts are inherently intertwined.

In an era of globalization, it is easy to forget that local businesses actually have been the economic norm for most of human history and, contrary to public perceptions, continue to account for most of the world's economy today. One distinguishing feature of very poor countries is that a large percentage of the population is engaged in subsistence agriculture—that is, local farming. As countries develop, farm families migrate to the cities for industrial jobs. But vast numbers remain jobless or underemployed and effectively wind up as microentrepreneurs in the informal sector. Even in an advanced industrial economy like the United States, roughly half the economy in terms of jobs and output comes from self-employed individuals or from small or medium-sized enterprises, nearly all of which are locally owned.[4]

So localization is neither new nor uncommon. But awareness of its potential power in promoting sustainability and prosperity is.

Localization and Sustainability

For a generation, "sustainability" has been defined as meeting this generation's needs without compromising the ability of future generations to meet their own needs. There is a growing appreciation, however, that this definition can be improved with a more nuanced understanding of place: a community should meet its current needs, present or future, without compromising the ability to meet the needs of future generations living in other communities, present or future. This new definition highlights the importance of every community maximizing its level of self-reliance, presumably through a diverse assortment of businesses behaving in a sustainable fashion. Localization, of course, does not guarantee sustainable behavior, but it increases its likelihood in at least four ways.[5]

First, an economy highly dependent on non-local businesses must continually make sustainability compromises to prevent its most important firms from exiting. For example, the state of Maryland is highly dependent on a poultry industry (dominated by two companies, Tyson and Perdue) that continually threatens to move to more "business-friendly" jurisdictions like Arkansas and Mississippi. Despite its impressive performance in other categories of sustainability like smart growth, the state has found it politically impossible to regulate the poultry industry's practice of dumping more than a billion pounds of manure into the Chesapeake Bay, the largest estuary in North America. Were the Maryland economy made up of locally owned businesses, officials could raise environmental standards with confidence that its enterprises would adapt rather than flee.[6]

The absence of local ownership means that non-local corporations can dictate the terms of sustainability in the communities in which they operate. Their ability to leave a community in a heartbeat means they can more easily leave environmental problems behind. The expansion strategy of Walmart, the largest chain retailer in the world, has included closing older stores (and resisting resale to competitors) while opening new superstores only a few miles away. As a result, some 350 empty Walmarts across the United States are causing serious environmental problems from runoff, flooding, and urban blight.[7]

Second, the presence of local business owners in a community can lead to greater environmental responsibility through accountability. A business owner can be shamed into thinking twice about polluting freely, for example, if the victims are attending the same church or going to the same schools. The responsibility that local owners feel to their own neighborhoods helps explain why U.S. locally owned businesses have been found to give 2.5 times as much money to local charities per employee as non-local businesses do.[8]

Courtesy Bellingham Farmers Market

Local farmers offer up their mixed greens at the Bellingham Farmers Market.

Third, because local businesses tend to use local materials and sell to local markets, their inputs and outputs require less shipping, consume less energy, and emit fewer pollutants, including greenhouse gases (GHGs). To be sure, a number of studies have argued that local food does not always minimize carbon emissions. Alaskans, for example, might find that growing bananas in their own greenhouses is more energy-intensive than transporting bananas from Guatemala.[9]

But the most widely publicized of these studies actually prove very little. For example, one report suggested that U.K. residents eating local lamb generated four times as many GHGs as they would have had they imported New Zealand lamb. But the study, whose funding by the New Zealand lamb export association went unnoticed, only compared energy-intensive, industrial-agriculture methods in the two countries, and it never even examined the GHG impacts of local production.[10]

Finally, every profitable green small-business model provides an invaluable jigsaw piece to the global puzzle of sustainability. A low-cost, Internet-based food distribution system—such as the Oklahoma Food Coop—can offer communities everywhere a model for greater food self-reliance. A successful local wind project, such as the subdivision-owned windmills in Hepburn Shire, just outside Melbourne, Australia, can help thousands of other windy communities worldwide see how to achieve energy self-reliance. According to localization advocates, a key to global sustainability and poverty alleviation (alongside Fair Trade and technology transfer programs) might be open-source platforms that spread without charge, particularly to poorer communities, start-of-the-art business models, technologies, and practices.[11]

Localization and Prosperity

The sustainability impacts of localization would be interesting but ultimately unconvincing if local businesses turned out to have few economic benefits for a community. In fact, a growing body of evidence suggests that localization, done properly, can increase prosperity for three reasons.

First, the immobility of local businesses means that economic development efforts focused on them are more likely to produce enduring results. An investigative report on the cost effectiveness of tax abatements in Lane County, Oregon, found that 95 percent of the tax abatement dollars given between 1990 and 2002 had gone to six non-local companies—three of which came, took the benefits, and then shut down and moved elsewhere. The rest went to about a hundred local companies. The public cost to the region of a non-local job, in tax-abatement terms, was about $23,800. The comparable cost of a local job was $2,100—the same per-job cost reported by several microenterprise organizations in the western United States. Thus non-local jobs were more than 10 times costlier. On a long-term, net jobs basis (taking into account the big firms' departures), non-local jobs were 33 times more expensive.[12]

Second, a local business tends to generate a higher economic multiplier than a comparable non-local business. In the summer of 2003, for instance, two economists studied the impact of a proposed Borders bookstore in Austin, Texas, compared with two local bookstores. They found that $100 spent at Borders would circulate $13 in the Austin economy, while $100 spent at the two local bookstores would circulate $45—translating to three times the jobs, earnings, and tax collections.[13]

Many other studies in the United States and the United Kingdom all point in the same direction, and for an obvious reason: local businesses spend more of their money locally. Unlike a chain book store, for example, a local bookstore has local management, uses local business services, advertises locally, and enjoys a stream of local profits.[14]

Third, the uniqueness of a local business fits hand-in-glove with other theories of economic development. For example, a community rich in local business creation attracts and retains entrepreneurs and entrepreneurial young people. As Richard Florida of the Creative Class Group argues, such "creative economies" succeed because they are tolerant, diverse, and fun, and in the end such economies depend on the ability to seed and expand local businesses.[15]

Most economists and economic developers are only dimly aware of these findings, since they are based on new studies and theories. But even as these ideas spread, resistance will run deep, because most economic developers know they will get more press, political kudos, and budgetary rewards for a single big-business deal creating 1,000 jobs than for 100 deals that each create 10 jobs. From an economic standpoint, however, the jury has returned with a clear and convincing verdict: locally owned businesses are significantly better bets for income, wealth, and jobs.

Localization and Efficiency

Skeptics of localization continue to assert that local businesses simply have poorer, more expensive goods and services that cannot possibly achieve the higher economies of scale inherent in global businesses. Yet at some point increasing scale brings diminishing returns and poorer performance. The recent global financial meltdown is a poignant reminder that many global corporations, not to mention the global financial institutions that have been their enablers, carry many more risks than people ever appreciated. In fact, what is becoming clear is that the global scale of business carries many profound dis-economies.

For example, even when nonlocal production can bring down costs by siting a factory in a jurisdiction with low-wage labor and high-pollution technologies, long-distance distribution is becoming increasingly inefficient. Consider food. Economist Stewart Smith of the University of Maine estimates that $1 spent on a typical U.S. foodstuff item in 1900 wound up yielding 40¢ for the farmer, with the other 60¢ split between inputs and distribution. Today, about 7¢ of every retail food dollar goes to the farmer, rancher, or grower, while 73¢ goes to distribution. Whenever the distribution costs tower over the production costs, there are huge opportunities for cost-effective localization. Food localization reduces the need for and expense of many components of distribution, such as refrigeration, packaging, advertising, and third parties. And as oil and energy prices rise in the years ahead, distributional inefficiencies like these will increase, opening up new opportunities for localization.[16]

Other trends also are making local businesses more competitive. For 50 years consumers in industrial countries have been shifting their expenditures from goods to services, which fuels localization because local services, where providers and clients have face-to-face relationships, have always been highly competitive. Homeland security concerns are nudging officials to promote self-reliance in commodities like food and energy. While the spread of the Internet is not unambiguously positive for localization (mass retailers like Amazon and eBay could not exist without it), it ultimately levels the playing field by providing local competitors with a low-cost tool for marketing themselves.

Even without these trends, small-scale businesses are already competitive in almost every business category. The North American Industrial Classification System, an important database produced by the U.S. Census Bureau, contains 1,100 such categories, and there are more small businesses—nearly all of which are locally owned—than large ones in all but 7 of them. The point is that even in very small communities, a smart economic developer can find exciting examples of small-scale success in almost every industry and replicate them.[17]

Fulfilling the Market Potential

Despite the market potential for more localization, formidable barriers stand in the way. Consumers are deluged with billions of dollars of global advertising and are often unaware of competitive local goods and services. Small-business owners—distrustful of their local competitors and overwhelmed by the daily work of keeping their firms alive—fail to forge natural business partnerships that might otherwise be beneficial. Investors are deterred from putting their money into profitable local businesses by obsolete security laws that make it unreasonably expensive. And public policymakers worldwide, despite all their positive rhetoric about small business, seem unable to break their addiction to subsidizing global businesses. The localization movement aims to dismantle these barriers.

To help consumers find and buy competitive local goods and services, Local First campaigns, like the one in Bellingham, are providing information about which businesses and products are in fact local and what their prices and quality are compared with the global competition. These initiatives are also nudging consumers to buy local through myriad tools. Local coupon books provide consumers with introductory discounts to local business. Local debit, credit, gift, and loyalty cards reward local purchases. Local barter and money systems induce participating consumers to use their credits exclusively with local businesses.[18]

To improve the competitive practices of local businesses, alliances like the Sustainable Business Network of Greater Philadelphia (a BALLE affiliate) are organizing conferences where they can showcase best business practices in everything from marketing strategy to energy-reduction technologies. Peer networks, especially those organized by sector (food, energy, retail, and so forth), are helping local businesses improve their competitiveness. Local businesses are learning that by working together, they can achieve most of the economies of larger scale that might otherwise give some global businesses a competitive advantage. Tucson Originals in Arizona, for example, enables participating local-food businesses to improve their bottom line through joint procurement and marketing.[19]

Perhaps the biggest obstacle to localization is the unavailability of capital. Complex securities laws governing capital markets make it unaffordable for small investors to place their savings in small businesses even in wealthy nations. In Australia, for instance, local businesses account for two thirds of the economy and have steadily improved their share of gross domestic product vis-à-vis global business, yet almost none of the 9 percent "superannuation" funds that citizens must put into their retirement accounts can be placed in local business. A growing mission of the localization movement is to deregulate grassroots participation in capital markets, help small businesses issue local stock inexpensively, provide liquidity to these markets through local stock exchanges, and create new investment professionals—advisors, broker dealers, traders, fund managers—who specialize in local investment.[20]

Changing investment rules is really a subset of a much larger policy reform agenda. Local business alliances are beginning to stake out policy positions dramatically at odds with the traditional business community. For example, while the U.S. Chamber of Commerce has been opposing "cap-and-trade" legislation to curb GHGs, a number of local business associations have been lobbying for the legislation. A similar split can be seen around proposals to eliminate tax loopholes for U.S. multinationals: the Chamber opposes these reforms, while local business networks support them.[21]

The biggest public policy change sought by localization advocates is to overhaul the priorities of economic development. Public dollars, they argue, should be focused exclusively on nurturing local business. Every economic development dollar and hour spent on attracting or retaining non-local business is a dollar and hour unavailable for the superior payoffs, in both sustainability and prosperity, for localization.[22]

The agenda for localization actually contains hundreds of action points for activists, businesses, and policymakers, many of whom never agree on much of anything. Localization is forging unlikely new alliances between green businesses and anti-business greens and between free-market conservatives and anti-globalization progressives. And this, in the end, might be the most compelling feature of localization and its most enduring contribution—a culture of sustainability rooted in deep democracy.

Government's Role in Design

In a sustainable society, eco-friendly choices should not be difficult to make. The sustainable choice in any situation, whether it be buying a new lightbulb or designing a suburban development, should be the default choice, the path of least resistance, even natural. This section confirms that governments—which set laws, create societal priorities, and design the cities and towns where people live—will be central players in nurturing such a culture of sustainability.

An important role of governments—one that is almost invisible when it is done well—is that of "choice editing." Michael Maniates of Allegheny College notes that editing citizens' options through laws, taxes, subsidies, and so on has been a long-standing role of governments. What is new today is that choice editing is now being used to make the sustainable choice the default one by design. From a plastic bag ban in Rwanda and the phaseout of incandescent bulbs in Canada to sweeping carbon taxes in Sweden and subsidies on solar power in China, many governments around the world are starting to try to make it effortless for people to live sustainable lives.[1]

Another concept that sorely needs to be reconsidered is national security. As human activities disrupt a growing number of ecological systems, it will become increasingly clear that the biggest threats to national security are not foreign armies or terrorist groups but the weakened state of the planet. Michael Renner of Worldwatch describes how to take the almost $1.5 trillion spent each year on militaries around the world and use it instead to heal environmental and social problems. This shift will do more to protect people than the largest nuclear arsenal ever could, and in the process it will create additional economic opportunities and new openings to improve diplomatic relations between countries.[2]

Where people live will also need to be redesigned in order to make it easy to live sustainably. Peter Newman of Curtin University of Technology outlines how and where this is happening already, so that cities and towns have smaller ecological footprints or even no footprint at all. Cities could become free of cars and could generate a significant portion of their energy and even their food by harnessing their rooftops and green spaces for solar arrays, wind turbines, and gardens. And by tapping into community networks, city dwellers can be mobilized as active participants in accelerating the shift to sustainable urban design.

Key social services like health care need to be overhauled as well, as Walter Bortz of the Stanford University School of Medicine notes. In many societies today, health care is focused too much on treating symptoms instead of on preventing disease and encouraging healthy and sustainable living. By shifting from "sick care" to health care, governments can prevent millions of unnecessary deaths and improve the lives of millions more. They can also save billions of dollars and, by reducing the need for resource-intensive treatments, cut the ecological impacts of keeping people healthy.

One other key redesign needed is that of the very system of law. Cormac Cullinan, an environmental attorney in Cape Town, describes how legal systems today fail to integrate the rights of Earth's systems and how this in turn allows the shortsighted conversion of ecosystems into resources at the expense of both human communities and the Earth community. Recognizing Earth's rights in law will help make it natural to consider the broader tradeoffs of development choices made today and will give citizens legal recourse when ecological degradation masquerades as economic development.

Within these articles there are also two Boxes: one on how other social services could be redesigned to provide more for less and in ecologically restorative ways and another on the international community's role in making global consumption and production patterns sustainable through the Marrakech Process of the United Nations.

The importance of government's role in creating sustainable societies cannot be overstated. If policymakers make sustainability their priority, bolstered by citizens' support, vast societal transformations can occur so that one day living sustainable lives will become natural—by design.

—*Erik Assadourian*

Editing Out Unsustainable Behavior

Michael Maniates

By late 2010, Australians are going to have a hard time finding an incandescent bulb for their nightstand lamps or desk lights. The Australian government, troubled by potential electricity shortages and global climate change, is the first to "ban the bulb" in favor of energy-sipping compact fluorescent lamps (CFLs) and LEDs. The impact will be significant: 4 million fewer tons of greenhouse gas emissions each year by 2012, together with sizable economic savings. And Australia is not alone. The European Union is slowly phasing out incandescents by 2012. Canada, Indonesia, and even the United States are next in line.[1]

Environmental analysts like Lester Brown of the Earth Policy Institute are delighted. Brown says that if everyone followed Australia's lead "the worldwide drop in electricity use would permit the closing of more than 270 coal-fired (500 megawatt) power plants. For the United States, this bulb switch would facilitate shutting down 80 coal-fired plants." But others are not so sure. Reports abound of people hoarding incandescent bulbs in Australia and Germany, among other countries, and some experts wonder if incandescents are being forced out too quickly. And then there is the prickly philosophical question at the heart of it all: Should products be removed from the menu of consumer choice because of their environmental or other socially objectionable qualities? Who decides what stays on the shelves and what goes? Shouldn't the consumer be allowed to choose freely? Is "lightbulb fascism" intruding into the marketplace?[2]

Choice Editing Is Nothing New

Welcome to the world of "choice editing," where the tussle over lightbulbs is but the opening salvo in a larger struggle to crowd out environmentally negative products in favor of more benign choices. Choice editing for sustainability is more than simply deleting what does not work. In the words of the U.K. Sustainable Development Council, it "is about shifting the field of choice for mainstream consumers: cutting out unnecessarily damaging products and getting real sustainable choices on the shelves." (See Box 16 for some initiatives on sustainable consumption at the international level.)[3]

Choice editors remove environmentally offensive products from commercial consid-

Michael Maniates is Professor of Political Science and Environmental Science at Allegheny College in Pennsylvania.

Box 16. The U.N. Marrakech Process on Sustainable Consumption and Production

In recognition of their disproportionate share of global consumption and the resulting impact on sustainability and equality, industrial countries agreed in 2002 to take the lead in accelerating the shift toward sustainable patterns of consumption and production.

To achieve this, a global informal multistakeholder expert process was launched in 2003 in Marrakech, Morocco, to support regional and national initiatives to accelerate the shift to sustainable consumption and production (SCP) and to elaborate a 10-year framework of programs on SCP, which will begin after its structure and content are negotiated at the U.N. Commission on Sustainable Development meeting in May 2011.

A key element of the Marrakech Process is its seven Task Forces, which are voluntary initiatives led by governments in cooperation with various partners:

- ***Sustainable Lifestyles*** (Sweden). Identifies and compares grassroots social innovations for sustainability from around the world, finds promising examples, and diffuses them. Develops train-the-trainer tools for sustainable consumption in youth, CD-roms on sustainability in marketing, and on-line galleries of sustainability communication. Projects implemented in more than 30 countries with materials in over 10 languages.
- ***Cooperation with Africa*** (Germany). Affirms Africa's own 10-year framework on SCP (the first region to have developed and launched such a program) by supporting an All Africa Eco-Labeling scheme, the establishment of a network of Life-Cycle Assessment experts in Africa, and initiatives to "leapfrog" straight into clean energy sources.
- ***Sustainable Public Procurement*** (Switzerland). Develops analysis and Web-based Status Assessment tools to support public-sector organizations' attempts to justify, develop, and gauge the success of sustainable procurement programs.
- ***Sustainable Products*** (United Kingdom). Catalyzes networks of experts in key product areas to upwardly revise standards, develop labels, work together on policy roadmaps,

eration, like smog-producing charcoal lighter fluid in Los Angeles or leaded gasoline in Europe and North America. Or they make such products expensive to use, like Ireland's levy on plastic shopping bags, which has reduced plastic bag use by 90 percent. But like any good editor, choice editors cannot just chop. They must offer options or, at the very least—in the words of environmental reporter Leo Hickman—a sufficiently compelling illusion of choice. In Los Angeles, backyard cooks denied their lighter fluid had the choice of chimney or electric briquette fire starters. In Ireland, shoppers can purchase any number of cloth bags, some trendy or stylish. And in Australia and the growing number of other countries looking to edit out incandescents, consumers will see more choice among CFLs, LEDs, and other innovative lighting technologies.[4]

If the idea of governmental choice editing rankles, perhaps because it sounds manipulative or too "Big Brother," remember that choice editing is neither new nor novel. Government has long been at it, in ways both obvious and obscure. (See Table 8.) Safety and performance standards for everything from the food people eat to the cars they drive constrain and shape choice. The same holds true for tax, tariff, and subsidy policies that heighten the desirability of some products while making others unattractive or unavailable. More subtly, government decisions about where to build roads and rail lines, what schools and hospitals are constructed or closed, and which research and development initiatives are supported or

Box 16. *continued*

and collaborate on compliance. Three product areas identified so far: lighting, home entertainment products, and electric motors.

- ***Sustainable Tourism*** (France). Creates demand for greener travel offerings with the *Green Passport Program* for citizens, fosters industry supply with the revised *Environmental and Sustainable Tourism Teaching Pack for the Hospitality Industry*, and encourages investment by convening a Sustainable Investment and Finance in Tourism Network.
- ***Sustainable Buildings and Construction*** (Finland). Works to move green building standards beyond the realm of the voluntary by developing policy recommendations and working in partnership with national governments and private firms participating in the U.N. Sustainable Buildings and Climate Initiative.
- ***Education for Sustainable Consumption*** (Italy). Focuses on integrating sustainable consumption into core curriculum in the Mediterranean region, while working with the UNESCO Associate Schools Network Project (a global network of 8,500 educational institutions in 179 countries founded in 1953) to disseminate best practices in sustainability education to teachers around the world.

By bringing consumption into the global dialogue on sustainability, the Marrakech Process raises questions of lifestyle, values, and progress, creating a unique space within national governments and regional forums for reforming the cultures and institutions at the basis of all socioeconomic systems, while bringing a suite of tools to the table for policymakers who are serious about greening the economy and improving human well-being.

Clearly more could be done with greater leverage and resources. Unfortunately, the low profile of the Marrakech Process means the effort suffers from a lack of serious attention by senior decisionmakers. In the run-up to the negotiations in May 2011, this fledgling but transformative U.N. process could be helped by the greater involvement of governments, the private sector, and the public.

—*Stefanie Bowles*

Source: See endnote 3.

starved converge to write the menu for housing, education, and jobs from which everyone must choose.

The real worry is not that government engages in choice editing. Rather, it is that for decades such editing has aided and abetted an especially narrow view of progress, one that imagines mass consumption as the foundation of human happiness, egalitarianism, and even democracy itself. As prize-winning historian Lizabeth Cohen writes in *Consumers' Republic*, "A strategy…emerged after the Second World War for reconstructing the [U.S.] economy and reaffirming its democratic values through promoting the expansion of mass consumption." A central plank of this strategy was to make energy-intensive, resource-depleting, mass-consuming choices appear natural and inevitable: witness the single-family, detached home to be filled with products, a family car to get to it, and dispersed and abundant shopping outlets. Other, more environmentally sustainable consumption options and patterns—efficient streetcar and intercity rails systems, for instance, or a returnable-bottle network for milk, soda, and other products—were cast as backward, were made more difficult to find or rely upon, and subsequently disappeared.[5]

Cohen's incisive gaze rests on the United States, but similar stories hold true for much of the industrial world, and parallel tales are now being told in developing countries, most notably India and China. They all point to a provocative question: if the rise of fundamen-

Table 8. Examples and Features of Choice Editing

Types of Choice Editing	Examples	Important Features
Eliminate offending choices	• Montreal Protocol and CFCs • Shift away from leaded petrol in the North America and Europe • Ban on incandescent bulbs in Australia • Compressed natural gas for public transportation in India • Walmart's decision to carry only MSC-certified wild-caught fresh and frozen fish	• Strong legislation, often supported by business interests • Requires new choices to offset the loss of previous choices • Demands a "phase-in" period that allows for adjustment
Slowly trim away the worst products and practices	• Japan's "top runner" program for energy efficiency • LEED building requirements in the United States, which gradually increase the standards for certifying a new building as "green" or "sustainable"	• The use of labeling to identify, over time, the most offending practices and products • Clear standards and methods of evaluation • Collaborations among government, industry, and consumer groups
Make offending choices less attractive or increasingly difficult	• Ireland's levy on plastic shopping bags • Shifting fatty and processed foods from eye level to higher or lower shelves	• Two primary instruments: taxation and product placement and positioning • Wide range of choice is retained, but incentives and positioning privilege sustainable choices over unsustainable ones
Change context for choices; alter "choice architecture"	• Creative use of defaults (for instance, consumers are subscribed to renewable forms of electricity and must intentionally refuse this option) • Focused changes to material flows; for university and corporate composting programs, for example, shift to all compostable dining ware (plates and utensils) in cafeterias to eliminate mixing of compostable and non-compostable waste by consumers • Embedded cues and drivers that encourage reduced consumption (for example, when trays in university cafeterias are removed, students take only what they need, reducing food waste, water use, and energy consumption) • Create real choice for trading leisure for income: four fifths work for four fifths pay as a viable work option	• Enduring question: How can consumer experience be structured so that doing the right thing is natural and requires little or no thought while doing the wrong thing is difficult and requires conscious thought and focused intent • Building a choice architecture to oppose consumerism often involves reintroducing meaningful choice: choices among varied transportation options, for example, or about work time and leisure

tally unsustainable consumer cultures was facilitated by choice editing—by an elite who intently shifted the field of choice for mainstream consumers—will transforming consumerism into something more sustainable require a similar degree of determination and sophistication by government and business?

The answer appears to be yes. In 2006, for example, the Sustainable Development Roundtable (SDR)—a project of the Sustainable Development Commission and the National Consumer Council in the United Kingdom—released an analysis of 19 promising transformations in consumer cultures, ranging from sustainable forestry products to fair-trade and organic food product lines. SDR concluded that "historically, the green consumer has not been the tipping point in driving green innovation. Instead, choice editing for quality and sustainability by government and business has been the critical driver in the majority of cases. Manufacturers, retailers and regulators have made decisions to edit out less sustainable products on behalf of consumers, raising the standard for all."[6]

A classic example of this is the Montreal Protocol's phaseout of ozone-destroying chlorofluorocarbons (CFCs). "Powerful economic, political, and technical factors combined to facilitate the phase-out of CFCs," write James Maxwell and Sanford Weiner of the Massachusetts Institute of Technology. They note that a critical factor was DuPont's desire to create new consumer demand for its CFC substitute while establishing a competitive advantage over its major global competitor, which had no such substitutes. The ozone layer is healthier today because consumers shifted to more ozone-friendly substitutes, but this shift came about largely because of methodical choice editing that pushed consumers in that direction.[7]

Of course, consumers still have an important role to play as they vote for sustainability with their purchases. But Tim Lang of City University London, who coined the idea of "food miles," speaks for many analysts of sustainable consumption when he asks "why should the consumer be the one left in the supermarket aisle to agonize over complex issues such as animal welfare, carbon footprints, workers' rights and excessive packaging, often without any meaningful data on the label to inform their decision-making?" Why, in other words, don't producers and governments shift their current choice-editing practices so that consumers choose only among a range of environmentally "good" products? That way, making the right choice is—as businessman and environmental writer Paul Hawken puts it—as "easy as falling off a log."[8]

One answer is that the favored alternative—labeling products as environmentally "good" or "bad" and letting consumers decide—is sometimes thought to be less controversial. Product labeling is an important component in the transformation of consumer societies to sustainable ones. Yet experience suggests that when product information is made available, perhaps as part of ecolabeling schemes, it influences no more than a minority of shoppers—and not nearly enough, not fast enough, and not consistently enough to drive the transformation of consumer life required by a planet under stress.[9]

At least three factors limit the effectiveness of labeling: the varying degree of environmental commitment among the general population; the complexity of consumer-choice decisions, which are structured by intricate sets of social processes and cultural influences; and a corrosive "choice architecture"—the potent context within which people make decisions. Nutrition labeling, for example, does not stand much of a chance in most supermarkets, given that products are positioned (or hidden) on shelves and at end-of-aisle displays to foster impulse purchases of fatty, sweet, and processed foods and that sugary products are shelved at a child's eye level. It is no surprise, then, that the Sustainable Development Commission found that information about the environmental and economic benefits of less environmentally destructive products "failed to get more than a minority of people buying" the best products. But the Commission also found that when labeling and other information efforts were part of choice-editing efforts by government, producers, and retailers, consumer practices changed across the board.[10]

N-O-M-A-D

New Delhi traffic jam: less pollution may be only half the battle.

Editing for Sustainability

If the goal is to move consumers toward less environmentally damaging patterns of consumption, contemporary experience says that choice editing delivers. At a growing number of colleges and universities across the United States, for instance, fair-trade coffee and renewably generated electricity are increasingly on the menu—and are often the only choice available on campus.[11]

In California, consumers can choose from a variety of electricity generation options, and the most environmentally dedicated customers can opt for rooftop solar arrays where site conditions and the ability to pay permit. Regardless of their preferences, 20 percent of their electricity will flow from renewable sources by 2010 due to Renewable Portfolio Standards imposed on electric utilities by the state government. These are driving the development of renewables faster than uncoordinated consumer demand ever could. California's proportion of renewable electricity will slowly grow, and 38 other states are following suit.[12]

In 2003, London implemented Europe's first "congestion pricing" program for its city core: drivers pay a fee to operate their car in central city areas during peak periods, with the revenue going to boost bus service and fund subway renovations. Initially treated with skepticism, the program now enjoys growing public support and is a model for major cities worldwide. And in India, in response to a Supreme Court public health order, the government has required all buses, taxis, and auto-rickshaws in major cities to switch from dirty fuels to cleaner burning compressed natural gas. Despite some initial protests, New Delhi has led the way, and commuters are now part of an ambitious effort to curb air pollution. These examples, and others like them, demonstrate the effectiveness and political viability of choice editing.[13]

Business offers its own set of examples, though whether these practices will endure and expand absent government regulation or persistent pressure by citizens' groups remains to be seen. Reacting to pressure from environmental groups, since 1999 Home Depot—the largest home improvement retailer in the United States—has sold lumber certified and labeled by the Forest Stewardship Council. But it also has quietly altered significant aspects of its wood-product supply chain; it is consequently harder today than 10 years ago for anyone to purchase environmentally "bad" lumber at Home Depot.[14]

B&Q, Home Depot's counterpart in the United Kingdom, pursued a similar strategy and has perhaps the most robust commercial system in place for certifying the sources of its timber supply, easily outpacing U.S. retailers. Interviewed in the late 1990s, Allen Knight, then Environmental Policy Coordinator for the company, explained that B&Q embarked on sustainable wood "even though there was no indication of consumer demand for certified products." He observed that "customers do not ask for certified products because they are unaware of them: Raising awareness and creating markets are the retailer's role."[15]

Not to be outdone, in early 2006 Walmart pledged to source all its wild-caught fresh and frozen fish from suppliers certified as sustainably harvested by the Marine Stewardship Council (MSC). Moreover, it required its suppliers to expand renewable fisheries rather than jockey for access to or ownership of existing suppliers. The blue MSC ecolabel figures prominently on Walmart wild-caught fish, but unlike other labeling schemes the certification is not meant to persuade buyers to choose sustainable wild-caught fish over less sustainable options, as the company has edited those out completely.[16]

Also in early 2006 Hannaford Supermarkets in the United States implemented its "guiding star" program in 270 stores, in which products identified as especially healthy or nutritious are given one to three stars. Some 28 percent of items in the stores receive the rating, with the remainder not being good enough to get a star at all. Dan Goleman, author of *Ecological Intelligence*, reports that "poorly rated brands dropped as much as 5 percent in sales," while sales of some three-star brands went up by 7 percent. "Brand managers started contacting Hannaford to ask what they needed to do to get higher ratings," Goleman noted.[17]

Hannaford's apparent success comes because they understood their program as more than a simple labeling exercise. It was about changing critical components of the "choice architecture" at its stores. "It includes signs, shelf tags, an advertising campaign, collateral materials, training materials, a web site, and community outreach, among other elements," explains Hannaford spokesperson Michael Norton. And it meant changing product placement and shelving strategies to reinforce healthier shopping habits.[18]

Obstacles to Change

There remains immense potential for choice editing to drive fundamental changes in consumption. But at least two obstacles stand in the way. One is the persistent belief that product labeling alone can drive necessary change. Even when logical and clear, labeling places the burden on consumers to drive needed social change with their purchasing decisions. It also reinforces what Thomas Princen at the University of Michigan calls one of the most disabling myths about political life: the notion of consumer sovereignty, which says that the decisions that producers and marketers make about what to produce and what to sell is driven solely by independent, uninfluenced consumer choices. The consumer decides, in other words, and the producer responds. This idea denies the power that government and business have over the menu and architecture of consumer choice. In doing so, it undermines the very rationale for choice editing.[19]

Japan has pioneered a better use of labeling, one that could move consumer cultures toward an ethos of sustainability. Since 1998 the government has divided products up into similar categories and classes, and then graded and labeled them on a 1–5 scale for energy efficiency. Tiers one and two are the standard set by the best-performing products—and it is the standard that the entire industry must meet within five years. As these "top runners" improve, the overall standard shifts upward, placing ongoing pressure on manufacturers to improve their product lines or face a ban on their products. In the short run, energy-conscious consumers are empowered: the top-runner label offers important information about the overall energy costs of a consumer choice. In the longer run, the field of choice changes: the label provides a regulatory platform for driving constant product innovation, increasing the range of choice among the higher-performing categories and editing out the worst products. Germany is considering a similar program. Advocates of choice editing hope that Walmart's recent commitment to environmental labeling will incorporate this "top runner policy."[20]

A second impediment to the power of choice editing is its prevailing focus on "consumption shifting" rather than "consumption reducing." Most choice editing has been about moving consumers to less environmentally damaging products. But genuinely sustainable patterns of consumption must also involve reductions in overall consumption. How can the context within which everyday people make consumption decisions be edited to encourage that? John de Graaf suggests one answer: make it attractive for people to trade work for leisure in ways that would lead to a voluntary reduction in income (but not health and other important benefits) for more free time, which in turn has known environmental benefits.[21]

Cornell economist Robert Frank offers another solution: shift taxes toward luxury consumption, reduce or eliminate taxes on income diverted to savings, and invest more government resources in public uses—parks, inviting pedestrian walkways, mass transit—that would reduce individual pressures to consume (thus supporting de Graaf's agenda for less work, less income, but more life satisfaction).[22]

In *Nudge*, economist Richard Thaler and legal scholar Cass Sunstein provide a suite of additional ideas for altering the "choice architecture" in service of sustainable consumption. These include the pervasive use of defaults to "nudge" consumers in environmentally appropriate directions. A person could opt out of these defaults, but the burden rests on the individual to choose the wrong behavior over the right one. Examples include automatic and certified carbon-offsets for all travel bookings, default savings plans, and pricier renewable energy automatically included in residential energy bills (so a customer would have to say explicitly "I want to use dirty, polluting coal to save a small amount of money").[23]

Choice editing has been with us a great long while, and it is here to stay. If that seems far-fetched, just bring an especially critical eye to the layout of products and displays in a supermarket. Which products draw customers' eyes? Which are easily reached? The question now is this: Will a primary focus on the promise of product labeling alone (and underlying notions of consumer sovereignty) continue to shape policy for sustainable consumption? Or will more-realistic assessments emerge about how and why people make consumer choices? Government and business, operating from a view that mass consumption means mass prosperity, have tightly held the reins of choice editing for too long. Now is the time for a more nuanced, more sustainable vision of choice and choice architecture to prevail.

Broadening the Understanding of Security

Michael Renner

In 1985, when the world was still trapped in the Cold War standoff, political scientist Daniel Deudney called for "large-scale cooperation between the United States and the Soviet Union in the manned exploitation of deep space and multilateral efforts to secure earth by making better use of space technologies." He argued that such a common, collaborative project could be "harnessed to transform the superpower relationship and to create a common security system."[1]

Whether space exploration was then or is ever the right vehicle for bringing about a more cooperative and peaceful world order is debatable. But the underlying argument is worth pondering: can humanity, in rallying around a common purpose, leave behind its costly history of conflicts and divisions? The Cold War is long over, but security concerns have hardly vanished. Nations around the world, and especially the poorest countries and communities, confront a multitude of interlinked challenges and pressures. These include rising competition for resources, environmental breakdown and the specter of severe climate disruptions, a resurgence of infectious diseases, demographic pressures, poverty and growing wealth disparities, and convulsive economic transformations that often translate into joblessness and livelihood insecurity.

Understanding how these social, economic, and environmental conditions can undermine human security and may even translate into conflicts and instability requires a broader definition of security, one that understands the influence of economic, demographic, and environmental pressures that cannot be resolved by force of arms. Recent years have indeed seen a growing recognition of such dynamics.

Key Challenges

A number of these conditions and dynamics can be seen as an outgrowth of the dominant economic model premised on essentially unlimited resource consumption. This model is not only putting humanity on a collision course with the planet's ecological limits, it has also led to tremendous social and economic inequality.

Nonrenewable resources. Throughout history the pursuit of resources such as fossil fuels, metals, and minerals has led to repeated outside interventions in resource-rich countries. The specter of peak oil and comparable contradictions between surging demand for and

Michael Renner is a senior researcher at the Worldwatch Institute who focuses on security and economics.

finite deposits of other resources raise the likelihood of intensifying geopolitical rivalries. But resource wealth has also fueled serious human rights violations, corrupt systems of governance, and even a series of civil wars. Revenues from mining and logging operations have mostly benefited a small minority, while the social and environmental burdens have typically been shouldered by poor and disadvantaged communities.[2]

Renewable resources. Water, arable land, forests, and fisheries are essential for all of human life, and the livelihoods of hundreds of millions of farmers, ranchers, and nomadic pastoralists depend directly on them. Distributional disputes may grow more pronounced with resource depletion and pollution. Almost one third of the world—estimates vary between 1.4 billion and 2 billion people—already live in water-scarce regions. Aside from population growth and poor management practices, climate change impacts could increase the affected number of people by anywhere from 60 million to 1 billion people by 2050. A recent study found that due to such impacts as rising temperatures and increased drought, half of the world's population could face severe food shortages by the end of this century.[3]

Disease burdens. Food shortages make affected populations more vulnerable to diseases. The world is experiencing a resurgence of infectious diseases, with the poor being the most vulnerable. Pathogens are crossing borders with increasing ease, facilitated by international travel and trade, migration, and social upheaval. In addition, logging, road-building, and dam construction bring humans close to new pathogens. And climate change enables vectors for diseases like malaria or dengue fever to spread. At the same time, an increasing number of societies are confronting an epidemic of obesity—a symptom of overconsumption and sedentary lifestyles.

Disasters. A combination of ecosystem destruction, population growth, and economic marginalization of the poor has led to more frequent and more devastating disaster events. The number of natural disasters (excluding geological events such as earthquakes and volcanic eruptions) has risen from 233 in the 1950s to more than 3,800 in the current decade, and the number of people affected has grown from nearly 20 million to 2 billion. The pace is likely to accelerate as climate change translates into more intense storms, flooding, and heat waves. Disasters can undermine human security by exacerbating poverty, deepening inequalities, and undermining the long-term habitability of some areas. The experiences of Haiti, Nicaragua, Bangladesh, India, and China suggest that unrest and political crisis can erupt where relief and reconstruction efforts are slow or incompetent.[4]

Unemployment. The global economic crisis that broke into full view in late 2008 sharpened concerns about unemployment, uncertain economic prospects, and the growing move toward the informal sector in the world economy. Almost half the world's workforce, some 1.5 billion people, is classified by the International Labour Organization as being in vulnerable employment arrangements; more than 1.2 billion workers are mired in poverty, earning less than $2 a day. Close to 190 million people were unemployed altogether in 2008, a number that was expected to rise by 30–60 million in 2009. North Africa, the Middle East, Eastern Europe, sub-Saharan Africa, and Latin America have particularly high rates of joblessness. The unemployment figure for young people, at 12 percent, was double the overall rate. When large numbers of young adults face bleak prospects for earning enough to establish and support a family, their discontent can translate into societal instability.[5]

Population movements. A range of factors contribute to population movements, and sometimes the boundaries between voluntary and involuntary flows are blurred. In addition to 42 million international refugees and internally displaced persons fleeing warfare and persecution, some 25 million people are thought to have been uprooted by natural disasters. As many as 105 million people have been made homeless by projects such as dams, mines, roads, and factories. Environmental degradation is behind at least some of these numbers. Projections of the number who may get displaced due to climate change by 2050 vary from a low of 25 million to a high of 1 billion. Refugees and migrants may be seen as unwelcome competitors for land, water, jobs, and social services, possibly leading to social unrest and violence.[6]

The Need for New Priorities

Over the years, academics and policymakers have come to accept the validity of a broader view of human security than just a military one. Roughly a dozen governments engage in an ongoing dialogue through the Human Security Network. A number of national government agencies and intergovernmental bodies have developed policy guidelines, commissioned research and strategy papers, and convened meetings to assess conflict prevention and post-conflict peacebuilding efforts in this broader context. In 2007, for the first time ever, the United Nations Security Council discussed the security implications of climate change, including border disputes, migration, societal stress, humanitarian crises, and shortages of energy, water, arable land, and fish stocks.[7]

These developments notwithstanding, government policies and budget priorities do not indicate any major shift. And much of this discussion remains solidly within the national security mindset, steeped in traditional perceptions of "threats" as opposed to common vulnerabilities. Reflecting the views of U.S. military and intelligence agencies, an August 2009 *New York Times* article warned of military interventions in response to climate-induced crises that "could topple governments, feed terrorist movements or destabilize entire regions." But instead of a militarization of environmental and human security challenges, a fundamental re-evaluation of security policies is needed.[8]

Traditional security perspectives remain dominant in most national budgets. In 2008, the world spent almost $1.5 trillion for military purposes—the largest amount since the end of World War II and many times more than is available for human security priorities. Western countries did increase their development aid to $120 billion in 2008, up from $52 billion (in current dollars) 10 years earlier. Aid by non-Western donors and multilateral agencies brings the total for development assistance to about $139 billion. That still leaves a military-to-aid budget ratio of more than 10 to 1.[9]

Budgets for climate change are also increasing but are still small compared with military budgets. In fiscal year 2010, the United States will spend $65 on the military for each $1 devoted to climate programs. The nuclear weapons budget—$9.9 billion—is more than four times the amount requested for renewable energy and energy efficiency programs. In Germany, the military-to-climate budget ratio was 9 to 1 in 2008, and in Japan it was 11 to 1.[10]

Bilateral and multilateral funds to assist developing countries with climate mitigation and adaptation tally about $20 billion over the next five years or so. On an annual basis, this is about a third of what the United States alone spends on military aid to other countries and less than a quarter of the value of global arms transfers to developing countries.[11]

Solutions

Policies that defuse conditions that may lead to grievances and disputes represent smart security policy. A robust and comprehensive approach to creating a more stable world entails measures designed to stop environmental decline, break the stranglehold of poverty, and reverse the trend toward growing inequity and social insecurity that all too often breeds despair. There are a number of concepts and initiatives that create goodwill and foster cooperation around shared needs and interests and thus contain the seeds for a recalibrated security policy.

USAID

In the Democratic Republic of Congo these men have found work manufacturing anti-malarial, insecticide-treated bednets.

Before discussing them, however, it is necessary to acknowledge that a fundamental insecurity in international relations will continue to cast a shadow until the establishment of global political institutions with the power to act as credible guarantors of a nation's security. Such institutions might rely on trade sanctions, diplomatic pressures, or even U.N.-sanctioned use of force. At present, U.N. peacekeeping forces are often hobbled by inadequate resources, while regional alliances like NATO lack global legitimacy. Narrow calculations of national interest hold sway. However, a variety of pragmatic and imaginative steps can be taken to lay the groundwork for a new culture of security.

Millennium Development Goals (MDGs). While poverty as such does not necessarily lead to violence, there is no doubt that the absence of equitable development breeds insecurity and discontent. A sustainable security policy will need to work to lessen human vulnerability and improve social and economic well-being. While not couched in the language of security, this goal finds expression in the Millennium Development Goals—some 21 targets of slashing poverty and hunger, combating health threats, and improving primary education to be achieved mostly by 2015. But progress toward these goals has been quite slow and uneven. The MDGs need a major boost in resources and commitment, especially in the face of the global economic crisis that threatens to reverse earlier progress on several goals.[12]

Curbing energy and materials appetites. An alternative energy policy geared toward developing renewable sources and boosting efficiency is not only essential for reducing environmental impacts and greenhouse gas emissions but can be a tool for peace in that it helps to lower the likelihood of resource conflicts. In this context, the creation of the new International Renewable Energy Agency in January 2009 is a welcome step forward. But a recent study finds that to bring about the transition to a low-carbon economy, public R&D for clean energy and energy efficiency will need to grow at least three- to fourfold. And complementary demand-side policies are essential—boosting efficiency and promoting sufficiency through less consumption-intensive lifestyles.[13]

Reducing materials throughput is similarly key to lessening the likelihood of resource conflicts. In the past decade or so, recognition of such conflicts has risen dramatically, in part because of effective nongovernmental groups' campaigns against "blood diamonds" and other resources from conflict zones. Governments and international agencies have responded by imposing embargoes on a number of governments and other actors profiteering from illicit resource exploitation and by promoting greater transparency. These measures will be far more effective if paired with efforts to critically examine, and curb, consumers' voracious resource appetite, which makes these commodities so lucrative in the first place.[14]

Environmental peacemaking. While environmental degradation can contribute to conflicts, environmental cooperation also holds great potential as a peacemaking tool. If well managed, cooperative efforts around shared ecosystems and natural resources can build trust and establish collaborative habits, especially if government contacts are augmented by vibrant civil society dialogue. Over time, such a dynamic may grow sufficiently strong to help overcome unresolved broader disputes. The notion of blending ecology and transboundary politics has been put into practice, to some extent, in two specific areas: river basin management among riparian nations and border-straddling peace parks.[15]

Cooperative water management efforts have been undertaken in international river basins such as the Nile, Danube, Indus, Jordan, and Mekong. These kinds of accords will be increasingly put to the test as populations and water consumption grow and as climate change heightens water scarcity in some parts of the world. Undoubtedly, the task of sustainable water stewardship is more challenging than a shared exploitation of plentiful water resources. These issues arise not just in transboundary settings but also within national borders where different communities and regions jockey for access to water.[16]

Peace parks are protected areas that straddle national borders and are dedicated to protecting biological diversity and promoting peace and cooperation. There are now 188 such areas worldwide. Though they can themselves be a source of conflict if they disregard the livelihoods of local communities, conservation zones can in principle facilitate cooperation and the resolution of territorial conflicts. Most peace parks to date have been established between countries that do not have active conflicts. But one notable case in which the creation of a conservation corridor helped with conflict resolution involved the 1995 border war between Ecuador and Peru. Proposals have been made to establish peace parks in such highly disputed areas as the Kuril Island (Russia-Japan), the Siachen Glacier (Pakistan-India), the Mesopotamian marshlands (Iran-Iraq), and on the Korean peninsula.[17]

Peacekeeping and environmental restoration. United Nations peacekeeping and post-conflict efforts increasingly take into account environmental dimensions. Some 11 peacekeeping missions in countries like the Democratic Republic of the Congo (DRC), Sudan (Darfur), Liberia, Georgia, Lebanon, and Timor-Leste have participated in tree planting efforts. These initiatives are seen as important both locally—countering deforestation and globally in the fight against climate change. U.N. officials recognize that traditional peacekeeping alone is unlikely to have lasting success without these and such environment-related efforts as rehabilitation, recycling, disaster relief, flood protection, and water quality.[18]

Since 1999, the U.N. Environment Programme has done a number of detailed post-crisis environmental assessments, identifying environmental risks to health, livelihoods, and security. Assessments have been carried out in the Balkans, Ukraine, Lebanon, Occupied Palestinian Territories, Sudan, Rwanda, Nigeria, the DRC, and Afghanistan. These help improve the understanding of environmental factors in conflict and pinpoint how environmental restoration can help stabilize war-torn societies.[19]

Disaster diplomacy. Disasters that strike in active or latent conflict zones may inflict suffering that cuts across the divides of conflict, often triggering goodwill and possibly jolting the political landscape. Common relief and reconstruction needs offer opportunities for collaboration, which in turn can build trust, break ingrained conflict dynamics, and perhaps facilitate reconciliation among adversaries. There have been attempts at disaster diplomacy in relations between Greece and Turkey, China and Taiwan, India and Pakistan, Ethiopia and Eritrea, and other nations.[20]

Still, there are no guaranteed outcomes. The aftermath of the 2004 tsunami disaster had diametrically opposite outcomes in two of the hardest-hit areas. In Indonesia's Aceh province, the disaster helped trigger a process that led to a successful peace agreement. But in Sri Lanka, a groundswell of popular-level goodwill did not reach to the elite political level, and post-tsunami aid became a divisive issue. Humanitarianism does not automatically create peace, but it can offer a window of opportunity for conflict transformation.[21]

Health diplomacy. Conceptually similar to disaster diplomacy, the notion of health diplomacy has been proposed as a way to generate goodwill by providing medical assistance to other countries, improving relations, and resolving conflicts, as well as advancing common public health objectives. This is especially important in the face of what some have called disease globalization (rapidly spreading epidemics like SARS or avian flu).[22]

Cuba has been a pioneer in this regard. It has engaged in vigorous "medical diplomacy" since the 1960s. It has invited thousands of students from many developing countries to be trained in its medical schools, sent thousands of its own doctors and nurses to provide care to poor communities abroad, and dispatched disaster-relief teams to several countries. In 2006, close to 29,000 Cubans served in 68 countries (though by far most of them were in Venezuela under an "oil for doctors" scheme). Achieving dramatic improvements in the health of assisted populations, Cuba's efforts have focused heavily on capacity building and preventive medicine. The programs have largely been free from political conditionality.[23]

Greening employment. Employment is affected by a multitude of factors, but disregarding environmental and resource constraints will be increasingly costly for businesses and workers. Yet "greening" technologies and workplaces through large-scale public and private investment—generating so-called green jobs—could inject a new positive dynamic into labor markets. The economic stimulus programs passed by many governments in response to the global economic crisis entailed substantial green spending. There have also been calls for a far more ambitious Global Green New Deal. Much of the green jobs discussion has focused on industrial countries and a handful of emerging economies with regard to high-tech sectors like wind and solar energy or electric vehicles. But green jobs also offer important opportunities for poverty reduction and livelihood promotion in developing countries. This involves support for recycling and composting efforts and investments to protect biodiversity, restore degraded farmland and watersheds, and make farming more organic and climate-resilient.[24]

The concepts and initiatives just described need to be replicated and scaled up. And close attention is warranted to ensuring that they are not undertaken in isolation but actually reinforce each other. In part this will take substantial investments—with resources channeled from outdated, adversarial security policies toward programs that can address the roots of insecurity and promote cooperative behavior. But a more fundamental need is institutional renewal and profound cultural change—moving away from a warrior culture that always sees new enemies lurking and toward an understanding that different nations and communities need to make peace not only with each other but also with nature.

Building the Cities of the Future

Peter Newman

Imagine a city that uses 100 percent renewable energy...where most transport is by electric light rail, biking, or walking...where the solar office block is filled with green businesses... where the local farmers' market sells fresh, bioregional produce...where parents meet in the parks and gardens while their children play without fear in streets that are car-free. This is a reality in Vauban, a new eco-city of 5,000 households within Frieburg, Germany. And in nearby Hanover, a city of 500,000 people has reduced its greenhouse gas emissions by 50 percent.[1]

How did these communities transform their cultures to make the transition that every city now faces? Vauban and Hanover took the opportunity to use every policy lever possible at every step of the way—from planning to delivery—to ensure that the goal of sustainability drove each decision.

Cities have always been places of economic and social opportunity. They emerged when hunter-gatherer societies were transformed into settled societies based on agriculture. Today's cities have grown large during the industrial era and still provide the main economic and social opportunities for the world's growing population. But cities are now having a significant environmental impact, as they are based around the consumption of fossil fuels and materials at increasing rates. They must continue to provide opportunities, but they must become more like Vauban and Hanover—sitting lighter on the planet. Indeed, the key question now is whether cities can not only reduce their impact on Earth but also contribute to its regeneration.[2]

Around the world, cities are becoming more sustainable through resilient buildings, alternative transportation systems, distributed and renewable energy systems, water-sensitive design, and zero-waste systems—with all the cleverness of a new industrial green revolution. From new cities like Masdar in Abu Dhabi to redeveloped areas like Treasure Island in the United States, Vauban and Hanover in Germany, and BedZED and the new Olympic village in London, these pioneers are dramatically reducing their ecological footprints.[3]

Helping Urban Residents Live Sustainably

BedZED is a carbon-neutral development and social housing experiment in inner London. It has many ecological innovations: it

Peter Newman is professor of sustainability at the Curtin University Sustainability Policy Institute in Perth, Australia.

used local and recycled materials; its energy-efficient design is combined with photovoltaics (PV) and biomass-fueled combined heat and power; it recycles gray water and harvests rainwater; it has local facilities to reduce the need for travel and is near a train station; and it has on-site permaculture gardens. When a detailed assessment of residents' ecological footprints was made, however, a huge variation was found in how people made use of the area's ecological features. The average footprint for some residents was around 4.4 hectares per person (still less than the average for London of 6.6 hectares), yet some residents were able to get their impact down to 1.9 hectares per person.[4]

Experiences in many early European experiments in urban ecology may hold the explanation for this. Buildings and neighborhoods that were not developed within a community can fail to achieve their design outcomes. If innovations are imposed on people who do not know how to use the new buildings as designed or why they should use less power or water or fuel, residents can simply transfer their old consumptive lifestyles to the new "eco" situations. The growth of sustainable cities will only be mainstreamed when the green transformation involves all elements of the policy process—especially the processes that help people want to change.[5]

Several key government policies can help cities move toward sustainability:

- Infrastructure to enable energy, water, transport, and waste to be managed with minimal ecological impact;
- A design to ensure that the infrastructure is efficiently available to all;
- Innovation through R&D and demonstrations to continually ensure the latest eco-technology becomes mainstream;
- Tax incentives to direct investment into these new technologies and provide people with the motivation to change their behavior;
- Regulations to set the standards high enough for sustainability technologies to cover their externalities; and
- Education to ensure households and communities want to make the changes needed.

Nowhere is this more evident than in policies about getting people out of cars.

Kicking the Car Habit

Car use is easily adopted as a way of life in cities, especially those that were developed in the past 50 years. U.S. cities use twice as much transport fuel per person as Australian cities, and those cities in turn use twice as much as European cities and five times as much as Singapore, Tokyo, and Hong Kong. Policymakers often claim that cities with a high dependence on cars are impossible to change. But with cars now being the largest single technology contributing to climate change and the one growing the fastest, it is time for decisionmakers everywhere to see how the policy changes just described can bring about a cultural transformation and get their cities to kick the car habit.[6]

A first priority is infrastructure. Cars are chosen for most destinations because they are quicker than other more-sustainable modes, and people do not like to commute more than an hour a day on average. Thus if a modern electric rail system or bus rapid transit can be installed down an urban corridor that is faster than the traffic, then people move quickly to use it. Perth's new Southern Rail meets this goal and now takes 55,000 people a day, compared with 14,000 who used to take the bus; this is the equivalent of eight lanes of traffic. Similarly, a good bicycle system and walkable urban environment means that in Copenhagen cars were used for only 27 percent of all work trips in 2003 compared with bicycles on 36 percent of such trips.[7]

The design of the city is totally enmeshed in its infrastructure priorities. When cities favor sustainable modes of transportation, then land use tends to cluster around it. But if a city only builds highways, it generally scatters in highly car-dependent patterns. Density and transport fuel use are closely linked. Planning cities to be much less car-dependent will be a key part of any plan to reduce a city's carbon footprint. For example, "transit-oriented developments" have been shown to cut residential car use in half, and residents save 20 percent on their household income by having one less car per household.[8]

New technology to make cities smarter and more sustainable is appearing and needs government assistance to be facilitated and tested. The new plug-in electric vehicles (for cars and for transit) need testing, along with the associated Smart Grids and renewable energy use that can allow cities to become 100 percent renewable. Green transit-oriented developments that can demonstrate the new technology would seem to be ideal sites for trials of such technology so that renewable transport can also mean less car use.

Every nation and city has its own way of making the adoption of more planetary lifestyles convenient and easy compared with more consumptive lifestyles. When it comes to cars, however, the more that a city is car-dependent, the harder it is to use tax incentives to change people's lifestyles. European cities have much higher gasoline taxes than American and Australian cities, and accordingly they use cars less.[9]

In the car-dominated cities of North America and Australia, the major public policy to reduce the global and local impacts has been through regulations on vehicles that have forced them to become cleaner. Following introduction of these, most urban atmospheres have become cleaner, although fuel use has continued to increase as vehicles became bigger and their use has continued to grow. Regulations also are applied to safety and congestion management, but this will continue to worsen if more and more car use is facilitated.

All these necessary policy approaches will be wasted without education on a changed role for the car and on climate change. For example, something known as the Jevons Paradox—increasing efficiency means increasing consumption—has been found to apply to car use. If people buy cars that use less fuel, they just drive them more—undermining most gains made possible through the new technology. Thus cultural change to help people to want to drive less needs to be part of any city's policy arsenal if it is to face up to the challenge of growing a sustainable city. One such program shows that this is indeed possible.[10]

German sociologist Werner Brög has developed an approach to travel demand management that is based on the belief that cultural change toward less car dependence can happen in any part of any city as long as it is community-based and household-oriented. After some trials in Europe, Brög's approach was adopted in large-scale projects in Perth, Western Australia. It has since spread across most Australian cities and to other European cities, especially in the United Kingdom, and has now been piloted in six American cities.[11]

Known as TravelSmart, the approach targets individual households directly (rather than through mass media) in a letter from the Mayor or State Minister (funds for the program are usually a partnership of the two), asking them to participate in the program. Follow-up phone calls elicit the residents' interest in receiving information and, for the few who need extra support, a potential visit from a TravelSmart officer. People select information materials to suit their individual needs and these are delivered by staff using bikes and trailers. The information is packaged in specially designed TravelSmart bags and includes walking and transit information, as well as pamphlets on why it is good for their health and the planet for people to get out of their cars more often. They encourage people to start with local trips, especially the school trip for children, which is now seen as an essential part of the healthy development of young people's sense of place and belonging in any community as well as a way to reduce obesity.

In communities where TravelSmart has been conducted, people have reduced the kilometers traveled by vehicle by around 12–14 percent—a result that seems to last for at least five years after the program ends. Where transit is not good and destinations are more spread out, the program may only reduce car use 8 percent, but where these are good it can rise to 15 percent. This is not a revolution, but it has many synergistic positive outcomes.[12]

Tom Chance, BioRegional

The BedZED development, Hackbridge, London, U.K.

People involved in TravelSmart become real advocates of sustainable transport—telling their friends how much better they feel after bicycling, walking, or taking the bus or train instead of driving. They show friends how much money it saves as well as making them feel they are doing their bit for climate change and oil vulnerability. There is evidence in Brisbane, Australia, that at least 50 percent more people than those involved in the initial household interviews were actually following the program when the surveys were done; in other words, people were spreading the message to their friends and colleagues.[13]

When people start to change their lifestyles and can see the benefits, they become advocates of sustainable transport policies in general. Governments find it easier to manage the politics of transformation to reduced car use and lower oil use when the communities they are serving have begun to change themselves.

The city of Perth has been rebuilding its rail system over the past 20 years following a strong social movement that demanded a better system. The extension of the rail system to far outer suburbs has been more positive and politically achievable than expected, with a massive 90 percent support for the last stage, the Southern Suburbs Railway. In parallel to this political process, Perth had some 200,000 households undergoing the TravelSmart program, which seems to have helped. Indeed, the Southern Suburbs Railway increased public transport patronage by 59 percent in areas without TravelSmart but by 83 percent in areas where TravelSmart was deployed to promote the new rail services. Patronage on the rail system has gone from 7 million a year to 110 million in 17 years, moving public transport from 5 to 10 percent of the work journey trips taken in the city. Perth has become a model across Australia for other cities that are now determined to upgrade their rail system funds to provide the needed infrastructure.[14]

The TravelSmart program recognizes a fundamental principle about cultural change: it works best when the change is supported by a community, when it is part of the development of social networks that support the changes in lifestyle. TravelSmart develops this social capital around sustainable transport modes rather than the dominant culture of the car. It does this through relationships established with the TravelSmart officer and with others in the local community who are making the same first steps to get out of their cars. In the workplace, TravelSmart is found to work well when a TS Club is formed that enables people to share experiences, bring in local speakers, and lobby for facilities like showers for bike riders and transit passes instead of parking spaces.

Planetary Lifestyles

The same approach to cultural change that TravelSmart uses can be applied to other aspects of sustainability at the household level—reducing energy, water, and waste. The program needs to provide infrastructure for the new technologies, an urban design that ensures the technologies are efficiently available for all residents, R&D on the best options available, regulations to set the energy and water use in buildings and appliances at the highest possible level, tax incentives to push people toward more "planetary lifestyles," and education to motivate people.

As with TravelSmart, the possibility of using educational programs to underpin these policy areas is critical to achieving the necessary planetary cultural change. In many cities, approaches to community-based planetary education are emerging as the politics of climate change becomes a major political force.[15]

Perth has built on its TravelSmart program to create a successful household education approach, known as LivingSmart, that brings sound and locally relevant material into people's homes. The eco-coaches who have worked with the first 15,000 households in a trial run have found enormous enthusiasm from people who have been looking for this targeted assistance. Using unsolicited phone calls to residents, the program is finding that 74 percent of households are interested in making changes to improve energy, water, waste, and travel sustainability. Half of the households contacted are signing up for ongoing coaching for special meters, advice on gardens, workshops, and home audits.[16]

Unlike TravelSmart, where change tends to occur slowly and incrementally, the LivingSmart program is receiving reports from households of instant and radical changes—replacing inefficient lights, for example, or ordering PV, solar hot water, and grey water recycling systems. The program is aiming to reduce carbon dioxide emissions 1.5 tons per household a year. (Australians on average are responsible for 14 tons per household.) This will save participants up to 10 percent in their gas, electric, water, and petroleum bills.[17]

The social capital being built up around these new technologies and lifestyles is also proving highly infectious and can become the basis of a major social movement if governments are prepared to adopt the approach more broadly.

The end result of household programs like these, combined with all the other policy initiatives, may be the beginning of a transformative sustainability process—not just in the actual savings in fossil fuels and other valuable materials, but in the growing sense that households and communities can achieve a transition to a more sustainable city. This hope is the currency of growth toward sustainable cities. It can enable people to begin to imagine a city that is more regenerative than destructive of Earth.[18]

Reinventing Health Care: From Panacea to Hygeia

Walter Bortz

According to Greek legend, Asclepius, Apollo's son, was charged with the oversight of human well-being. He in turn delegated his responsibilities to his two daughters, Hygeia and Panacea. Hygeia was entrusted with the health aspects of humanity's life course, and Panacea with the disease and illness elements. This dichotomy of health and disease has pervaded medicine's brief history.[1]

Since the discovery of penicillin in 1865, modern medicine has focused the bulk of its efforts on the treatment and repair of infectious disease, with many positive results. The development of antibiotics and the embrace of antisepsis (the prevention of infection) have unburdened humanity from many historically persistent scourges. Smallpox was eradicated in 1979, polio has been eliminated from much of the world, and infections such as guinea worm disease, measles, and rubella are no longer present in many regions. In several industrial countries, life expectancy rose 30 years in the space of a century, an event that has not been rivaled since.[2]

In the last 60 years or so, however, new disease conditions have emerged that are not caused by bacteria, viruses, or other microorganisms. Instead, they are triggered by environmental pollution and by lifestyle factors such as poor diet and a lack of exercise. In many countries, obesity has become the "norm," with health implications such as diabetes, hypertension, and arthritis. For the first time in history, Africa is now home to more people who are overweight than are underfed. Rather than focusing on isolated disease components and individual events, medicine has entered the era of multiple causes and diagnoses.[3]

Facing New Health Challenges

The major contributors to global mortality today are for the most part preventable. According to the World Health Organization (WHO), childhood and maternal malnutrition cause an estimated 200 million "years of life lost" annually, followed by physical inactivity and obesity (150 million years), unsafe sex (80 million years), and tobacco (50 million years). A study of the "actual causes of death" in the United States in 2000 lists tobacco as the number one killer, with poor diet and physical inactivity coming in a close second.[4]

The global community has made important progress in responding to these chal-

Walter Bortz is clinical associate professor of medicine at Stanford University School of Medicine.

lenges—from improvements in water quality to the treatment of infant diarrhea. Yet the collective response by the medical system has been primarily to alleviate symptoms. This is because it is rarely possible to "cure" the major killers of today. The two principal therapies in medicine's black bag—surgery and pharmacy—are largely irrelevant to the new disorders of aging and poor lifestyle choices. The medical system can treat symptoms, but heart attacks, stroke, diabetes, emphysema, arthritis, and neurologic disorders remain resistant to curative effort.[5]

These ailments are, however, notably open to modifications in lifestyle—from improved diets and exercise to efforts to reduce the use of tobacco and alcohol. But improving health literacy remains a significant challenge. Hygeia's product of "health" has effectively played second fiddle to Panacea's product of disease.[6]

Overhauling Global Health Care

From a financial perspective, prevention pays poorly, while sickness pays. In the United States, health care spending accounts for over 15 percent of gross domestic product (GDP)—a figure that is projected to reach 20 percent by 2015. Yet the current U.S. health system, addicted to high payments from surgery and pharmacy, does not address the job requirements of medicine. In 2000, WHO ranked the United States first in the cost and responsiveness of its health system, but thirty-seventh in performance and seventy-second in overall health. U.S. infant mortality is the highest among industrial countries (see Table 9), and studies suggest that as the obesity epidemic spreads, today's children may be the first in U.S. history who will not live as long as their parents. In the southeastern part of the country, life expectancy is falling to levels that approach those in Russia.[7]

Nearly all industrial countries provide some form of mandated universal health insurance coverage, but the United States is notably absent from this list. The degree of privatization in the U.S. medical system exceeds that in nearly all other countries, eroding much of the "locus of control" within communities. Nobel laureate economist Kenneth Arrow has observed that medical care cannot function like a standard competitive market because of inherent uncertainties and the gross imbalance of skills between physician and patient. Such "market failure" creates unlimited opportunities for perverse incentives such as rewarding medical procedures instead of health outcomes.[8]

Worldwide, the primary emphasis of med-

Table 9. Health Care Performance, Selected Countries, 2006

Country	Health Expenditures as Share of GDP	Infant Mortality	Life Expectancy	Healthy Active Life Expectancy*
	(percent)	(number per thousand live births)	(years)	(years)
Cuba	7.1	5	78	68
France	11.1	4	81	72
Japan	7.9	3	83	75
Sweden	7.9	3	81	73
United States	15.3	7	78	69

**Data for 2003.*
Source: See endnote 7.

icine needs to be on health, not disease, and on prevention instead of repair. (See Box 17.) The medical structure must serve its function, which is the assurance of the human potential. Panacea must be downsized and Hygeia reconstituted in place of disease medicine.[9]

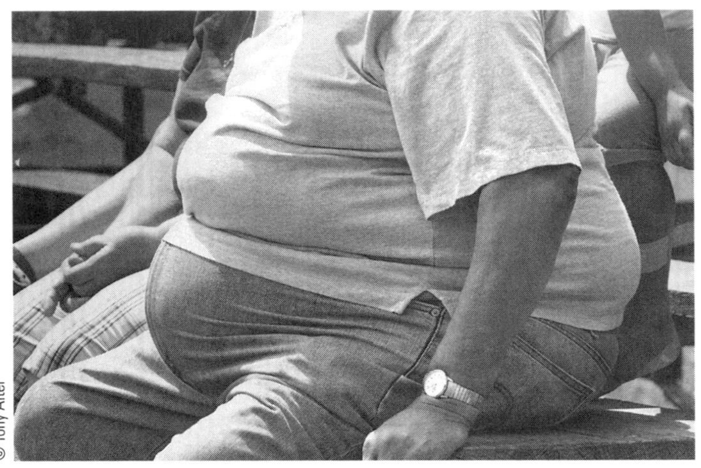
© Tony Alter

An example of obesity in the U.S.

Emphasizing Health over Disease

There are hints of a transition in this direction. The region of North Karelia in Finland was notorious for having one of the highest rates of heart disease in the world, affecting 855 out of every 100,000 residents. But since 1972, the North Karelia Project—an integrated, population-based preventive effort—has registered a 68-percent reduction in heart-related deaths and a 49-percent reduction in overall mortality. WHO has since replicated the experience in other communities.[10]

France, credited as having one of the best-performing health care systems in the world, has always maintained a prominent role for its village doctors. According to a WHO report, these physicians provide the "personal, one-to-one, empathy, trust and intimate knowledge of individual case histories" that are often lacking in more "advanced" health care systems.[11]

And Cuba, perhaps more than any country, has emphasized primary care. The country's "neighborhood health system" serves as an intimate antenna to the health circumstances of the people. Health care spending in Cuba represents only 7.1 percent of GDP, yet the country's average life expectancy is 78 years and its infant mortality rate is lower than the U.S. rate. Interestingly, the rates of diabetes and obesity in Cuba fell precipitously after the U.S. trade embargo was imposed in the 1960s, as access to total calories, unhealthy foods, and mechanized transport all declined.[12]

The global health budget needs to include a massively increased commitment to health education. Health illiteracy is the biggest killer everywhere, and it deserves significant attention. People need to learn, for example, that when medical costs are included, fast food is in reality not all that cheap.

Research and training institutions should be redesigned to increase knowledge of the environment as well as determinants of human behavior. Medical schools must realign their teaching to the personal requirements of health medicine. And health educators must assume prominence over disease technologists. If something like half of all illness is preventable or lifestyle in origin, then half of health care costs are approachable—suggesting that a new emphasis must be placed on knowing how to age and how to die.[13]

Greening Health Care

The medical care system needs to be "greened" as well. The current health care sector is characterized by high usage of energy and other resources, rising emissions of greenhouse gases, and the release of toxins such as mercury and pharmaceutical chemicals into the environment. Medical institutions are also notorious

Box 17. Making Social Welfare Programs Sustainable

Around the world, but especially in industrial countries, social welfare programs take up a large share of government budgets. In 2005, social expenditures accounted for 16 percent of gross domestic product in the United States, 19 percent in Japan, 27 percent in Germany, and more than 29 percent in France and Sweden. The fundamental goal of these publicly mandated programs is to guarantee a certain minimum standard of living to every citizen, based on the premise that it is the state's responsibility to provide for the general well-being of its citizens.

In an ecologically constrained future where sustainability will need to take precedence over economic growth, governments will have to find ways to make social programs as effective as possible while minimizing cost and environmental impact. Fortunately, newer, more effective, more efficient programs are continually being tried, and many small programs that have proved effective are being scaled up in industrial and developing countries alike.

"Social farming," for example, is the use of agriculture for social purposes. More specifically, it involves the use of agricultural resources for rehabilitation, social inclusion, and education. It serves several purposes in that it generates agricultural products while creating jobs, providing social services, and building social capital. Thus far the popularity of social farming has been greatest in Europe, where roughly 700 social farms have been initiated in the Netherlands, and more than 1,200 have been started in France.

Norway has implemented an innovative twist on social farming on an island 75 kilometers south of Oslo at Bastøy Prison, which has been converted into the world's first "ecological prison." Inmates who have been transferred from maximum security prisons now live in unlocked houses, care for livestock, practice organic agriculture, and run a forestry operation. The prison's "human ecological philosophy" effectively rehabilitates prisoners by teaching useful skills while offering them responsibility and a crucial level of control over their own affairs. The prison also uses less energy, generates less waste through its composting program, and produces food and wood products for consumption and sale, thus reducing its ecological impact while saving money for the Norwegian government.

"Conditional cash transfer" (CCT) programs represent another innovative, low-cost, and highly effective social service. These involve giving money directly to poor households based on certain agreements, such as sending children to school and providing for their health and nutrition. The concept stems from the idea that the conditions on which transfers are made, especially those related to health and education, will generate human capital that will provide returns significantly greater than the value of the cash transfers themselves. *Oportunidades* in Mexico and *Bolsa Familia* in Brazil are two of the more well-known and successful CCT programs. Studies in Mexico have found that *Oportunidades* has significantly increased student enrollment and considerably reduced both maternal and infant mortality, among other outcomes.

Social programs of this type are proving very effective because they are low-cost, low-impact, and highly targeted. As these and other programs are scaled up and expanded to reach more people, governments will continue to discover more and better ways to meet their responsibilities to both people and the planet in more effective and less costly ways.

—*Kevin Green and Erik Assadourian*

Source: See endnote 9.

A severly malnourished child, said to be 9 years old, in a village in India.

users of products made from polyvinyl chloride, which results in the production of dioxin, a known endocrine disruptor, when the wastes are incinerated.[14]

Hospitals are supplanting steel mills and oil refineries as major polluters. According to the U.S. Environmental Protection Agency, hospitals in the United States use more than twice as much energy per unit area as office buildings. And a 2001 study of Lion's Gate Hospital in Vancouver, Canada, calculated the "ecological footprint" of the facility to be 2,841 hectares, corresponding to a land area about 719 times larger than the hospital's actual area. This was significantly higher than the estimated footprint for the City of Vancouver, which was about 180 times its actual area.[15]

Health Care Without Harm, a global coalition of 473 organizations in 52 countries, is working to limit the environmental footprint of industrial medicine by addressing everything from toxics reduction to the purchasing of more environmentally friendly products by medical institutions. In addition to reducing their ecological impact, green hospitals can also have a positive impact on patient health.[16]

Ultimately, hospitals and health care systems need to be redesigned in ways that do not make patients or other people sicker, such as through the production of toxic waste. Even more important, health care needs to be redesigned so that people are not getting sick in the first place and therefore are less likely to end up in the hospital to suffer needlessly and to indirectly cost societies and the environment significant resources. This will require not just a subtle retuning of health care policies, but a complete reinvention of health care as it is practiced today.

Earth Jurisprudence: From Colonization to Participation

Cormac Cullinan

"If we have our land and clean air and water, our communities can have *sumak kawsay*—the good life," the indigenous leader said with calm conviction. "I don't know why you are calling this a new development model—we have always lived this way. The duty of the state is to ensure that these fundamental rights are protected in order to safeguard the well-being of our people."[1]

The leader was speaking to legislators, politicians, lawyers, and activists gathered in Quito in November 2008 to discuss how best to implement the provisions in Ecuador's new constitution that recognize that nature has rights that must be enforced by law. The constitution sets the achievement of well-being in harmony with nature (*el buen vivir* or *sumak kawsay*) as a fundamental societal goal. Inclusion of these provisions was achieved in a remarkably short time by the collective efforts of indigenous peoples' representatives and environmental nongovernmental organizations (NGOs) supported by lawyers from the Community Environmental Legal Defense Fund (CELDF) of the United States.[2]

In a world where almost all legal systems define nature as property and "natural resources" as available for state-sanctioned exploitation and where the highest goal of government is the pursuit of an ever-growing gross domestic product, Ecuador's constitution is a strong indicator that a centuries-old logjam in legal and political thinking and practice is beginning to break up. Legislators are starting to recognize that human well-being is a consequence of the well-being of the Earth systems that sustain us.

From Colonial Law to Earth Jurisprudence

Almost all of the "environmental crises" that threaten contemporary industrialized civilization are caused by ecologically unsustainable and harmful human practices. Since these practices reduce the prospects of our descendants surviving and thriving, from an evolutionary perspective—as well as from ethical, spiritual, and pragmatic perspectives—they are contrary to the interests of the species. The fact that many of these practices are allowed to continue and even receive incentives indicates that today's governance systems are dysfunctional.

Legal systems are failing to protect the

Cormac Cullinan is a practicing environmental attorney, author, and honorary research associate of the Department of Public Law, University of Cape Town.

Earth community in part because they reflect an underlying belief that humans are separate from and superior to all other members of the community, and that the primary role of Earth is to serve as "natural resources" for humans to consume. These beliefs are demonstrably false. Humans are, of course, but one of many species that have co-evolved within a system they are wholly dependent on. In the long term humans cannot thrive in a degraded environment anymore than fish can survive in polluted water.

Just as colonial laws did not recognize the rights of indigenous peoples and facilitated the exploitation of them and their land, most contemporary legal systems do not recognize that any indigenous inhabitants other than humans are capable of having rights. The law defines land, water, other species, and even genetic material and information as "property," which entrenches an exploitative relationship between the owner (a legal subject with rights) and the property (legally speaking, a "thing" incapable of holding rights). Most legal systems also grant human beings legal rights to exploit all aspects of the Earth community (through mining, fishing, and logging concessions, for example), with predictably dire consequences for the integrity and functioning of indigenous communities.

One of the most exciting developments in law today is the emergence on several continents of initiatives to bring about a fundamental change in human legal systems. These all share the belief that a primary cause of environmental destruction is the fact that current legal systems are designed to perpetuate human domination of nature instead of fostering mutually beneficial relationships between humans and other members of the Earth community. They all advocate an approach known as Earth jurisprudence. (See Box 18.) According to this philosophy, human societies will only be viable and flourish if they regulate themselves as part of the wider Earth community

Box 18. Principles of Earth Jurisprudence

- The universe is the primary law-giver, not human legal systems.
- The Earth community and all the beings that constitute it have fundamental "rights," including the right to exist, to have a habitat or a place to be, and to participate in the evolution of the community.
- The rights of each being are limited by the rights of other beings to the extent necessary to maintain the integrity, balance, and health of the communities within which it exists.
- Human acts or laws that infringe these fundamental rights violate the fundamental relationships and principles that constitute the Earth community and are consequently illegitimate and "unlawful."
- Humans must adapt their legal, political, economic, and social systems to be consistent with the fundamental laws or principles that govern how the universe functions and to guide humans to live in accordance with these, which means that human governance systems at all times must take account of the interests of the whole Earth community and must:
  - determine the lawfulness of human conduct by whether or not it strengthens or weakens the relationships that constitute the Earth community;
  - maintain a dynamic balance between the rights of humans and those of other members of the Earth community on the basis of what is best for Earth as a whole;
  - promote restorative justice (which focuses on restoring damaged relationships) rather than punishment (retribution); and
  - recognize all members of the Earth community as subjects before the law, with the right to the protection of the law and to an effective remedy for human acts that violate their fundamental rights.

and do so in a way that is consistent with the fundamental laws or principles that govern how the universe functions.[3]

This approach requires looking at law from the perspective of the whole Earth community and balancing all rights against one another (as is done between humans) so that fundamental rights like the right to life take precedence over less important ones such as the right to conduct business. Currently the rights of humans, and particularly of corporations, automatically trump the rights of all others. This also means that while a fox eating a rabbit could be seen as a violation of the rabbit's right to life, it does not violate the laws that govern the universe because the maintenance of predator-prey relationships is fundamental to preserving the integrity of the whole community. Killing to survive serves the greater good in a way that killing for sport does not.

The Evolution of Earth Jurisprudence

A few prescient commentators have for several decades drawn attention to the need for legal systems to take an evolutionary leap forward by recognizing legally enforceable rights for nature and other-than-human beings. One of the best-known articulations of this position is that of Christopher Stone, who in 1972 published a seminal article entitled "Should Trees Have Standing? Towards Rights for Natural Objects." He pointed out that the widening of society's "circle of concern" had led to the recognition of more extensive legal rights for women, children, Native Americans, and African Americans. There was no good reason, he argued, why increasing public concern for the protection of nature could not lead to the recognition of nature's rights. This would allow legal suits on behalf of trees and other "natural objects" and would mean that damages could be recovered and applied for their benefit.[4]

As Chilean lawyer Godofredo Stutzin pointed out in 2002, one practical advantage of recognizing rights for nature is that anyone seeking to alter or destroy any aspect of it would have to put forward reasons to justify why this should be permitted, instead of making people who wish to prevent destruction prove why nature should be conserved.[5]

Perhaps the clearest calls for the development of a new jurisprudence have come from Thomas Berry, eminent American cultural historian, religious scholar, and philosopher. He maintains that the legal systems in countries such as the United States legitimized and facilitated the exploitation and destruction of Earth. Berry has argued that "we need a jurisprudence that would provide for the legal rights of geological and biological as well as human components of the Earth community. A legal system exclusively for humans is not realistic. Habitat of all species, for instance, must be given legal status as sacred and inviolable."[6]

In April 2001 the Gaia Foundation of London convened a meeting of lawyers, eco-psychologists, wilderness experts, anthropologists, and environmentalists to begin the process of developing this new jurisprudence. And immediately before the World Summit on Sustainable Development in 2002, *Wild Law: A Manifesto for Earth Justice* was published, articulating an Earth-centric approach to law and governance. The term "wild law" refers to laws that articulate and give effect to Earth jurisprudence by fostering mutually beneficial instead of exploitative relationships between human beings and other members of the Earth community.[7]

Wild Law proposed that the primary purpose of legal and political systems should be to ensure that human beings act as "good citizens" of an Earth community rather than merely defining antisocial behavior in relation to other human beings. This would require recognizing that the other members of the Earth community also have rights that must be balanced against human rights. The precise nature and way in which Earth jurisprudence was expressed would vary according to the particular context, but all would be consistent with the fundamental principles on which the Earth community is ordered.[8]

In some cases the alignment of laws with the fundamental principles of the natural system of order has been happening for pragmatic reasons as lawmakers and officials seek to create more effective governance systems. For example, the widespread adoption of the "ecosystem approach" in relation to fisheries and the conservation of wild species and places can be seen as pragmatic recognition that it is impossible to manage human impacts on an ecosystem successfully by looking only at a part of that system, such as a particular fish stock. Similarly, concepts like intergenerational equity recognize the need to align human legal systems with the far longer time scales on which nature operates, while moves toward bioregional planning reflect a growing acceptance of the fundamental natural principle of diversity and the benefits of shortening feedback loops by allow for more local decisionmaking.[9]

BK59

Energy landscape: Open-pit coal mining in western Germany.

Helping Local Communities Change the Rules

In the United States, much of the pioneering work in Earth jurisprudence has been undertaken by the Community Environmental Legal Defense Fund, founded and led by Thomas Linzey. For many years the CELDF successfully represented communities that wished to prevent or challenge authorizations for corporations to undertake a range of environmentally destructive activities, such as the disposal of sewage sludge on land, the establishment of massive pig farms, or mining. Initially the CELDF used the conventional legal strategy of attacking deficiencies in the authorization processes. Despite initial successes, however, Linzey soon realized that the victories were short-lived because the corporations simply repeated the process in a manner that complied with all legal requirements—and eventually triumphed.

Communities could not protect themselves and the ecosystems within which they lived because the rules of the legal system as a whole were skewed in favor of both corporations and property owners. In fact, environmental laws mainly regulate how quickly natural communities are destroyed rather than preventing the destruction. A fundamentally new approach was required.[10]

The first step was to expose the limitations of existing regulatory systems and how corporations have shaped the law so that it allows commercial interests to override the interests of local communities and facilitates the lawful degradation of nature. To do this, CELDF and Richard Grossman (co-founder of the Programme on Corporations, Law and Democracy) established the Daniel Pennock Democracy School, which runs intensive short courses around the United States for communities that organize themselves to resist environmentally and socially harmful activities in their areas.[11]

The second step was to empower local communities to use legal systems proactively to support the establishment of sustainable, local economies. Realizing that local communities could not secure their own well-being without protecting the integrity and functioning of the ecological communities within which they lived, the CELDF developed a strategy of assisting communities to draft local ordinances that:

- re-assert their right to prohibit activities harmful to their well-being,
- recognize rights for natural communities,
- enable local governments and individuals to sue for damages to be used for the restoration of any damage to ecological communities, and
- strip away the legal personality of corporations who contravened the ordinances (and hence their right to benefit from the civil rights in the U.S. Constitution).[12]

If corporations and state governments take legal action to challenge the validity of these ordinances, they simply further expose the extent to which the legal system has been hijacked by vested interests.

The CELDF has helped more than 100 local governments in the United States pass local ordinances with one or more of these features. In the process of drafting "home-rule" charters, local communities such as Spokane in Washington State and Blaine Township in Pennsylvania are accepting that the only way to fulfill their role as trustees of natural communities is to create legal mechanisms that will enable local people and communities to enforce the inalienable and fundamental rights of natural communities as well as their own rights to a healthy environment.[13]

In South Africa, the 2008 National Environmental Management: Integrated Coastal Management Act now requires decisions about the coastal zone (which includes the 200-nautical-mile exclusive economic zone) to be made in the interests of the "whole community," which includes more than just humans.[14]

Teaching Wild Lawyers and Civil Servants

Conventional law schools are being challenged to identify how natural systems function and how the interests of other-than-human members of natural communities should be taken into account in decisionmaking. The Center for Earth Jurisprudence (CEJ) was established in 2006 by two Catholic Universities in Florida in order to re-envision law and governance in ways that support and protect the health and well-being of the Earth community as a whole. Inspired primarily by the works of Thomas Berry, the CEJ adopts a multidisciplinary approach and seeks to train a new breed of lawyers who are equipped to deal with the reality of regulating human behavior in a highly interdependent Earth community. In the United Kingdom, the UK Environmental Law Association has established a standing Wild Law working group and hosts annual Wild Law Weekends in the countryside at which members explore and develop these concepts.[15]

In Africa, when Mellese Damtie introduced his students at the Ethiopian Civil Service College to the book *Wild Law*, they were particularly enthused by the suggestion that African customary law—long ignored as "primitive"—could be a source of inspiration for contemporary governance systems. Field research by government administrators studying at the college revealed that a rich heritage of customary laws and cultural practices designed to ensure respect for nature had survived among rural communities in Ethiopia. For example, in areas where reverence for rivers means that people remain silent or speak only in hushed tones when crossing them, the watercourses are in a far better condition that elsewhere.[16]

Future Prospects

The view that the long-term viability of human societies cannot be attained at the expense of the Earth community is supported both by the teachings of many ancient traditions and religions and by the findings of physics and ecology—all of which point to the interconnectedness of everything and the futility of attempting to understand any part of a system without reference to its context. Achieving widespread acceptance of this perspective in a consumerist world presents a major challenge, particularly in the face of corporations and persons with a vested interest in maintaining the exploitative status quo.

The rapidly intensifying challenge of climate change has exposed how ineffective international and national governance regimes are in dealing with the side effects of consumerism and the excessive use of fossil fuels. But there are still major differences regarding how best to respond. Most governments today favor a combination of new technology and better application of existing regulatory systems. Ecuador is exceptional in opting to make a fundamental change to the architecture of its governance system by recognizing the rights of nature and redefining its concept of development. There, the existence of a large number of people who had not wholly adopted western consumerist values appears to have been a crucial factor in securing the recognition of the rights of nature in the constitution. And in a speech to the U.N. General Assembly in April 2009, President Evo Morales of Bolivia called for a Universal Declaration of the Rights of Mother Earth, indicating the potential for these ideas to spread rapidly.[17]

At present the most promising prospects for promoting "eco-centric" law and governance appear to be at the local level, where appeals to traditional values and cultures of resistance have increasing resonance. The CELDF's democracy schools in the United States reconnect people with activist movements of the past, including abolitionists and suffragists. In India, Navdanya—an organization founded by environmental activist Vandana Shiva—is a prime mover in the Earth democracy movement that has succeeded by building on existing cultural understandings of the sacred dimensions of seeds, food, water, and land and on traditions of resistance to colonial authority.[18]

In Africa and Colombia, the Gaia Foundation and local organizations have been working with traditional communities and elders to develop a similar approach, which they term "community ecological governance." Reconnection with elders and the rediscovery of the wisdom in customary law systems has also inspired Kenyan lawyers and activists from the NGO Porini to go to court to win the right for local communities to assume custodianship of sacred hills and groves and to begin restoring them.[19]

The speed and the extent to which existing environmental and social justice organizations and networks adopt this perspective is likely to be a crucial factor in determining the impact of eco-centric governance initiatives. If these organizations realize that they could greatly enhance their effectiveness by collaborating on the basis of the common understanding that sustaining human well-being requires protecting the whole Earth community, this eco-centric approach would spread rapidly through the web of relationships that already connects them. This could foster a rapid uptake of this approach of Earth jurisprudence.

Media: Broadcasting Sustainability

The media can be a highly effective tool to shape cultures—painting pictures of how people live, broadcasting social norms, modeling behaviors, acting as a vehicle of marketing, and distributing news and information. These important roles can be used to spread either a cultural pattern of consumerism or one that questions consumerism and promotes sustainability. Although the vast majority of media today reinforce the former—through advertising, product placement, and much of the content—there are efforts worldwide to tap media's vast reach and power to promote sustainable cultures, as described in this section.

Considering the commanding role that marketing plays in stimulating consumerism, redirecting it to promote sustainable behaviors will be essential. Jonah Sachs and Susan Finkelpearl of Free Range Studios describe "social marketing"—marketing to encourage socially positive behaviors like avoiding smoking, wearing seatbelts, practicing safe sex, or consuming less stuff—which can play an important role in redirecting how people live. Granted, at the moment just a tiny percentage of marketing budgets promotes such social goods.

While social marketing is encouraged, governments will need to limit or tax overall marketing pressures. A few governments are working to tackle advertising directly, such as the Spanish government, which voted to ban commercials on its public television stations starting in 2010. Yet with advertisers' influence over policymakers, these efforts have been few and far between. Robin Andersen and Pamela Miller of Fordham University point out that media literacy can help limit the effectiveness of the romantic visions of consumption created by marketing—and unlike regulation, it can be easier to introduce across societies.[1]

Beyond the mass media, the arts also play an important part in inspiring people to better understand the effects of consumerism and to live sustainably. For example, the cover of *State of the World 2010* by artist Chris Jordan is a recreation of a famous woodprint by Japanese artist Katsushika Hokusai—except Jordan's version is made out of 1.2 million bits of plastic trash. This vast number, representing the pounds of plastic that enter the world's oceans every hour, has a visual power that can represent the destructive nature of consumerism far better than yet another statistic would. Music, as Amy Han of Worldwatch describes, can also be a useful educational tool, inspiring people to live more sustainably and mobilizing them to join political efforts to help drive change.[2]

Two Boxes in this section expand on the

role of the arts: one describes the power of film and the other considers the potential for all individuals to become artists rather than consumers. Finally, a Box on the importance of journalism in effectively educating people about the environment and their role in it rounds out the section.

People spend significant portions of their lives interacting with media. Today they have the potential to create their own programming, music, art, films, and news and to distribute these farther than ever before—not just through formal channels but through YouTube, Facebook, local radio broadcasts, Web sites, even posters and self-published books. The more that this content can promote sustainability and redirect people away from consumerism, the more likely it is that humanity will avoid a future conjured up by movies like *Soylent Green* or *WALL-E* and instead create a future of high-quality lives for all.

—Erik Assadourian

From Selling Soap to Selling Sustainability: Social Marketing

Jonah Sachs and Susan Finkelpearl

Sixty years ago, Americans greeted the postwar era with a thrift-based value system that had gotten them through two decades of war and economic depression. Corporate industry, meanwhile, exited the war capable of producing more goods than ever. But with the soldiers they once supplied now back home, they needed a new customer base. If only industry could reverse the thrift-based values of the American people, then their ramped-up infrastructure could continue pumping out goods, which would be readily bought by willing consumers.

Enter Madison Avenue. Marketers responded to industry's challenge decisively, taking a dramatic leap forward in marketing sophistication. They rejected the typical fact-based approach of advertisements in favor of an identity-based, storytelling construct. The result? They created a radical reversal of thrift values and an explosion of consumerism that ignited in the United States in the 1950s and spread around the world. This became the era when people met the Marlboro man and came to believe that the cigarette someone smoked said a lot about who the person was. They embraced the idea of perceived obsolescence, accepting that owning this year's model television was a sign of virtue even if last year's model was still working perfectly well. Before long, even cultural resistance had requisite consumable products, such as the Volkswagen (VW) Beetle.

As is clear today, Madison Avenue's success has had deep, unintended consequences, and sophisticated story-based marketing continues to drive its relentless growth. Yet the seeds of the current consumption crisis may also contain powerful solutions. If marketers were able to motivate a massive reorientation of cultural values and behaviors in relatively little time 60 years ago, can they do it again? Could a revolution in social marketing, where marketing principles are used to change social behavior rather than sell a product, drive a new set of values that would lead to the lifestyles and political changes necessary to confront today's ecological crises?

Certainly, social marketing faces major hurdles. In 2008, spending on advertising was estimated at over $271 billion in the United States and $643 billion worldwide. Today approximately only one in every thousand marketing dollars is spent on broadcast pub-

Jonah Sachs is co-founder and creative director of Free Range Studios, a design and communications firm. **Susan Finkelpearl** is online strategy director at Free Range Studios.

lic service announcements that market for the public good—and only a tiny fraction of that is spent on sustainability issues.[1]

But there are also enormous opportunities. Social marketing has a 40-year history of experience to draw upon, plus there are vast lessons to learn by observing traditional consumer marketing. The Internet has rapidly leveled the playing field in the media marketplace by reducing distribution costs and removing the barriers of traditional corporate gatekeepers who limited the broadcast of messages that ran counter to consumerism's values. And the emergence of social media has spawned a "viral" distribution model through which an inspiring message can move almost instantly and at nearly no cost through networks of mutual trust.

For social marketers to play a role in the transition from consumerism to sustainability, they will need to draw on the main lesson learned by consumer marketing in the 1950s: facts alone do not sell behavior change. Instead, people working to foster sustainable behavior must use storytelling to reach audiences on a human, personal scale.

Stories Change Behaviors

As social marketers craft a strategy for this critical next decade, understanding and harnessing the power of emotional storytelling may be their most important task. Table 10 outlines a few of the most successful product and social marketing efforts since the 1950s and describes how human-scale character and stories, as opposed to facts and product attributes, have built the most powerful brands and behavior change.[2]

Iconic, story-based campaigns do not simply shift the perception of a product or activity. To change behavior on the scale they do, such campaigns have to shift how millions of people see themselves and how they are defined by, for example, their choice of cigarette, car, computer, or social behavior. But is it storytelling per se that makes these campaigns so successful?

Writer and philosopher Joseph Campbell offers a compelling reason to believe that human-scale storytelling is key to opening people up to changing instinctive "tribal" identities and altering behavior. Campbell's views even imply that social marketing may have an advantage over product marketing in this arena.[3]

In his seminal work, *Hero with a Thousand Faces*, Campbell presents a survey of mythology across broad cultural contexts and millennia and finds strong commonalities. He hypothesizes that human beings are, in fact, genetically hardwired to see their world in terms of stories. And what's more, these stories are strikingly similar. They share certain archetypal characters like the hero, the nemesis, and the mentor, and they follow a plot of invitation to adventure, acceptance of that invitation, battle with the nemesis, and then return.[4]

What is of particular interest to social marketers about Campbell's theories is that the setting for these adventures is often a broken world in need of healing. What's more, the return involves the hero coming back to society with the wisdom to heal it. Seen through this lens, stories of a societal shift from consumerism to sustainability fit perfectly into humanity's pre-formed ideas of what a hero's journey is all about. A hero is someone who helps to heal society's ills.

Campbell's theories do not stop at saying that people respond to stories. He believed that stories motivate behavior and identity, which might explain the success of storytelling marketing efforts to change consumer activity. "The myth is the public dream and the dream is the private myth," wrote Campbell in describing how deeply people internalize stories and seek to place themselves as the heroes within them.[5]

In the field of public health, the power of

Table 10. Selected Successful Product and Social Marketing Campaigns

Product or Cause	Story-Based Campaign	Result
Marlboro cigarettes	In a series of windows into the life of a fictional American hero, the Marlboro Man, the campaign focuses on the man. The product is merely an accessory.	The Marlboro Man is one of the most familiar faces in the world and solidified Marlboro as the top cigarette brand for the past 40 years.
Volkswagen Beetle	A campaign that began in 1959 spoke frankly about consumer frustration with planned obsolescence and the Big Three car companies' branding puffery. Instead of targeting consumers' impulse to buy a car, it targeted their impulse for cultural resistance.	The campaign completely reversed Americans' perception of what had been seen as a "Nazi car." The VW beetle became the symbol of cultural resistance and 1960s culture. It is still one of the most analyzed and admired campaigns in advertising history.
Seat belt use	In 1985, the "You can learn a lot from a dummy" campaign introduced two charming crash-test dummies, Vince and Larry. The dummies showed viewers exactly what it looked and felt like to be in a car accident.	In 1986, 39 percent of drivers in 19 U.S. cities reported using their safety belts, compared with 23 percent in a 1985 study. The campaign was a significant factor among several that influenced this increase. The campaign also created political cover for mandatory seat belt legislation that eventually pushed compliance nationwide past 85 percent.
Apple computers	Apple's "1984" ad said nothing about computers and ran only once on television during the 1984 Super Bowl. It simply showed a lone rebel smashing through the Orwellian dominance of its PC competitor, laying the groundwork for Apple users to identify heavily with the brand.	*Adweek* called 1984 "the best ad ever created"; Apple II sales accounted for 15 percent of the market share in its first year. It was the beginning of a string of story-based campaigns that would make Apple one of the most identifiable lifestyle brands in history. Apple's more recent "Get a Mac" campaign has millions of Americans identifying so much with the brand that they repeat the mantra "I'm a Mac."
Raising awareness about over-consumption	*The Story of Stuff* took users into the 10-year journey of activist Annie Leonard as she explored where "stuff" comes from and where it winds up when it gets thrown away. Leonard's high-level analysis of the materials economy was boiled down to simple stories told on the human scale.	This movie, by Free Range Studios, quickly went "viral" on the Internet when it was released in 2007. Since then, it has been seen by more than 7 million people in 224 countries, translated into 10 languages, and featured in hundreds of U.S. classrooms.
Reduction in obesity in the United States	Morgan Sperlock's film *Supersize Me* showed viewers the disconcerting health and appearance effects on one man of eating nothing but McDonald's meals for 30 days.	The film was an enormous critical and commercial success. Shortly after the film's release, McDonald's removed the Supersize option from its menu.

Source: See endnote 2.

archetypal storytelling has gone well beyond theory and has proved to be effective worldwide. Beginning in the 1970s, Mexican television executive Miguel Sabido began to practice Entertainment-Education (E-E), which spread public health messages by embedding them into soap operas. Sabido's shows influenced audiences by encoding health behaviors into the interpersonal dramas of three types of role models: positive, negative, and transitional. These models map closely to Campbell's archetypes of the mentor (the source of wise behavior), the nemesis (the antithesis of the mentor), and the hero (the initiate who must choose the correct behavior).[6]

Viewers of Sabido's E-E shows were expected to identify closely with a transitional character and, by seeing that person make good choices about sex, marriage, and family planning, believe that they too could make positive behavior changes.

Courtesy Free Range Studios

The online movie *The Story of Stuff* reminds viewers how marketers use emotion to sell their goods.

In the years since its launch, E-E has been adopted into radio plays, animations, reality dramas, and even mobile phone programming with consistently demonstrated success well above other forms of public health education. For example, in South Africa the weekly drama *Tsha-Tsha* drew an audience of 1.8 million. People exposed to the show and with good recall of its plot reported significantly higher rates of HIV prevention practices, such as abstinence and safe sex. And a study in Tanzania found that 40 percent of new family planning users at government clinics came in because of hearing the radio drama *Twende na Wakati*. Similar results have been documented in an analysis of 39 family planning communications worldwide between 1986 and 2001.[7]

Few Stories Address Climate Change

Although social marketers have had some stunning successes in harnessing the power of stories, when it comes to the most pressing environmental sustainability issues, the lesson has not been applied adequately.

A survey of the Web communications of the "environmental G8," the foremost international nongovernmental organizations (NGOs) addressing climate change, reveals an approach that is still heavily devoted to the facts of the climate crisis, its dire consequences, and current policy proposals to address it. Emotional appeals that aptly reflect the reality of visitors' lives and concerns, as well as the frames through which they receive and evaluate information about the crisis, are sorely lacking.[8]

A recent study by the Yale Project on Climate Change and George Mason University's Center for Climate Change Communication signals that the time for a fact-alone approach has past. Seventy percent of Americans already believe climate change is a problem and 51 percent view it as a serious problem. With the public recognizing the need to address climate change, NGOs must shift gears to inspire action, not merely persuade people that climate change exists through a barrage of facts.[9]

Moving beyond facts and information alone is critical because when it comes to taking action, humans tend not to be rational actors. In the wake of the 1970s energy crisis, researcher Scott Geller demonstrated this when he exposed research participants to three hours of slide shows, lectures, and other educational materials about residential energy consumption. The result? Participants were more aware of energy issues, understood more about how they could save energy in their homes, but failed to change their behavior.[10]

Fortunately, there is a dawning realization among social marketers and the scientists whose work they support that facts alone are not enough. This was captured perfectly by activist Bill McKibben in describing the work of NASA scientist James Hansen: "I think [Hansen] thought, as did I, if we get this set of facts in front of everybody, they're so powerful—overwhelming—people will do what needs to be done," McKibben told the *New Yorker*. "Of course, that was naïve on both of our parts."[11]

Today, McKibben and Hansen are key evangelists of the Internet-savvy, story-based campaign known as 350.org, which seeks to cast the climate crisis in terms of the health of a single organism. As its Web site explains: "We're like the patient that goes to the doctor and learns he's overweight, or his cholesterol is too high. He doesn't die immediately—but until he changes his lifestyle and gets back down to the safe zone, he's at more risk for heart attack or stroke."[12]

Based on the patterns of success seen on Madison Avenue and Mexican soap operas and on the predictions of Joseph Campbell, this shift to campaigns like 350 is desperately needed in order to see the mass behavior shifts required for a sustainable future.

Social Marketing Meets Social Media

For most of the past 40 years, social marketing distribution has occurred in a uniform way. Whether messages were made available through radio, television, or print, the dominant approach until a decade ago was the one-to-many broadcast model.

Today that model is quickly being overtaken by a many-to-many "narrowcast" model that is made possible by the Internet. In this new world, messages travel through personalized social networks. As each audience member handles the message, he or she may comment on it or even alter it. Effective social marketing has become not just about creating great stories but about sparking great conversations out of which great social change stories can arise.

To understand how powerful social marketing efforts might move around in this new media landscape, it is important to first understand the basics of social media today:

- Social media refers to a new crop of Internet tools and content, where anyone with an Internet connection can publish text, images, and video easily through Web sites such as Facebook, Twitter, YouTube, and Flickr or with tools such as blogging and podcast software. Once published, others can interact with the content by commenting on it, integrating it with other content, sharing it, or rating it.[13]
- Social media tools and users are growing exponentially, so that today online forums are no longer only for ardent Internet users. Facebook alone boasts 250 million active users. About 70 percent of these individuals live outside the United States, and the fastest-growing Facebook demographic is people 35 or older.[14]
- Social media are redefining people's core social networks. A recent Pew study found that people's networks are more geographically dispersed, mobile, and varied thanks to the Internet. The study goes so far as to say that social media are changing the traditional orientation of human behavior.[15]
- Social media content is among the most trusted sources of information for Americans today. Sixty million Americans said information shared on the Internet has helped them make a major life decision, and 90 per-

cent say that they trust the recommendations of their networks over any other form of communication (such as advertising).[16]

What are the inherent opportunities here and how will this enhance or diminish the power of stories to create social change?

First, social media amplify the public's appetite for and access to human-scale stories. For instance, after the 2008 Sichuan earthquake and the 2009 Iranian presidential elections, Twitter allowed thousands of authentic individual stories to flood out of countries that previously would have repressed or controlled the message. In the past, China's government had buried stories of natural disaster, leaving little space for public response. After the Sichuan tragedy, the unfiltered stories of heartbreak generated 1.5 billion yuan ($208 million) in relief donations from Chinese citizens alone.[17]

Similarly, after the Iranian elections, marchers in Tehran were joined in solidarity around the world by demonstrators in Washington, London, Islamabad, Sydney, São Paulo, and dozens of other cities. These story-based social marketing efforts harnessed social media both to spontaneously disseminate key information and to create dramatic results that would not have been possible using the broadcast model.[18]

Second, social media do not remove the need for traditional "tribal" identities; they create an even deeper need for them. The Pew study showed that all this incredible new technology has not fundamentally changed the size of social networks. People still tend to interact in small "tribes" of about 35 "close ties." These close-knit communities, however, are no longer necessarily held together by geographic proximity or traditional markers of social status. Thus the tribes need new identity-forming concepts and behaviors to hold them together.[19]

The group 350.org has taken advantage of this by organizing a global protest at the micro-social network level. By early September 2009 its highly successful social marketing campaign had signed up over 1,700 groups in 79 countries to create actions before the Copenhagen climate talks at the end of the year. The organization did not provide top-down instructions for how these networks should behave. Instead, it offered a sort of social and identity glue that the networks eagerly embraced and used to further the organization's cause.[20]

Third, social media can offer a natural advantage to social marketing over product marketing. Because these networks are made up of permission-based communications, it is difficult for people to "advertise" to each other without breaching natural social taboos. On the other hand, social groups tend to welcome education and values-based messages. Thus, despite having smaller budgets, social marketing campaigns will likely move more quickly through social media.

Now Is the Time

Return for a moment to the 1950s, a turning point in the evolution of the consumer-based society. The marketing revolution that helped reverse cultural norms so swiftly can be seen as a small miracle—a miracle to learn from and perhaps repeat. It is true, of course, that the stakes are much greater and the hurdles to cross in terms of behavior and political change seem much higher. But this is not the 1950s, when television was new and a handful of players dominated the media landscape. This is 2010, a time of exponentially greater connectivity, free information flow, and dramatically lower distribution costs. By combining the key lessons of marketing's past with the opportunities of today's social media revolution, social marketers armed with the power of storytelling have the chance to create another great shift and move the world toward a sustainable future.

Media Literacy, Citizenship, and Sustainability

Robin Andersen and Pamela Miller

A series of advertisements for Italian-based Diesel brand clothing features alluring young people in suggestive poses wearing jeans, swimsuits, and other clothes while enjoying luxury, popularity, and admiration for their perfect bodies and good looks. Although the ads use common popular-cultural themes and marketing strategies that tie beauty, belonging, and happiness to a line of clothing, the models in them lounge on no ordinary beach. In the water stands a partially submerged Mount Rushmore. In other ads, models appear in a rainforest in Paris amid palm trees and lizards surrounding the Eiffel Tower, a couple sprawls on a rooftop in Manhattan while New York City is almost completely engulfed in water, and the Great Wall of China is surrounded by a vast and empty desert. Thus Diesel's 2007 Global Warming Ready campaign created scenes of consumer bliss in a future world that has been drastically altered by rising temperatures and seas.[1]

Commercial messages that assert consumption equals happiness even as the negative environmental consequences of industrial production occur illustrate the challenges and necessity for media literacy as a cornerstone in the transition to sustainable cultural practices. Understanding visual language and revealing the false promises implied in such carefully choreographed ads are important tasks.

Info-literacy challenges audiences to become sophisticated "readers" of media text, especially with regard to visual images. Consumers are rarely aware that pictures are routinely "touched up," nor do they regularly consider why emotional gratifications are not easily fulfilled in the realm of consumption. Photographs create associations and implied meanings that are fundamental to the strategies of persuasion. A picture of a group of friends all wearing Diesel clothes or drinking the same soda confers a sense of group identity and belonging. But if such messages were stated more bluntly—"wear these jeans and you will have the friends you want" or "people who drink Coke are thin, popular, and always happy"—the assertions would be hardly credible.

Learning how to critically engage with television, magazines, films, and the Internet is essential in a sprawling media landscape where users are exposed to more and more media every year. Increasingly this landscape is dominated by advertising, and gaining immunity to its persuasions is an important step along the

Robin Andersen is professor of communication and media studies and director of graduate studies and **Pamela Miller** is a graduate student in public communications at Fordham University.

Diesel

The Mount Rushmore ad in Diesel's 2007 campaign.

path to sustainable cultural practices. But a deeper critique of consumerism is required in order to build a more sustainable culture—one that goes to the heart of consumption as a social practice.

Diesel's ads claim that the company and its brand of clothing are Global Warming Ready, but no mention is made of the environmental impact of producing the clothes. Clever ad campaigns may cause consumers to feel clever by association, but they often encourage them to think uncritically about whether the company behind the campaign follows sustainable business practices. Does it use alternative energy sources in production or distribution to reduce its carbon footprint, pay its workers adequately, or use organic fibers in any way? What industrial by-products are created, and how are they treated?

The Diesel images speak of inevitability and acquiescence to a global crisis, and their wide circulation in popular culture in place of narratives about the urgency and necessity for citizen action reinforces defeatist and apathetic attitudes to global warming. This cultural attitude complements a larger media context that offers little real information about the causes of and solutions to climate change.

Take, for example, a segment televised in 2006 by WTOK-11 in Meridian, Mississippi. It featured two "top weather and ocean scientists" who asserted that a link between the recent severe hurricane season and climate change was "all hot air." The channel made no effort to inform viewers that this was a re-edited video news release produced by a public relations firm, Medialink Worldwide. Nor were viewers told that the client behind the video—Tech Central Station Science Roundtable—was run by the lobbying firm DCI Group, whose client list includes ExxonMobil, a corporation that has made a sizable contribution to the Tech Central Science Foundation for "climate change support." Few members of WTOK-11's audience could have recognized that this "news" segment did not contain a particular scientific argument but instead served the political and economic interests of the oil company lobbyists who wrote it and paid for it.[2]

James Hansen of the National Aeronautics and Space Administration identifies a lack of public knowledge as a main obstacle to reversing climate change, pointing to the gap between what the scientific community understands and what the public and policymakers know. He argues that public understanding of the effectiveness of reducing fossil fuel use and carbon dioxide emissions is thwarted by "intensive efforts by special interest groups to prevent the public from becoming well-informed." One study of press reporting on this issue found that the practice of journalistic balance serves to amplify a small group of global warming skeptics, many of whom, it has been revealed, are indirectly funded by special interest groups.[3]

How Critical Should Media Literacy Be?

While the broader role of media literacy to create sophisticated critical "readers" of media texts is clear, disagreements about the degree and levels of criticism have emerged over the years. Some advocates want to expand analysis into other realms, including corporate media practices and policy reform. Rejecting this approach, in 2000 the Alliance for a Media Literate America (AMLA, now rechristened the National Association for Media Education) stated that it was not an "anti-media movement" but one dedicated to finding a "more enlightened way to understand our media environment." Not interested in "media bashing," the AMLA created controversy by accepting funding from media conglomerate Time-Warner. That deal led to the formation of a more critical group, the Action Coalition for Media Education (ACME) in 2002. ACME seeks to broaden the concept of "literacy," which focuses on messages, to include "education," which includes messages, structures, and reform activism. Writer Bill Yousman identifies the central question that divides the media literacy community in the United States: "Is media literacy aimed at creating more sophisticated consumers of media, or is it about nurturing engaged citizens?"[4]

The consequences of these varying approaches are significant. As Yousman explains, "It is one thing to teach children how to decode an advertisement for fast food, for example, so that they may see how the image of a hamburger is artificially constructed, and doesn't actually resemble the actual product that you purchase at the counter. It is another thing entirely to encourage an understanding of fast food as a mega-billion dollar global industry that is spreading particular industrial practices and ways of thinking about food, labor, the environment, and the like, throughout the world."[5]

Sophisticated consumers make better choices about what to buy, but the potential for media literacy as a force for sustainability will depend on people around the world creating and supporting alternative choices, not the ones offered by the unsustainable practices of current global manufacturing. As media literacy develops, these issues and concerns will remain at the forefront of debates over curriculum. In the United States, media literacy is offered in many schools in 50 states. And a dynamic media literacy movement is growing worldwide, which includes community activists, grassroots practitioners, media reformers, and policymakers as well as educators.[6]

Media Literacy and Global Organizations

Media literacy has become an important item in the global educational curriculum, with the support and promotion of key world bodies. (See Table 11.) Educators are no longer isolated in a few schools or regions. Indeed, UNESCO has worked for 26 years to extend the reach of media education worldwide. The agency works within the framework of the Grunwald Declaration of 1982, which enjoined global educational systems to "promote citizens' critical understanding of 'the phenomenon of communication' and their participation in media." In 2007, the Paris Agenda identified key components for media education, and UNESCO followed with the development of a Media Education Kit the same year.[7]

UNESCO's current initiative, Training the Trainer on Media and Information Literacy Curricula, promotes teacher-training programs in developing countries. The agency also seeks to foster a global environment that encourages free, pluralistic, and independent media as a fundamental component of media education, extending education into adult communities. Many in global organizations realize that receiving and creating media content and having full access to new media technology will allow global citizens to reap the full benefits of Article 19 of the Universal Declaration of Human Rights, on freedom of opinion and expression.[8]

UNESCO is partnering with the UN-Alliance of Civilization, which also identifies

Table 11. Efforts to Promote Media Literacy, Selected Countries

Country	Programs
Argentina	The School and Media Program became a nationwide initiative in 2000. One effort involves distribution to high schools of a free monthly magazine with notable online/print news articles.
Australia	The Australian Communications and Media Authority is currently pursuing a Digital Media Literacy Research Program that aims to improve knowledge about digital media literacy levels and to aid development of consumer education and protection.
Austria	The Ministry of Education, Science, and Culture distributes a quarterly journal on media education to all schools. Ministry-evaluated resources for educators are accessible online and other teaching materials are available for order.
Canada	In 2006–07, Ontario's Ministry of Education instituted a policy mandating instruction in "four program strands"—reading, writing, oral communication, and media literacy—for all schoolchildren.
Finland	Government policy for 2007–11 includes specific initiatives encouraging media literacy, especially in younger citizens. The Citizen Participation Policy Programme emphasizes the cultivation of "information society skills" as a catalyst for citizenship.
France	The Ministry of Education's Centre for Liaison Between Teaching and Information Media produces teaching tools, trains educators in the process of analyzing and using news media messages, and connects teachers and students with media professionals during an annual Press and Media Week.
Hong Kong, China	The Education Department recently introduced the New Senior Secondary Curriculum, emphasizing the ability "to make critical analyses and to judge the reliability of the news and the suitability of ways of reporting used by the mass media."
Russia	Since the early 2000s, the Russian Academy of Education Laboratory has worked to incorporate media literacy into national arts and culture-related curricula.
South Korea	Newly reformed national curriculum, mandatory for students aged 5–16, encourages media literacy practices in Ethics, Social Studies, and Practical Studies courses.
Sweden	Nordicom's International Clearinghouse on Children, Youth and Media continues to promote media literacy in youth, incite constructive public debate, and inspire research and policymaking.
Turkey	School systems first introduced media literacy programs into the curriculum as elective courses in 2006. The government's regulatory bodies have begun proactive collaboration with nongovernmental groups and educators to promote media literacy.
United Kingdom	Under the 2003 Communications Act, duties of the Office of Communications (Ofcom) include "furthering the interests of citizens, in relation to communications matters, and of consumers, by promoting competition in relevant markets."

Source: See endnote 7.

media literacy as an indispensable tool for global citizenship. Understanding that institutional media are key generators that circulate symbols in social and political life, media literacy is no longer an option for global citizenship but a necessity for social development and civic engagement and for sustainable societies. In the words of Divina Frau-Meigs and Jordi Torrent, Project Managers for the Media Literacy Program at the Alliance, "A threshold has been reached, where the body of knowledge concerning media literacy has matured, where the different stakeholders implicated in education, in media and in civil society are aware of the new challenges developed by the so-called 'Information Society,' and the new learning cultures it requires for the well-being of its citizens, the peaceful development of civic societies, the preservation of native cultures, the growth of sustainable economies and the enrichment of contemporary social diversity."[9]

Media Literacy Education and Global Citizenship

A main goal of media literacy education is to find ways to encourage media users to actively engage through critical awareness and creative media skills. Citizen participation is especially crucial for addressing global issues and finding collective solutions to environmental problems. Writing about "critical citizenship," Costas Criticos of the University of Natal in South Africa argues that "a citizen or a society unable or unwilling to be critical will militate against the growth and maintenance of a healthy civil society." Many nations are presently affected by the influence of "global nodes of information power and practice" that contribute greatly to the marginalization of regional voices. Teaching media literacy facilitates critical citizenship, and encourages marginal voices to produce counter-discourses. Creative counter-narratives that embody the wisdom of regional sustainable practices will be key to envisioning a sustainable future.[10]

Fackson Banda of the School of Journalism and Media Studies at Rhodes University in South Africa advocates a mode of media training embedded within the concept of citizenship. His proposal is rooted in postcolonial theory, with the primary aim of recovering "lost historical and contemporary voices of the marginalized, the oppressed, and the dominated, through a radical reconstruction of history and knowledge production." Banda calls on African media educators to reconceptualize media structures to "improve the relevance of local media to civic life, encouraging informed use of and participation in media." Such participation can help with local community witnesses and information dissemination about local conservation issues throughout the world. (See Box 19 on environmental journalism in India.)[11]

The Development Through Radio project, piloted by Panos Southern Africa in Zambia and Malawi, is "aimed at cultivating engaged and engaging citizenship." The women involved in the project were given the skills to produce radio programs and to make sense of the context of media production. The groups made audio recordings about a mutually agreed upon topic and then coordinated getting their tapes to central studios in their respective major cities. Producers at the Zambia National Broadcasting Corporation and the Malawi Broadcasting Corporation recorded responses to the women's concerns by relevant urban-based policymakers or leaders of nongovernmental groups. They then edited the recordings into a single program for broadcast, promoting further discussion and creating an empowering, cyclic dialogue.[12]

While this global movement continues to grow, the full implementation of media literacy programs and the addition of citizens' voices to public dialogue face challenges on many levels. Blocks to full participation in the information society arise every day, yet with the convergence of new media—including wireless telephones, the Internet, satellite broadcasting, and digital technologies of all sorts—virtually anyone can create media content. Only about one fifth of humanity has access to the Internet, however.[13]

Often because of dire financial concerns, efforts to promote media literacy must at times

Box 19. The Evolving Role of Environmental Journalism in India

After the historic Earth Summit in Rio in 1992, unfortunately environmental journalism in developing countries like India went into a steep decline. Part of the reason was the overkill in coverage of the summit and the subsequent failure of major powers to live up to the promises made at Rio. The other major reason was that by the mid-1990s, thanks to economic reforms, India's economy boomed. Business publications sprouted, and within a couple of years business TV journalism followed suit. This spawned a massive boom in a niche area of journalism and offered hundreds of jobs with good salaries to young journalists. Business development, as opposed to sustainable development, was now attracting talent.

Suddenly environmental activism was viewed as a major obstacle to industrial development. The middle class, whose huge constituency supported environmental activism, seemed more focused on securing good jobs and building houses. This does not mean that there were no brilliant journalistic works on environment during this period—there were, but they were few and far between.

Then nature began to strike back with an unprecedented fury. The deluge that paralyzed Mumbai and the great drought of 2002 signaled to Indians that all was not well with the weather. There were alarming studies of rapid glacial melt in the Himalayas. This coincided with a continuous stream of reports by the Intergovernmental Panel on Climate Change—coverage of these was helped in no small measure by the fact that the chairman, Dr R. K. Pachauri, is Indian. Al Gore's film "An Inconvenient Truth" boosted awareness among urban Indians. There was a dramatic revival on interest in these issues, and environment was on the front-burner again.

Yet most journalists were stymied by the new challenges of covering climate change issues. People did major reports, to be sure, but they were not pushing hard enough to understand the subject. Environment had become far more complex than just forests and wildlife. Now journalists needed to understand issue ranging from economics to science and development. In retrospect, getting readers or viewers to understand the link between carbon emissions and climate change may have been the easy part.

When it came to the hard questions—such as which sectors in the industry are the highest emitters or what technologies could make a difference and whether companies were using them—journalists' efforts to find answers were by and large missing. There has been no independent media investigation of the claims of success by government and industry. There has been no great double-checking of government data on India's emission levels. Nor has there been a really hard look at the viability of the renewable energies being pushed. There have been no great guides for the public on how they can reduce their carbon footprint, for example.

It is not too late though. Following the crucial Copenhagen conference, the baton to do business-as-radical rather than as usual will fall on public initiative driven by perceptive media coverage. Just as in the United States, where a recalcitrant government was forced to act when states like California passed their own legislation on climate issues, public pressure will build to enormous levels as natural catastrophes strike India. Journalists can play a constructive role in channeling their anger and their desire for change by exploring ways out and offering solutions. With political pressure on environmental issues inadequate, journalists will have to play the role of both torchbearer and public watchdog.

—*Raj Chengappa*
Managing Editor, India Today

rely too heavily on corporate entities, as is the case of Argentina, where media literacy courses are sponsored by Telecom and Microsoft but also by Coca-Cola and Adidas. Such funding could no doubt result in the exclusion of critical discussion of corporate practices, such as Coca-Cola's many environmentally and socially irresponsible practices. Stakeholder negotiations are key, as are media regulatory bodies able to address ethical and content issues without suspicion of censorship.[14]

Media Literacy Is the Literacy of Our Time

As media engagement is understood as a global necessity for sustainability and citizenship, collaborations across cultures and borders become essential. Robin Blake of the U.K. Office of Communications, which regulates media in the United Kingdom, has identified a research framework for shared knowledge that includes four key research areas for media literacy: social, political, regulatory, and commercial. Another essential component is documentation of the role played by long-standing grassroots practitioners who often spearhead citizen media across the globe. The work of such independent media producers is being documented in the series *Waves of Change*, which features examples of grassroots radio in Bolivia, El Salvador, South Africa, and the United States and of groups producing community video and television in India, Brazil, and Mexico. Information about past and present grassroots efforts to promote sustainability and media literacy throughout the global village is also made available online.[15]

David Gauntlett, a U.K. media literacy educator who works with children and video production, discovered that youthful audiences have internalized environmental problems and their solutions in a one-dimensional "narrative": the problem has been created by individuals and must be solved by individuals. Gauntlett's analysis reveals an increasingly narrow range of acceptable environmental content and an important "absent narrative" within television coverage. Left unaddressed is how to account for political and economic forces and instances of polluting industries that fall within government and legal regulations. Moreover, how should society address institutional practices such as the car-centered transportation system through an individual problem/solution framework when alternatives such as affordable and efficient public transportation are not available?[16]

Challenging environmental suppositions based in media stories is only a starting point for Gauntlett. He moves from the negative critique into positive creative solutions. The larger project is to overcome "passive paralysis," a consequence of the "sit back and be told" culture. Believing that the media literacy paradigm must include a fundamental transformation of engagement with media, Gauntlett encourages students to create alternatives, not to "sit around watching as the world gets worse." Or as media educator DeeDee Halleck puts it: "Don't watch TV. Make it."[17]

Such educators envision a transformed relationship to media in a new "making and doing culture," one that demands an expansive perspective able to envision positive proposals for a better future. By connecting to the world and seeking solutions to its problems, people reveal their presence in the world. Promoting counter-narratives that creatively address issues such as climate change is a powerful antidote to the cynical agreements often embodied in media, such as the Diesel ad campaign. Media are the means by which people communicate and share knowledge and creativity with the global public. Increasing access to media, learning how to use them, and creating public and legal structures that democratize them will allow people to cope with the challenges of finding sustainable cultures based on human and environmental priorities.

Music: Using Education and Entertainment to Motivate Change

Amy Han

Music has traditionally been cherished in society for its artistic beauty and its raw expression of life and spirit, and it continues to be enjoyed today. The songs of birds inspired Mozart and other great classical composers to recreate the elegance of nature's sounds, while folk music, passed down through generations, has served as an influential base for many other forms of storied expression—from country music and gospel to blues and jazz.[1]

Along with its emotional and creative elements, music has played a critical role in encouraging social engagement. Historically, the power of music to communicate and create connections has helped unite people around a common identity or purpose. In the Soviet Union, traditional Kazak folk songs celebrating birth, death, and other life stages were adapted into modern operas and literature supporting worker ideals, sovereignty, and nationalism. In the United States, the traditional hymn "I'll Overcome Someday" was taken up by the black Tobacco Worker's Union in the 1940s as the collective labor song "We Will Overcome"—and in the 1960s was adapted as the civil rights classic "We Shall Overcome."[2]

Music continues to be used as a way to connect with people's values, heritage, and cultural preferences in order to encourage behavioral change. For example, songs from Marvin Gaye's 1971 album "What's Going On"—which catalogued the Vietnam War, pollution, and economic hardship—are being revisited today in light of the current recession, climate change, and environmental decline. In August 2009, U.S. Environmental Protection Agency Administrator Lisa Jackson invoked Gaye's songs "Inner City Blues" and "Mercy, Mercy Me (the Ecology)" in a speech announcing a Greening the Block initiative to empower climate-vulnerable and economically disadvantaged communities in the United States.[3]

In the current age of digital media, opportunities for remembering, sharing, and using music for mobilization are expanding. Technology has not only preserved music for future generations, it has facilitated people's access to it, enabling independent artists to post their work on the Internet, fans to share files and lyrics, and virtual communities to come together through social networking sites such as Facebook and Twitter. Although music has morphed, mixed, diversified, and

Amy Han is a *State of the World 2010* project assistant at the Worldwatch Institute.

globalized over the centuries, it remains a potent force in society, with a significant part to play in inspiring sustainability through education as well as entertainment. (See Boxes 20 and 21 for the similar roles played by other artists and by movies.)[4]

Music as Education

From conception, humans are exposed to music. Babies in the womb are lulled by the rhythmic beats of the heart, and young children are introduced to music through song and

Box 20. Lights, Camera, Ecological Consciousness

Cinema is a powerful visual and auditory medium that contributes to people's understanding of the world and their role in it. In its most direct form, a documentary film can raise awareness of an issue and generate public dialogue. In recent years, the documentary as a genre has seen a resurgence, and many have been related to sustainability—including *March of the Penguins* (2005), *An Inconvenient Truth* (2006), *the 11th Hour* (2007), *Blue Gold* (2008), and *Home* (2009).

Home is an effort to illustrate humanity's impact on the planet using all aerial footage in a feature-length documentary. Within a few weeks of its release on 5 June 2009, which was World Environment Day, some 200 million people had watched it in more than 120 countries, and it was dubbed or subtitled in 33 languages. Despite this success and the success of other eco-documentaries, documentaries tend to attract audiences already sympathetic to the issues, thereby limiting their transformative potential.

Fictional movies, which for many people are easier to watch, are uniquely positioned to stimulate cultural change for sustainability. They can depict challenging future scenarios, such as *WALL-E* (2008) and *The Day After Tomorrow* (2004), and give voice to the struggles faced by communities, such as *Erin Brockovich* (2000). As they are less overtly educational or political, they appeal to audiences on a human level by personifying what typically are perceived as abstract global-scale ecological issues. Although dramatic films do not directly prescribe actions as documentaries often do, they have the power to normalize sustainable lifestyle choices through the actions of characters on-screen and sometimes through the actions of celebrities off-screen.

While a few notable sustainability-themed documentaries have been able to transcend a niche audience and reach across the world—in the case of *An Inconvenient Truth* grossing $50 million in the process—most eco-conscious filmmakers will have to use creative tools to ensure broad distribution of their films. *Home* could never have reached the audience it did had it needed to make a profit; the PPR group provided a generous grant to enable a broad distribution, and the film can be watched for free at www.youtube.com/homeproject. Other filmmakers are using innovative tools like crowd funding, in which many people invest small amounts to finance a film's production and distribution. This allowed *The Age of Stupid* (2009) to maintain creative control over the film and its distribution and to be launched in over 60 countries.

Both fictional and nonfictional cinema can play an important role in drawing attention to environmental topics and in creating space for sustainability values. If people are to come together to solve the sustainability crisis, they must make and demand films that not only inform audiences and generate public dialogue but also exemplify and project sustainable lifestyles.

—*Yann Arthus-Bertrand*
Director, Home
Source: See endnote 4.

Box 21. Art for Earth's Sake

The dominant thinking in western society is that of separation: the separation of mind from matter, science from spirituality, art from daily life. From the Renaissance onwards, artists worked as individuals, in their studios, separating themselves from their fellow craftsmen and women. They practiced art as a way of self-expression. Their art produced mostly items of luxury and status. Thus art became disconnected from the natural world, from living communities, and from life itself. For centuries, art was practiced only by those with special talent, purchased only by those with great wealth, and seen mostly in churches, museums, and art galleries.

But the exclusive practice of art is now being challenged by people with ecological and social sensibility. Joseph Beuys, one of the founders of the Green Party in Germany, said "Everyone is an artist" and began the process of reclaiming art from galleries and museums. He began to reconnect art with ecology, politics, and everyday life. Similarly, Sri Lankan art historian A. K. Coomaraswamy said "the artist is not a special kind of man, but every man is a special kind of artist." When artists let go of their egos and their wish for celebrity status and personal glory, then art becomes truly boundless.

Art is a force for transformation and self-realization. As a potter transforms an ordinary lump of clay into a work of beauty, that clay transforms the potter into an artist and craftsperson of his or her community. This transformative power of the arts gives us a sense of belonging and unlocks the doors of optimism and hope.

Unfortunately, at the moment a scenario of environmental doom and gloom is expounded by experts and activists alike. Book after book tells us that we have passed the tipping point and have reached the point of no return. Artists are some of the few people who sow the seeds of hope and empower the disempowered.

Of course no one should doubt the severity of the climate crisis. Our present way of life, so dependent on the use of fossil fuels, is hanging on a cliff edge. If we go any further we will fall into the abyss. Yet artists go beyond fear, beyond doom and gloom. Their work is rooted in love of life. The potential of growth and progress in the sphere of arts and crafts is immense, and this can occur with little damage to planet Earth.

To meet the challenge of this environmental, social, and spiritual crisis, we need to change from being consumers to being artists. As the British architect, textile designer, and artist William Morris pointed out long ago, arts and crafts ignite our imagination, stimulate our creativity, and bring us a sense of fulfillment. Poetry, painting, pottery, music, meditation, gardening, sculpting, and many other forms of arts and crafts can produce beautiful objects to use—objects that do not require the use of fossil fuels.

The climate crisis and the economic downturn offer us an opportunity to change our direction from gross to subtle, from glamorous to gracious, from hedonism to healing, from the conquest of Earth to the conservation of nature, from quantities of possessions to quality of life. This will transform us from being mere consumers of goods and services to genuine makers of arts and artifacts. In the present state of the world and under the influence of unsustainable consumerism, human beings are reduced to the condition of passive recipients of factory-made objects. This must change. We need to move toward a state where humans are active participants in the process of life and in the making of things that are beautiful, useful, and durable.

—*Satish Kumar*, Resurgence

Source: See endnote 4.

dance. Music and rhythm aid in intellectual development, as research has pointed to the value of music in developing cognitive skills as well as in helping individuals develop a sense of organization, self-awareness, and self-confidence. This educational contribution has been taken so seriously that music has been regarded as its own language and is even believed by some to have a powerful effect on a person's moral character.[5]

Increasingly, children's music contains not only civil themes such as friendship and sharing but also educational messages about the environment and sustainability. For 15 years, the Japanese Ministry of the Environment has supported the television program "Ecogainder," which features a group of environmental superheroes who serve as role models for children around the country, and strengthened this message with a catchy theme song. In North America, the popular musician Raffi has entertained young people for decades with songs about the environment and respect for the natural world. Raffi positions his music as a call to action and challenges "Beluga grads," or people who grew up listening to his songs such as "Baby Beluga" in the 1970s and 1980s, to both embrace sustainability in their own lives and pass these teachings on to their own children.[6]

The appeal and relevance of music as a tool for environmental education is not limited to youth. Irthlingz, an art-based educational group, uses music to inform both children and adults about issues that affect the planet. In 2007, students performed the organization's musical revue "Penguins on Thin Ice," which includes songs about energy and climate issues, before an audience of civil society leaders at the United Nations Commission for Sustainable Development meetings in New York.[7]

In Mozambique, musical and theater traditions are proving integral to larger efforts to address the challenges of rural sanitation, waterborne disease, and environmental health. The band Massukos tours the country and combines traditional rhythms with modern lyrics to teach people about handwashing and sanitation, at times drawing an entire village to hear their messages. Often accompanying the band are practical projects that promote community sanitation, sustainable agriculture, and reforestation, and the government has set up related forums to teach about hygiene and environment-related illness. Massukos has signed with a British recording label, and lead band member Feliciano dos Santos has won the prestigous Goldman Environmental Prize. The band is now expanding its audience and musical messages through international performances.[8]

Festivals, Activism, and Entertainment

Music is also being used to educate audiences through less explicit means. In the 1980s, as popular music culture continued its global spread, musicians began bringing attention to wider humanitarian causes by organizing large-scale, widely publicized entertainment events. In 1985, Irish singer Bob Geldof and Scottish singer Midge Ure organized "Live Aid," the world's first multi-venue super-concert, which was broadcast live to some 400 million viewers in 60 countries. That same year "We Are the World," written by pop icons Michael Jackson and Lionel Richie, united 45 recording artists from around the world for famine relief, helping to create the charity USA (United Support of Artists) for Africa. As of mid-2009, an estimated 20 million copies of the song had been sold, raising more than $63 million for humanitarian aid.[9]

More recently, the Internet has enabled such events to have an even broader international reach. In 2007, the concert extravaganza Live Earth, started by producer Kevin Wall and former U.S. Vice President Al Gore, was broadcast for 24 hours across seven continents, featuring an all-star lineup of artists that included Madonna, The Police, and Snoop Dogg. Live Earth has since become a "multiyear campaign to drive individuals, corporations and governments to take action to solve the climate crisis." The event is now partnering with other climate protection groups, such as the Together campaign, which offers tips and consumer products online to help people lower their ecological footprints.[10]

Some artists have stepped beyond their musical boundaries to become well-known activists in their own right. U2's lead singer, Bono, known for his efforts to eradicate global poverty, has co-founded several organizing movements and engaged in extensive discussions with public and private leaders ranging from former President Bill Clinton to Pope John Paul II. Bono is also a spokesperson for the ONE campaign, founded in 2004 to rally grassroots support for international aid to fight extreme poverty and preventable disease.[11]

Concerts have become increasingly important opportunities for musicians and event organizers to demonstrate their commitment to environmental action. Large tours in particular can be resource-consumptive and responsible for high levels of greenhouse gas emissions. According to one estimate, the carbon footprint generated by U2's 44 international concerts in 2009 is equivalent to the waste produced by 6,500 Britons over a year or "the carbon created by the four band members traveling the 34.125 million miles from Earth to Mars in a passenger plane."[12]

To minimize their carbon footprint, many music venues now use renewable energy, such as solar power or biodiesel, to run their events, or they purchase third-party certified carbon offsets to ensure that the activities are "carbon-neutral." Roskilde Festival, which calls itself as the largest North European culture and music festival, has a Green Footsteps campaign that in 2009 completely ran the festival on wind energy and changed 90 percent of the lighting equipment to low-energy LED equipment. Glastonbury Festival is encouraging public transport use and the planting of tree hedges (over 10,000 since 2000), while incorporating the use of solar power in its festivals. It also plans to use tractors capable of running on 100 percent biodiesel, all steps taken to lower its own carbon emissions.[13]

Waste reduction is another key feature among event organizers who are encouraging fans to tread more lightly. In accordance with its "Love the Farm, Leave No Trace" principles, the Glastonbury festival asks attendees to bring fewer items that would normally become waste, has replaced plastic bags with 100 percent cotton bags, has required wood cutlery and compostable cups and plates at stalls, and in 2008 recycled just over 863 tons of waste. Smaller venues have taken similar greening actions. Seattle's annual music and arts festival, Bumbershoot, bans vendors from using Styrofoam and also reuses the previous year's signage. The High Sierra Music Festival's "Red, White, Blue and Green Campsite Challenge" makes the "Leave No Trace" outdoor ethic into a competition, rewarding participants with the least amount of impact with prizes.[14]

At the Ojai Music Festival in Ventura County, California, classical music fans are encouraged to help preserve the natural beauty of the area with a free bike valet area for alternative transport, water stations to refill reusable containers, and Zero Waste Stations to help sort trash. And the U.S.-based Dave Matthews Band, through its So Much to Save program, encourages fans to take actions to reduce their ecological footprint whether at the concert or outside of it, such as recycling or conducting energy efficiency audits in exchange for free downloaded music. During the first two months of the 2009 campaign, participants recycled an estimated 19 tons of waste, diverting more than

84 cubic meters of waste from landfills.[15]

Musicians continue to carry on the tradition of delivering important messages through their lyrics. Joni Mitchell's 1970 song "Big Yellow Taxi," which laments the conversion of the natural world to a "paved paradise," has been covered by multiple artists, including Bob Dylan and more recently the Counting Crows. Tracy Chapman's 1995 song "Rape of the World," which observes that Mother Earth "has been clear-cut, she has been dumped on, she has been poisoned and beaten up," is another example of an artist lending her creative talent to raise awareness about environmental devastation. Some musicians point specifically to the importance of activism: in their songs "Up to Us" and "We Must Act Now," the California "eco-rock" band the Depavers encourages listeners to stand up for their beliefs.[16]

Some artists are especially concerned about practicing what they preach. Blues musician Bonnie Raitt promoted her "Silver Lining" album with a Green Highway Festival and "an eco-partnership promoting BioDiesel fuel, the environment, and alternative energy solutions at shows and benefits along the way." Along with other artists, she founded Musicians United for Safe Energy, formed after a nuclear accident at Three Mile Island in March 1979; the group organized No Nukes concerts at Madison Square Garden in New York that same year. Her current tour allows concertgoers with VIP packages to choose a cause they want to support—energy, environmental protection, and human rights, among others.[17]

Country singer Willie Nelson also expressed his mood and hope for a "Peaceful Solution" by making a song that protests social injustice available to other artists for replay. Outside of his music, Nelson leads Farm Aid, an organization dedicated to stopping the loss of U.S. family farms and advocating change in current U.S. food and agricultural policy. Willie Nelson has even promoted his version of biodiesel, Bio Willie, to help reduce dependence on foreign oil.[18]

Conclusion: Engaging through Education and Entertainment

Beyond the individual efforts of artists, some people are working to constructively engage the music and broader artistic community in support of sustainable change. Organizations such as Tipping Point are holding roundtable conversations, discussions, and debates among creative artists to increase their engagement with the complex issue of climate change and to help catalyze societal shifts in thinking and behavior.[19]

The Judith Marcuse Projects, a nonprofit arts company, is using an "EARTH=home" stage production to give voice to youth, create connections between different sectors of society, and reach the broader community through "post-show talk-backs, presentations, workshops, community events, web-based resources, and media activities" on the environment. In addition, its International Centre of Art for Social Change is a collaboration with Simon Fraser University in Vancouver, British Columbia, to "house learning and dialogue programs, networking events and research projects designed to nurture and support the growing global community of arts for social change."[20]

While music can be a potent tool for mobilization, its power lies within the people who create, promote, and use it within a meaningful, proactive movement for sustainability. As Together campaign founder Steve Howard has observed, "When the music stops, we must all start to act."[21]

The Power of Social Movements

Throughout history, social movements have played a powerful part in stimulating rapid periods of cultural evolution, where new sets of ideas, values, policies, or norms are rapidly adopted by large groups of people and subsequently embedded firmly into a culture. From abolishing slavery and ensuring civil rights for all to securing women's suffrage and liberating states nonviolently from colonial rulers, social movements have dramatically redirected societal paths in just an eye blink of human history.

For sustainable societies to take root quickly in the decades to come, the power of social movements will need to be fully tapped. Already, interconnected environmental and social movements have emerged across the world that under the right circumstances could catalyze into just the force needed to accelerate this cultural shift. Yet it will be important to find ways to frame the sustainability movement to make it not just possible but attractive. This will increase the likelihood that the changes will spread beyond the pioneers and excite vast populations.[1]

This section looks at some ways this is happening already. John de Graaf of the Take Back Your Time movement describes one way to "sell" sustainability that is likely to appeal to many people: working fewer hours. Many employees are working longer hours even as gains in productivity would allow shorter workdays and longer vacations. Taking back time will help lower stress, allow healthier lifestyles, better distribute work, and even help the environment. This last effect will be due not just to less consumption thanks to lower discretionary incomes but also to people having enough free time to choose the more rewarding and often more sustainable choice—cooking at home with friends instead of eating fast food, for example, making more careful consumer decisions, even taking slower but more active and relaxing modes of transport.

Closely connected to Take Back Your Time is the voluntary simplicity movement, as Cecile Andrews, co-editor of *Less is More*, and Wanda Urbanska, producer and host of *Simple Living with Wanda Urbanska*, discuss. This encourages people to simplify their lives and focus on inner well-being instead of material wealth. It can help inspire people to shift away from the consumer dream and instead rebuild personal ties, spend more time with family and on leisure activities, and find space in their lives for being engaged citizens. Through educational efforts, storytelling, and community organizing, the benefits of the lost wisdom of living simply can be rediscovered and spread, transforming not just personal lifestyles but broader societal priorities.

A third movement that could help redirect broader cultural norms, traditions, and values is the fairly recent development of ecovillages. Sustainability educator Jonathan Dawson of the Findhorn ecovillage paints a picture of the exciting role that these are playing around the world. These sustainability incubators are reinventing what is natural and spreading these ideas to broader society—not just through modeling these new norms but through training and courses in ecovillage living, permaculture, and local economics. Similar ideas are also spreading through cohousing communities, Transition Towns, and even green commercial developments like Dockside Green in Canada and Hammarby Sjöstad in Sweden.[2]

Two Boxes in this section describe some other exciting initiatives. One provides an overview of a new political movement called *décroissance* (in English, "degrowth"), which is an important effort to remind people that not only can growth be detrimental, but sometimes a sustainable decline is actually optimal. And a Box on the Slow Food movement describes the succulent power of organizing people through their taste buds. Across cultures and time, food has played an important role in helping to define people's realities. Mobilizing food producers as well as consumers to clamor for healthy, fair, tasty, sustainable cuisines can be a shrewd strategy to shift food systems and, through them, broader social and economic systems.

These are just a few of the dozens and dozens of social movements that could have been examined. It is just our imaginations that limit how we can present sustainability in ways that inspire people to turn off their televisions and join the movement. Only then, with millions of people rallying to confront political and economic systems and working to shift perceptions of what should feel "natural" and what should not, will we be able to transform our cultures into something that will withstand the test of time.

—Erik Assadourian

Reducing Work Time as a Path to Sustainability

John de Graaf

There is a silver lining on the cloud of recession that hangs over the industrial world. Contrary to popular expectations, in some countries—particularly the United States—health outcomes are actually improving. Christopher Ruhm at the University of North Carolina finds a decline in mortality of half a percent for each 1 percent increase in U.S. unemployment. How is this happening? Many of the newly jobless suffer acute stress, and suicides are up. But some are using the time off to improve the rest of their lives—learning to save, finding time to exercise, bonding more closely to family and friends.[1]

More important, the crisis has meant a reduction in working hours for most Americans for the first time in decades. Some companies and public agencies have chosen to cut hours through shorter workweeks or furloughs instead of laying employees off. With more time and less money, people are smoking and drinking less, eating fewer calorie-laden restaurant meals, and walking or bicycling more. While auto sales have plunged, bicycle sales are on the upswing. As Americans drive less, they die less often in accidents—U.S. traffic deaths declined by 10 percent from 2007 to 2008. Air pollution from cars and factories (as they produce less) is also down, resulting in fewer deaths, especially among children.[2]

In time, workers may find that the increased family time, improved health, and other benefits of more leisure outweigh the income losses. This should inspire more efforts to trade productivity for time instead of greater purchasing power.

But we need to do this for another reason: preserving the biosphere for future generations.

The Need to Limit Consumption

Data from the Global Footprint Network suggest that if people in the developing world were to suddenly achieve American lifestyles, the world would need four more planets to provide the resources for their products and absorb their wastes. Already—and with half the world's people living in real poverty—Earth's carrying capacity is being overshot by some 40 percent.[3]

Some environmentalists suggest that the world can have its cake (expanded production) and eat it too simply by improving technologies and investing in clean energy. Too

---

John de Graaf is a documentary filmmaker, co-author of *Affluenza: The All-Consuming Epidemic*, and executive director of Take Back Your Time.

often, however, technological improvements such as greater fuel efficiency merely lead to greater consumption of a product—people drive more, for example. As Gus Speth, former dean of the Yale School of Forestry, puts it: "The eco-efficiency of the economy is improving through 'dematerialization,' the increased productivity of resource inputs, and the reduction of wastes discharged per unit of output. However, eco-efficiency is not improving fast enough to prevent impacts from rising."[4]

Speth spells out clearly the cost of current trends in resources, pollution, and equity: disappearing rainforests and fisheries, exhaustion of fossil fuels, increasing hunger, a rapidly widening gap between rich and poor. Despite the faith of many in "super" cars and order-of-magnitude technical advances, the burden of evidence is clearly on those who think the economy and human activities can continue to grow exponentially without increasingly severe environmental consequences.[5]

Industrial countries cannot deny the rights of developing nations to greater economic prosperity while others continue to consume at current levels. That would be asking them to sacrifice so that the rest of the world can binge awhile longer.

Is There an Answer?

The current situation cannot continue, but people in industrial countries are reluctant to reduce their "standard of living." Is there a solution to this stand-off? Yes: the rich nations of the world must immediately begin to trade advances in labor productivity for free time instead of additional purchasing power.

And people must understand that doing so will not be a sacrifice. Rather it will mean substantial improvements in the quality of life.

There is a simple economic law that might be called the growth imperative. Technical progress consistently makes it possible to produce more product per hour of labor expended. For example, hourly labor productivity in rich countries has more than doubled since 1970. The point is simple: to keep everyone employed at the current number of hours while productivity increases, it is necessary to simply produce and consume more. It is unlikely that scientific progress and increases in labor productivity are going to stop. Therefore in order to limit consumption to current levels (or lower), it will be necessary either to lay off a portion of the workforce or to reduce everyone's working hours.[6]

Since 1970, the United States has chosen to keep working hours stable—in fact, there is some evidence that U.S. working hours have even increased during the past 40 years. By contrast, most other industrial countries, especially in Europe, have used shorter workweeks, longer vacations, and other strategies to reduce working hours—sometimes significantly. Today, the average American puts in 200–300 more hours at work each year than the average European does. Europeans have made a better choice.[7]

The Benefits of Shorter Hours

Shorter working hours allow more time for connection with friends and family, exercise and healthy eating, citizen and community engagement, attention to hobbies and educational advancement, appreciation of the natural world, personal emotional and spiritual growth, conscientious consumer habits, and proper environmental stewardship. The positive impact of greater free time can be seen by comparing quality of life indices for European nations and the United States.

Since 1980, for example, the United States has fallen from eleventh place in life expectancy to fiftieth. West Europeans now live longer than Americans. On average—although this varies by country—they are also only a little more than half as likely to suffer from such chronic illnesses as heart disease, hypertension, and type 2 diabetes after the age of 50. The United States now lags behind Western Europe in virtually every health outcome, despite spending about twice as much per capita for health care. Moreover, Americans, with their more stressful and hurried lives, are nearly twice as likely to suffer from anxiety, depression, and other abnormalities of mental health.[8]

Seattle Municipal Archives Photograph Collection

Bored in Seattle: the production line of a bottle factory.

Happiness is also affected. While the United States ranks a respectable eleventh in the world in life satisfaction, a recent study found that the four happiest countries in the world—Denmark, the Netherlands, Finland, and Sweden—were all characterized by their remarkable attentiveness to "work-life balance."[9]

The environmental benefits of reduced work time are myriad and include:

- Less need for convenience products. Fast food, for example, is in part a response to an increasingly pressured way of life. Highly packaged and processed foods and other products, including throwaway products, also appeal to those who feel time is short.
- More time to reuse and recycle. Separating wastes into paper, plastics, metals, compost, or trash takes time. People often skip this if they are feeling rushed or overwhelmed.
- Time to make other behavioral choices, such as drying one's clothes on a clothesline rather than in a dryer. When pressed for time, "convenience" tends to take priority.
- Time to choose slower and more energy-friendly forms of transport, including walking, cycling, or public transit rather than driving, or to take trains rather than planes.
- Time to make careful consumer choices, including for certified products like Fair Trade, organic, and songbird-friendly coffee or Forest Stewardship Council lumber.

Moreover, reductions in work time translate rapidly into reductions in energy use, carbon footprints, and pollution (as already seen in the current recession). A study conducted by the Center for Economic and Policy Research, a prominent Washington think tank, concluded that if Americans were to reduce their working hours to European levels, they would almost automatically reduce their energy/carbon impacts by 20–30 percent.[10]

Rushing Through the Environment

Finally, for many people environmental awareness is enhanced by exposure to the natural world, particularly in childhood. From John Muir to Aldo Leopold to Rachel Carson to David Brower, prominent environmentalists have written of the impact of their experiences in natural settings on their later commitment to Earth. A love of nature often results in less desire for material things. Aware of this, Muir was one of the first to call for a law mandating vacation time; he called it a "law of rest." In 1876, on the one-hundredth anniversary of the Declaration of Independence, Muir argued for "Centennial Freedom" that would allow everyone, rich or poor, of whatever race or origin, time to get out into nature. "We work too much and rest too little," Muir declared. "Compulsory education may be good; compulsory recreation may be better."[11]

All Europeans enjoy at least four weeks of paid vacation by law. So do citizens of many African and Latin American nations. Yet the United States still has no law providing vacation time, and half of all American workers now get only one week or less off each year. Consequently, children are now only half as likely to spend unstructured time outdoors as they were in 1970, and visitors to Yosemite National Park—which is more than 300,000 hectares in size—spend on average less than five hours there. People rush through, snapping quick photographs of the granite cliffs and waterfalls, checking their watches, answering their cell phones, and dashing on. There is no time to appreciate the rhythms of Earth or experience a connection to other species, no sense of loss as they pass into extinction, no quiet time to reflect on the wondrous world that now is threatened with humanity's insatiable material demands.[12]

Trading Stuff for Time

What might people do to begin trading gains in productivity for time instead of stuff? The organization Take Back Your Time has been exploring the possibilities of this for the past eight years, encouraged by such developments as the Hours Adjustment Act in the Netherlands and France's 35-hour week.[13]

Dutch working hours are among the shortest in the world, and the Netherlands has the highest percentage of part-time workers. In part, this is a direct response to policy initiatives. European Union law already requires pay and benefit parity for part-time workers who do the same work as full-timers. Moreover, in the Netherlands the Work and Care Act and the Hours Adjustment Act encourage parents to share 1.5 jobs, each working three-quarters time, by requiring that employers allow workers to reduce their hours while keeping the same hourly rate of pay and prorating the benefits. While the right is used primarily by parents of young children, it applies to all employees. Those who choose this option also commonly fall into lower tax brackets; thus the economic penalty for working less is further reduced.[14]

In other European countries, innovative laws allow for such things as regular sabbaticals, phased-in retirement, and guaranteed days of rest, while sharply restricting long hours and overtime work. Europeans would do well to resist calls by corporate leaders to drop restraints on work time and follow the Anglo-American model, as their shorter work time has brought them a higher quality of life than in the United States.

In the United States and other long-hours nations, change must start with a sober assessment of the costs of the higher production/higher consumption lifestyles—what some now call "affluenza." Americans have the farthest to go in this and therefore perhaps the best opportunity to make quick progress. The United States stands alone among industrial nations and most other countries in its lack of laws guaranteeing such rights to time as paid maternity or family leave, paid sick days, or paid vacations. Paid maternity leave, for example, is now guaranteed everywhere except the United States, Swaziland, Liberia, and Papua New Guinea. Many immigrants to the United States are shocked at how few protections American workers have, particularly where the right to time is concerned. Bills currently being considered in the United States Congress would correct some of these deficiencies, but powerful forces are arrayed against them. Business lobbies resolutely oppose all "mandates" that would restrict their absolute control of working hours.[15]

On the other hand, there is some reason for

optimism. The voluntary simplicity movement has helped many Americans choose time over money where the choice was actually theirs to make and not the sole prerogative of their employers. The leaders of that movement understand that making these changes is not only a matter of voluntary action, and it can be helped by progressive policies. Strong organizations that advocate a better work-life balance, like the 1-million member group MomsRising, have emerged in recent years. And the great debate over national health care offers a chance to make points about the health implications of shorter work time.[16]

Since 2002, the Take Back Your Time campaign has worked to increase American awareness of the benefits of shorter working hours. These efforts have included celebrations of Take Back Your Time Day (October 24th) in about 200 U.S. municipalities, coverage of the issue in hundreds of media outlets, and campaigns for legislation such as the Paid Vacation Act of 2009 introduced by Representative Alan Grayson of Florida. His proposed law is modest by international standards—offering only one to two weeks vacation time for workers in firms of 50 employees or more. But it would be a "down payment" on further improvements and would enhance exposure of the issue in the media. Discussion of paid vacation—the epitome of leisure legislation—can help raise the broader issue of Americans' time poverty and its social and ecological impacts.[17]

In his inaugural address, President Barack Obama honored workers who accepted shorter hours rather than see their colleagues fired. But more can be done. Economist Dean Baker proposes that any further government stimulus packages include tax credits for companies that reduce working hours through shorter workweeks, family or sick leave, or extended vacation time without commensurate reductions in workers' pay and benefits. While temporary, such transition funds, which reduce short-term economic sacrifice, would make it possible for workers to see the value of increased leisure and reduced work time.[18]

Re-Visioning the Future

Clearly the world is at a crossroads. For all the remarkable benefits that investments in "green jobs" and new energy technologies will surely provide, they are only part of what's needed for long-run sustainability—necessary change, but not sufficient. To survive and to let people in developing countries somehow achieve secure and modest comfort, material economic growth in rich nations simply must be limited. Yet this must be done without stopping the progress of science and the advance of productivity and without casting millions into the hell of unemployment.

Ultimately, it can only be done by trading gains in productivity for time, by reducing the hours of labor and sharing them equitably. All of this means limiting greed, understanding that a life less rich materially but more rich temporally is not a sacrifice, finding new indices of success to supplant the gross domestic product (which is more a measure of the churn of money in the economy than of true value), and providing real freedom to workers so that their choice to limit their hours of labor does not come at the cost of being fired and losing their livelihoods and health care. It is time to take stock of the "best practices" already being implemented in some countries, expanding them and applying them throughout the world. This way lies hope, sustainability, and greater joy as well.

Inspiring People to See That Less Is More

Cecile Andrews and Wanda Urbanska

Voluntary simplicity is an age-old philosophy that advocates turning away from the pursuit of money, possessions, and greed in order to live more deeply and fully—limiting outer wealth for a greater inner wealth. Philosophers have seen simplicity as a central component of the "good life," arguing that the pursuit of wealth distracts people from more important things, and for much of human history it has also been a religious and spiritual ideal personified by people like St. Francis of Assisi and Gandhi. Today voluntary simplicity has become a movement for sustainability and happiness in a post-consumer society.[1]

Environmentalists have established the harm caused to the planet by consumerism. Voluntary simplicity builds on these facts to create a movement to change behaviors. It is a critique of the values of consumerism: the belief that money is the measure of all things; the practice of using people and the planet for personal benefit; the competitiveness that pits people against each other; and the acceptance of impersonal, sterile, authoritarian, and irresponsible values. In place of these, voluntary simplicity advocates caring and community. Above all, it is a challenge to the dominant philosophy about money found in most societies. As theologian Abraham Heschel puts it: "The most urgent task is to destroy the myth that accumulation of wealth and the achievement of comfort are the chief vocations of man."[2]

Levels of Simplicity

The subject of simplicity is a complex one, with at least three levels—practical, philosophical, and public policy. First is the practical level: cutting back and consuming less. People limit consumption for a variety of reasons—to clear out clutter, to reduce or avoid debt, to secure savings, to afford to work less, or to protect the planet. But focusing only on frugality does not work in the long run. It's like a diet: sooner or later people begin to indulge themselves again. So for deeper engagement, people need to understand that less consumption can lead to more fulfillment: more time for connection to others; more time spent in nature; more satisfaction, security, and balance.

Thus an enduring simplicity must move to

Cecile Andrews is the author of *Less is More, Slow is Beautiful*, and *Circle of Simplicity*. **Wanda Urbanska** is an author and the producer/host of *Simple Living with Wanda Urbanska*, the first nationally syndicated series dedicated to promoting simple, sustainable living.

a second level, a philosophical approach that asks what is important and what matters. At this level, voluntary simplicity becomes a way of living that asks about the consequence of behaviors for the well-being of people and the planet. In fact, it can be argued that consuming becomes a habit for people because they do not take the time to think and make choices based on their own best interests. In a rushed society, people do what is easiest—which is often the things the corporations want them to do.

At the philosophical level of simplicity, people strip away the inessential so that they have time for the essential. In particular, they explore the idea of the "good life" and the nature of happiness. As researchers like Tim Kasser, author of *The High Price of Materialism*, have found, after a certain point more money does not make people happier. Yes, people need a certain level of money, but the lust for more causes people to ignore the important things like friends, family, and community. Having supportive relationships is what makes people happy. Thus the public must come to understand that voluntary simplicity is not a sacrifice. It is about increased personal benefit, about greater life satisfaction and fulfillment—all with a smaller ecological footprint. It's about "less is more"—more security, more tranquility, more joy, more happiness.[3]

Finally, at the public policy level the issue is "less is more" for all people. Although individuals can make changes in their own behavior and live more simply, very few people can live a truly simple life in western industrial societies. For too long, the simplicity movement has focused primarily on individual change. It is time to move to a greater advocacy of public policy change. In order to enable all to live simply, society needs public policies that provide health care, vacations, parental leave, and reduced work hours.

Perhaps the most essential policy change has to do with wealth inequality. The biggest predictor of the health of a nation, as measured in terms of longevity, is the gap between the rich and the poor. It's not just that the health of the poor brings down the average—everyone is affected because inequality undermines social cohesion. Richard Wilkinson, author of numerous books on the wealth gap, shows how the stress of inequality undermines health and promotes consumerism. It is highly stressful when someone is denied respect and dignity in a status-conscious society, and stress makes people sick. Further, inequality contributes to consumerism: in an unequal society, people use material possessions to fight their way up the ladder of status. In *The Spirit Level: Why More Equal Societies Almost Always Do Better*, Wilkinson and Kate Pickett show how wealth inequality affects community life, mental health, and violence, among others—all factors making it hard to live simply and attain happiness.[4]

Motivating Change

How do we motivate people to begin to reduce their consumerism and to work for change? First, some will respond to information about the issues. Knowing the dire facts about climate change will motivate them to change. But for others, more is needed. Berkeley linguist George Lakoff says that too often change agents rely only on information and facts—but that is not enough. It is important to evoke empathy and caring. The simplicity movement does this through the vision and the experience of the joyful community.[5]

The voluntary simplicity movement holds out a vision of the good life, a life based on connection, caring, and the common good. Environmental author Bill McKibben, writing in *The Nation*, said: "In fact, the only way to endure the transition will be with a renewed sense of community. The real poison of the past few decades has been the hyper-individualism that we've let dominate our political life—the idea that everything works best if we think not a whit about the common interest. In the end, that has damaged our society, our climate and our private lives. The final hope we have is resurgence of a politics that calls on us to work together."[6]

This focus on community takes several forms in the simplicity movement: the study circle, cohousing, ecovillages, and the relocalization or Transition Town movement. Certainly not all of these efforts label themselves as "simplicity," but most people involved in them are trying to live more simply.

The simplicity study circle is a small-group method of community education and social change that has its roots in European history. Study circles originated in Sweden and the Danish folk education movement, although folk education has a long heritage in the United States as well. The Highlander Center in Tennessee, for example, began as a folk school after its founder Myles Horton visited Denmark in the 1920s. Highlander was also influenced by the popular education movement in Latin America, in particular the theories developed by Brazil's Paulo Friere, author of *The Pedagogy of the Oppressed*. The focus of folk education, popular education, and the study circle is the belief that if people come together to talk, they will find the answers to their own problems—that the wisdom is in the people. Whereas the purveyors of consumerism put a great deal of effort into manipulating people and their emotions, the community education approach restores people's abilities to think for themselves—an approach that can break the manipulation of advertising in consumer societies.[7]

While the study circle is a small group of six to eight people, other forms of community are larger. Efforts such as cohousing and ecovillages, for example, ask people to move into a new setting. At the same time, more and more people are working to transform their own neighborhoods into places that encourage sustainability and community. What began in the United States as the relocalization movement is now joined by the Transition Town movement, which originated in the United Kingdom and is spreading around the world. (See Box 22.) As of August 2009, almost 200 communities were recognized as official Transition Towns in the United Kingdom, Ireland, Canada, Australia, New Zealand, Italy, and Chile.[8]

The Transition Town movement focuses on reducing the use of oil by building "resilience" in a community—first by helping people to see that almost everything they buy involves oil, whether in its manufacture, transport, or marketing, and then by teaching people ways to reduce their use of oil by gardening and other traditional skills such as canning and knitting. Transition Town leaders work with neighbors to shop and eat locally. They encourage people to share through projects like local currencies, tool exchanges, car sharing, community gardens, community-supported agriculture, and farmers' markets. All these projects involve working with others in a collaborative and cooperative way, undermining the competitiveness of corporate consumerism.[9]

People involved in these movements may not even realize they are practicing "voluntary simplicity," and indeed the label is not important. For instance, the Slow movement has gained attention in Europe—particularly the Slow Food and Slow Cities movements as described in Carl Honore's *In Praise of Slowness*. The Slow Food movement, founded in Italy, encourages people to support local, organic food. (See Box 23.) It supports farming that nurtures the planet as well as promoting social justice—focusing on the practices of corporations. The *Cittaslow* movement states that its purpose is to resist "the fast-lane, homogenized world so often seen in other cities throughout the world"; it supports local food and artisans, less use of cars, and places for people to linger and enjoy.[10]

In the United States, the Slow movement

Box 22. Growing a Degrowth Movement

Today many people believe that economic growth will lead to perpetual improvements in well-being, even as growth has increasingly taxed Earth's ecosystems, exploits the poor, and threatens the security of future generations. To proactively address the current environmental, financial, social, and ethical crisis, a radically different societal model is necessary: a degrowth society. The movement in support of this—tailored for countries that have grown beyond their fair share of Earth's bounty—has developed a political platform that envisions degrowth societies centered on sustainability and proximity, where, for example, they relocalize production and consumption. Degrowth societies promote human relations instead of consumerism and reduce waste and polluting transport through the use of ecotaxes. All of this is done so that these societies will have sustainable ecological footprints and be in balance with nature.

Today there are degrowth political parties in France and Italy. The publication *La Decroissance* (*Degrowth*) can be found in newsstands across France and has readers in the rest of the francophone world as well. In Spain, Temps de Re-voltes organized a "degrowth publicity tour" of more than 30 small municipalities in 2008. In cooperation with local authorities, the group organized panels and discussed future energy crises and degrowth visions while celebrating local culture and traditions.

To have the impact needed to stabilize ecological systems, degrowth will need to be pursued at a variety of levels. Cities, towns, and villages will need to relocalize agricultural and energy systems, introducing community and backyard vegetable gardens as well as locally generated renewable energy to promote resilience. Local currencies such as Totnes Pounds or Ithaca Hours can help wealth remain in the hands of individuals and small local businesses as opposed to multinational corporations and financial institutions. There also need to be broader societal efforts such as reducing working hours and improving regulation of international institutions that only promote destructive growth.

Currently, the initiatives that best put into practice the values set forth by degrowth are the Transition Towns that are found mainly in the United Kingdom, but also in Australia, the United States, Japan, and Chile, among other countries. Transition Towns are based on preparing for resource scarcity and climate change by building communities that are both socially and economically resilient, where the focus is on improving quality of life for the inhabitants while living sustainably. The "showcase" of Transition Towns is Totnes in England.

The movement for a "degrowth society" is radically different from the recession that is widespread today. Degrowth does not mean the decay or suffering often imagined by those new to this concept. Instead, degrowth can be compared to a healthy diet voluntarily undertaken to improve a person's well-being, while negative economic growth can be compared to starvation. In a degrowth world, people will spend less time working and more time living. They will consume less but better, produce less waste, reuse and recycle more, understand the impacts of human behavior, and have ecological footprints that can be sustained. People will find happiness in human relationships and conviviality rather than the never-ending pressure to accumulate more and more goods. All this implies a serious rethinking of people's current concepts of reality and significant imagination, but the shifting ecological realities are sure to provide the necessary inspiration.

—*Serge Latouche*
Professor Emeritus of Economics
University of Orsay, France
Source: See endnote 8.

Box 23. The Slow Food Movement

The international Slow Food Movement started in 1986 as a protest against the opening of a McDonalds near the Spanish Steps in Rome. The restaurant became a physical representation of the erosion of the local, sustainable, and healthy food culture in Italy. Since then, Slow Food has become a global organization with chapters, or *convivia*, in 132 countries, and with more than 100,000 members. They work to promote "good, clean and fair food," thereby transforming cultures via food. This is done through a wide range of activities that both educate and inspire.

Slow Food aims to reconnect producers and consumers (or co-producers, as they prefer to call the educated consumer who supports slow food) and promote culinary diversity and healthy, tasty food on a local scale, while also seeing the "bigger picture," promoting biodiversity as well as international networking among artisanal producers, and enhancing traditional production in order to make it economically viable. Throughout the movement's activities, the emphasis is on making gastronomic pleasure and ecological responsibility inseparable.

The movement works to educate the public through a variety of initiatives. Many books now teach the art of "slow cooking." Lectures, articles, and Web sites describe the grim realities of agribusiness and fast food, as well as the benefits of buying local and Fair Trade products. The movement also uses events to educate and mobilize people. During one such event, Slow Fish 2009 in Genoa, Italy, 55,000 local and international guests learned about sustainable fish harvesting, met artisanal fishers, and got gastronomic education through food and wine tasting.

The Slow Food Foundation for Biodiversity—a division of Slow Food—works to "defend local food traditions, protect local biodiversity and promote small-scale quality products." Small artisan producers around the world are organized in 300 Presidia that focus on improving production techniques while preserving traditional products and methods and finding new markets for these. The Foundation also has an Ark of Taste that is a register of food products they hope to reintroduce in the marketplace but that are in danger of being forgotten because traditional production methods are no longer in use or certain ingredients are scarce.

Communities are also setting up Earth Markets and Slow Food cafés and restaurants—both of which help food producers and co-producers interact while promoting local food and helping customers. Earth Markets and cafés are now located in Delhi, Tel Aviv, Beirut, and Bucharest, among other cities.

Slow Food members also do a lot of lobbying on behalf of their causes, particularly on issues regarding agricultural and trade policy in the European Union. Recently, a Time for Lunch campaign was organized by Slow Food USA; it encourages Congress to improve the Child Nutrition Act, which sets the standards for school meals in the United States. The whole Slow Food movement—through its role in promoting good, clean, and fair food—is playing an important role in facilitating a shift to sustainable cultures.

—*Helene Gallis*

Source: See endnote 10.

has become a part of the simplicity movement, encouraging people to live deeply by exploring and reclaiming the ancient vision of leisure. People are beginning to find ways to take back their time in order to walk more, talk with their neighbors, and spend more time in local neighborhoods. Advocates of the slow life are involved in the Take Back Your Time campaign, a project to bring some European labor management polices to the United States. Certainly, without shorter work hours, more vacations, and parental and sick leave, it is difficult to live simply.[11]

Creating community is central to inspiring people to live more simply. And it is important to see that this approach has implications for democracy, which is the only way to wrest power from corporations, the force behind consumerism. Robert Wuthnow, in *American Mythos*, calls for more "reflective democracy," opportunities to talk about basic values and ideals. He argues that in the usual democratic discourse, people fail to move beyond the idea of "the informed citizen"—someone who engages in discussions about current events—when they also need to reflect on basic values and assumptions. Certainly voluntary simplicity is the "examined life," helping people determine what's important and what matters.[12]

As Robert Putnam, author of *Bowling Alone*, notes, the culture in which people talk over the back fence is the culture in which people vote. When people are involved in their local communities, they are talking with each other and are usually more involved with public policy—often trying to stop intrusive development in their neighborhoods. So the neighborhood movement is important in many ways. Ultimately, conversations and engagement with others help people transform the "lone wolf" culture and realize that true security lies not in material wealth but in people.[13]

In their new book, *Meeting Environmental Challenges: The Role of Human Identity*, Tim Kasser and Tom Crompton argue that it is important to focus on strategies that inspire people to move away from materialistic values. They maintain that focusing on fear—through dire warnings about the environment—can drive people to consume as a compensatory behavior. They cite simplicity circles as a way to offer social support that evokes more transcendent values of caring and concern.[14]

Another way to elicit feelings that evoke hope rather than fear is through people's stories. People become interested in voluntary simplicity when they read or hear a story they identify with. The story usually goes something like this: a corporate employee is stressed, sick, and depressed, so he consumes less, quits his job, moves to a smaller house, finds work that is more satisfying, reduces his work hours, plants a garden, and begins to work with his local community center. People see themselves in these stories. They begin to see that their desire for more—more money, more status—will not make them happy. They awaken from the spell of the false promises of consumerism, and they begin to search for a better way. One story about downshifting can perhaps do as much to change people's consumer behavior as 10 facts about climate change.

One effort to use stories like these is the U.S. public television series *Simple Living with Wanda Urbanska*. Rather than being shamed into change by finger-wagging and images of environmental degradation and calamity, the audience learns from the stories of real people. The series illustrates different approaches to simplicity, allowing all kinds of people to relate. For instance, it documents one family's commitment to bringing Great Plains bison back from the brink of extinction in Bozeman, Montana. It shows a Massachusetts church encouraging people to "roll or stroll for your soul" by asking congregants to ride bikes, walk, or carpool to service, followed by the minister's dramatic "blessing of the bikes."[15]

The series challenges viewers to keep possessions in service past their date of planned obsolescence in a playful, recurring feature called "The Thing That Refused to Die." One such "thing" was a 1930s Fireboat put out to pasture by the New York City Fire Department that was a knight in shining water when its ancient hoses were pressed into service to fight the fires at the Twin Towers on 9-11. The moral of this segment was simple: "That thing you put out to pasture may be the most valuable thing you own." Stories like these can motivate people to reexamine their consumption choices, to make change.[16]

Finally, a sense of joyful community is evoked by new kinds of social inventions: experiences that bring people together in new and creative ways that challenge consumerism. For instance, "Buy Nothing Day" led by *Adbusters Magazine* was originally the day after Thanksgiving in the United States, reportedly the biggest shopping day of the year. Now more than 65 nations participate with different activities—the zanier the better. Volunteers stand in shopping malls with scissors and a sign offering to cut up people's credit cards. Others sponsor a "zombie walk" through malls, mirroring the blank looks on the faces of shoppers. People have fun driving their shopping carts around in long conga lines in places like Walmart or filling their carts and leaving them without buying anything. *Adbusters* encourages "culture jamming" to fight consumerism, and it stages events like giving fake tickets to SUVs or sponsoring a "detox week," encouraging people to "unplug" from video games and computers.[17]

Another creative idea is The Compact, an initiative in which people agree to go a year without buying anything new. Some become involved in "freeganism," which can include "dumpster diving" for food and other items that have been thrown away but are perfectly good, gleaning, wild foraging, urban gardens, or squatting in empty buildings. Another movement that appeals particularly to young people is "couch surfing," where they travel cheaply by finding homes on-line to stay in for free. In a similar vein is "wwoofing" (World Wide Opportunities on Organic Farms), where people work on organic farms in return for room and board.[18]

Creating Post-Consumer Cultures

All over the world people are developing ways to challenge consumerism and create post-consumer cultures. At the Barefoot College in India, local impoverished people are encouraged to maintain their sustainable ways of living. Gaviotas, a village in Colombia, has reclaimed barren savannas and regenerated forests using innovative techniques such as solar and wind power and children's seesaws to drive a water pump. Europeans engage in "placemaking," where "spaces are turned into places" by, for example, Denmark's Jan Gehl, who has transformed urban spaces into experiences of community and conviviality by expanding the cafe society and the bicycle and pedestrian culture. A similar movement called "city repair" started in Portland, Oregon, and involves people "taking back" streets in their neighborhoods by painting designs in the intersections, moving lawn chairs into the streets, and creating straw bale benches and bulletin boards at the corners—all to bring people together.[19]

The goal is not only to get people to consume less but to create a new society, to inspire and motivate them to become more involved in social change efforts by evoking empathy, caring, and connection. When people get involved with others, they lose their desire to consume because they encounter a new, more satisfying way of life. Voluntary simplicity, then, is at the same time a practice, a philosophy, and a method of social change that can help transform consumer cultures by helping people understand that "less is more."

Ecovillages and the Transformation of Values

Jonathan Dawson

Tsewang Lden and Dolma Tsering, elderly Ladakhi women, are caught on film in an old people's home in London, incongruous in their fine and colorful traditional costume. They look on in shock at an old English woman, alone in a sterile white-painted room and so absorbed in watching television that she barely noticed the other women's entry. The Ladakhi women had never seen anything like this before. In the north Indian province that is their home, old people are integrated into the family, considered wise elders and honored.[1]

Lden and Tsering were participating in a "Reality Tour" organized by the Ladakh Project to enable small groups of Ladakhi women to visit western countries, where they see for themselves the reality of life in the West—good and bad—including community breakdown, loneliness, and violence. The organizers hope this will reinforce cultural self-confidence, help Ladakhis appreciate the many positive features of their culture, and show the dark side of today's globally dominant cultural orientation—consumerism—that is so rarely presented in the global media.

What is happening here is one small example of a much wider questioning of the values base underlying the consumerist culture and an exploration of what could replace it. The Ladakh Project is a founding member of the Global Ecovillage Network (GEN), an umbrella organization for ecovillages that includes some of the innovative experiments in post-consumerist, community-based living that are at the forefront of this wave of exploration.[2]

The commonly accepted definition of ecovillages, provided in 1991 by *In Context* editor Robert Gilman, is "human-scale, full-featured settlements in which human activities are harmlessly integrated into the natural world in a way that is supportive of healthy human development and can be successfully continued into the indefinite future."[3]

Today this global network contains an interesting and innovative alliance between intentional communities with a strong focus on sustainability (generally though not exclusively located in the industrial world) and networks of traditional communities in developing countries. Intentional communities are ones that have been formed consciously around specific values and objectives, most of which today have a strong focus on some dimension of sustainability and call themselves ecovillages. The communities in developing countries that are members of GEN seek to maintain their

Jonathan Dawson is a sustainability educator and author based at the Findhorn ecovillage in Scotland.

traditional values and cultural distinctiveness and to win back greater control of their economic destinies in the face of pressures unleashed by economic globalization.

The most visible and tangible projects within ecovillages tend to be those related to technology and the development of alternative systems of various kinds. Most first-time visitors to ecovillages are there to find out about ecological housing, biological wastewater treatment systems, renewable energy technologies, community currencies, and the like.

Less immediately obvious, but arguably even more significant, is the contribution of ecovillages to a radical transformation of values and consciousness. Ecovillages are engaged in the transformation of values in four ways that may make the transition to sustainability easier and more graceful:

- delinking growth from well-being,
- reconnecting people with the place where they live,
- affirming indigenous values and practices, and
- offering a holistic and experiential educational ethic.

Delinking Growth from Well-being

There has been growing awareness in recent years of the inadequacy of gross domestic product as a measure of true wealth, with its exclusive focus on economic capital formation but with no reference to other forms of capital—the health and biodiversity of the natural environs, the strength of communities, the well-being and happiness of people. What would a society look like that consciously developed its various forms of capital in a more balanced and integrated way? Could communities—indeed, whole societies—learn to substitute other forms of capital for economic wealth, demonstrating how quality of life could be maintained or even enhanced while significantly reducing consumption and material throughput? Ecovillages serve as research, training, and demonstration sites for such a proposition.

The attempt to delink growth and the accumulation of material goods from well-being lies at the heart of the ecovillage concept. The low levels of consumption that typically prevail within ecovillages result partly from the design of their systems so as to reduce energy and materials intensity and partly because, by opting out of the global economy to varying degrees, they forgo opportunities to maximize income.

Several recent studies confirm that the ecological impact of ecovillages is markedly lower than for average conventional communities. A 2003 study by the University of Kassel looked at carbon dioxide emissions associated with two ecovillages in Germany. It found that per capita emissions in the Sieben Linden and Kommune Niederkufungen ecovillages were 28 and 42 percent, respectively, of the German average. Sieben Linden scored especially well in the fields of heating and housing: as a result of renewable energy generation and the use of highly energy-efficient building materials and insulation, the community recorded emission levels just 10 and 6 percent, respectively, of the national average.[4]

Two studies of energy consumption at Ecovillage at Ithaca in upstate New York—one by Cornell University, another by the Massachusetts Institute of Technology—found that the community's consumption was more than 40 percent lower than the U.S. average. And a study undertaken by the Stockholm Environment Institute found that the Findhorn ecovillage in Scotland has a per person ecological footprint a bit over half of the U.K. average, the lowest footprint recorded for any settlement in the industrial world. Findhorn residents achieved an especially low footprint in the areas of home heating and food—21.5 and 37 percent per person, respectively, of the national average.[5]

There is substantial anecdotal evidence that the quality of life within ecovillages is generally high—certainly much higher than would be expected for communities that operate on low levels of income. The anecdotal evidence has been reinforced by a 2006 study comparing the contribution of built (economic), human, social, and natural capital to quality of life in 30 intentional communities with that in the town of Burlington, Vermont. The study found that the quality of life was slightly higher in the intentional communities despite the fact that average incomes were significantly lower because of a greater cultivation and appreciation of other forms of capital, especially social capital. Of special importance in determining quality of life, the study identified the strong social bonds that develop within intentional communities, their "ownership provisions as well as...process for allocating work and rewarding contributions," and the "emphasis the community placed upon the preservation of natural areas."[6]

The authors concluded: "Results of this study represent an existence proof: it is possible to achieve a high (and probably more sustainable) quality of life while consuming at rates much less than the U.S. average.... We have much to learn from intentional communities around the world that have been actively experimenting with issues related to quality of life and sustainability."[7]

It is especially interesting that many of the activities and design features that are responsible for low energy and resource use within ecovillages are also among the most important in contributing to a better quality of life. The decision by many ecovillages to grow a significant amount of their own food, for example, involves community members working cooperatively together in a way that strengthens relationships and builds a strong and nurturing sense of connection with the land.

Many of the other footprint-shrinking design features—preparing and eating meals together, car clubs, community-owned renewable energy facilities, community currencies and investment, and so on—similarly engender a spirit of cooperation that builds community and contributes to strong feelings of well-being.

Courtesy Findhorn Foundation

Harvesting organic vegetables at Findhorn ecovillage, Scotland.

This ethic extends into the economic life of ecovillages, where cooperation and solidarity are promoted and the relationship to work is transformed. The Twin Oaks ecovillage in the state of Virginia declares: "We use a trust-based labor system in which all work is valued equally. Its purpose is to organize work and share it equitably, giving each member as much flexibility and choice as possible. Work is not seen as just a means to an end; we try to make it an enjoyable part of our lives."[8]

Reconnecting People with the Place Where They Live

One of the more pernicious impacts of today's globalized economy is the weaker connections that people feel to the place where they live.

There has been a progressive homogenization across the world over the last 50 years or so of foodstuffs, clothing, farming technologies, building materials, styles, and so on. As a part of this trend, increasingly diets no longer reflect the changing seasons.

This disconnect is enormously important in providing a seed bed for alienation and consumerism. When resources are pulled in from all over the world, people lose all sense of the carrying capacity of the bioregions they live in—and thus of any obligation to attempt to live within such limits. People's natural propensity to love the web of life that all humans are part of becomes lost in a fog of ignorance of what that web looks and feels like in specific places.

Reestablishing a keener appreciation of the qualities, patterns, and rhythms of home places and what they can sustainably yield is fundamental to refinding a balanced and respectful place within them. Nurturing just such an enhanced appreciation is of central importance to the ecovillage ethic.

In part, this manifests in attempts to increase levels of self-sufficiency. Ecovillages typically seek to develop an enhanced understanding of ecological building techniques using local materials, local medicinal herbs, wild food foraging, organic food production and processing, energy generation with locally available renewable resources, and so on. They are seeking to deepen their connections in their own bioregions, to increase resilience in a period of energy transition, and to reduce dependence on money and the global economy.

Similarly, many ecovillages are engaged in initiatives to restore the health of their surrounding ecosystems. Over the last 40 years, to cite but one example, the Auroville ecovillage in southern India has planted nearly 3 million trees and engaged in widespread earth restoration projects that have simultaneously enriched the diversity of local natural systems and woven people more deeply into the web of life. Moreover, the way in which this was done—early efforts involved the planting of non-native species that created other environmental hazards, but these were progressively replaced with more diverse and native species—demonstrates how ecovillages are able to learn and be flexible with their efforts, catering to the needs of the environment as discovered along the way. Similar efforts at large-scale tree planting and earth restoration can be found at Sólheimar in Iceland, The Farm in Tennessee, and many other ecovillages.[9]

The journey toward being more rooted in bioregions is also a cultural one. Many ecovillages engage in rituals to mark and celebrate the turning of the seasons—building on, though generally not slavishly adhering to, traditional practices. Grishino ecovillage in Russia, for instance, has become an important center for the celebration of and training in traditional Russian song, dance, arts, and storytelling. In Findhorn, the turning of the year is marked through celebration of the Celtic festivals in song, dance, storytelling, and bonfires.[10]

Affirmation of Indigenous Values and Practices

The corporate marketing and advertising industries have played a central role in shaping the values underlying today's consumerist culture. They have played an especially devastating role in undermining the cultural self-confidence of groups falling outside of the global consumer class. Consequently, an important dimension of the value shift required in the transition to a sustainable global society lies in celebration of the diversity of human cultures, encouraging each to value and take pride in their distinctiveness.

Ecovillage networks in developing countries tend to be very active on this front. Activities with new groups generally focus on building cultural self-confidence and celebrating the communities' strengths and achievements.

The Sri Lankan nongovernmental group Sarvodaya, a founding member of GEN, works with over 15,000 communities island-wide. It has developed a methodology for community assistance that begins with an empowerment program. This includes a strong element of social and spiritual empowerment, including meditation, cultural validation, peacemaking, and conflict facilitation. Only when this foundation has been built does the more tangible work of economic empowerment and physical infrastructure development begin.[11]

The Ladakh Project in India similarly places great weight on building cultural self-confidence. It has helped to establish the Women's Alliance of Ladakh (WAL), a network of over 6,000 women from almost 100 different villages, with the twin goals of raising the status of rural women and strengthening local culture and agriculture. Some of the more creative programs initiated by WAL are No TV weeks, aimed at encouraging people to resist the consumerist ethic; annual festivals celebrating local knowledge and skills, including traditional spinning, weaving, and dyeing and the preparation of indigenous food; and the Reality Tours that brought Tsewang Lden and Dolma Tsering face to face with the reality of old people's lives in an industrial country.[12]

A Holistic and Experiential Educational Ethic

Something extraordinary has happened over the last decade or so in the relationship between ecovillages and the mainstream society that they were created to be an alternative to. As interlocking economic, ecological, and social crises have deepened, the various experiments that ecovillages have been engaged in are becoming recognized as of growing relevance far beyond the ranks of radical outsiders. One of the principal ways that the values and models they have developed are being shared more widely is through education.

Courtesy Kibbutz Lotan

Straw-bale, earth-plastered domes being built on geodesic frames will become student housing at Kibbutz Lotan's Center for Creative Ecology, Israel.

The various educational packages developed within ecovillages reflect the core ethics of the communities themselves in that they are holistic—exploring interdependence and the relationships between issues and subjects that are generally considered independently in more conventional settings—and experiential, in that they engage all of the learner's faculties—head, heart, and hands.

In this regard, ecovillage education can be seen as part of the wider trend toward environmental education based on systems thinking. What is distinctive in the ecovillage educational model is that the learning experience unfolds in the context of a live experiment in the translation of post-consumerist values into the fabric of a sustainable community. Immersion in such living laboratories can be a profound transformation for students as they experience in a very tangible way the dynamic relationship between values, lifestyle, and community structures.[13]

A number of ecovillage-based educational initiatives have sprung up over the last decade or so. The Ecovillage Training Center at The Farm in Tennessee, the Center for Creative Ecology at Kibbutz Lotan in Israel, and Ecological Solutions at Crystal Waters in Australia are three among many centers worldwide whose courses in the various dimensions of sustainability now attract participants from across the social spectrum.[14]

Numerous educational partnerships have also developed between ecovillages and more mainstream institutions that aid the diffusion of ecovillage values and models into wider society. A United Nations CIFAL training center, one in a network of 11 centers worldwide that provide training in sustainability to local authorities and other local actors, opened in 2007 at Findhorn in Scotland. This draws on expertise developed within and beyond ecovillages to build the planning and implementation capacity of local agencies in Scotland and, increasingly, in northern Europe.[15]

Meanwhile, the Findhorn College, an educational institution within the ecovillage, regularly hosts the University of St. Andrews undergraduate program in sustainable development. And as of September 2009, Heriot-Watt University in Edinburgh offers the first Master of Science degree in Sustainable Community Design—with two compulsory sections on Ecovillage Practice and Community Design Practice taught by Findhorn College staff at the ecovillage.[16]

A major new ecovillage-based educational initiative, Gaia Education, has developed a curriculum derived from good practice within ecovillages that has been endorsed by UNITAR and welcomed as a valuable contribution to the UN Decade of Education for Sustainable Development. The curriculum is now being taught in ecovillages and universities on every continent.[17]

An undergraduate study-abroad program, Living Routes, offers students at U.S. universities the opportunity to do formally accredited semesters at ecovillages on every continent, while Ecovillage at Ithaca, in New York, is engaged in an ambitious alliance with Cornell University and Ithaca College to enhance university-based sustainability curricula in the United States.[18]

These developments on the educational front represent an opportunity to spread ecovillage values and models into the wider society. As the world seeks to make the transition to a rich, diverse, and sustainable global society, the lessons learned by ecovillages are likely to be an important source of information and inspiration.

Notes

State of the World: A Year in Review

October 2008. International Union for Conservation of Nature, "IUCN Red List Reveals World's Mammals in Crisis," press release (Gland, Switzerland: 6 October 2008); Peter Smith, "The Green Commandment," *Courier- Journal* (Louisville, KY), 22 October 2008; *The Green Bible* (New York: HarperCollins, 7 October 2008); J. Timothy Wootton, Catherine A. Pfister, and James D. Forester, "Dynamic Patterns and Ecological Impacts of Declining Ocean pH in a High-resolution Multi-year Dataset," *Proceedings of the National Academy of Sciences*, 8 October 2008; James Kanter, "Indonesia Officials Unveil a Deal to Protect Forests," *New York Times*, 9 October 2008; "New Treaty Aims to Protect Shared Transboundary Aquifers," *Environment News Service*, 23 October 2008; Nathan P. Gillet et al., "Attribution of Polar Warming to Human Influence," *Nature Geoscience*, 30 October 2008.

November 2008. Jacqueline Alder et al., "Forage Fish: From Ecosystems to Markets," *Annual Review of Environment and Resources*, November 2008; United Nations Environment Programme (UNEP), "Wide Spread and Complex Climatic Changes Outlined in New UNEP Project Atmospheric Brown Cloud Report," press release (Nairobi: 13 November 2008); Philip Pullella, "Vatican Set to Go Green with Huge Solar Panel Roof," *Reuters*, 25 November 2008; Raymond Colitt, "Brazil Cracks Down on Amazon Loggers After Riot," *Reuters*, 28 November 2008.

December 2008. "Venice Is Hit by Serious Flooding," *BBC News*, 1 December 2008; Clive Wilkinson, ed., *Status of Coral Reefs of the World 2008* (Townsville, Australia: Global Coral Reef Monitoring Network, 2008); "California Air Resources Board, "ARB Says Yes to Climate Action Plan," press release (Sacramento: 11 December 2008); European Parliament, "European Parliament Seals Climate Change Package," press release (Brussels: 17 December 2008); "Greenland's Glaciers Losing Ice Faster This Year Than Last Year," *TerraDaily*, 23 December 2008; Munich Re Group, "Catastrophe Figures for 2008 Confirm that Climate Agreement Is Urgently Needed," press release (Munich: 29 December 2008).

January 2009. Gábor Horváth et al., "Polarized Light Pollution: A New Kind of Ecological Photopollution," *Frontiers in Ecology and the Environment*, eView, 7 January 2009; Plantlife International, *Medicinal Plants in Conservation and Development* (Salisbury, UK: November 2008); Environment News Service, "Japan Launches World's First Greenhouse Gas Observing Satellite," 23 January 2009.

February 2009. Andrew Torchia, "China Drought Deprives Millions of Drinking Water," *Reuters*, 7 February 2009; Susan Solomon et al., "Irreversible Climate Change Due to Carbon Dioxide Emissions," *Proceedings of the National Academy of Sciences*, 10 February 2009; UNEP, "Historic Treaty to Tackle Toxic Heavy Metal Mercury Gets Green Light," press release (Nairobi: 20 February 2009); U.S. National Aeronautics and Space Administration

(NASA), "NASA's Launch of Carbon-Seeking Satellite Is Unsuccessful," press release (Washington, DC: 24 February 2009).

March 2009. Marian Burros, "Obamas to Plant Vegetable Garden at White House," *Washington Post*, 19 March 2009; Drew Shindell and Greg Faluvegi, "Climate Response to Regional Radiative Forcing During the Twentieth Century," *Nature Geoscience*, 22 March 2009; "Rain-soaked Southern Africa Hit by Worst Floods in Years," TerraDaily.com, 27 March 2009; Jeff Mason and Thomas Ferraro, "Obama Signs Landmark U.S. Conservation Bill," *Reuters*, 30 March 2009.

April 2009. Henry Chu, "Beneath the G-8 Summit, A Valley of Misery for Italy Quake Victims," *Los Angeles Times*, 10 July 2009; David Randall, "Pandemic Fears as Flu Kills 68," *The Independent* (London), 26 April 2009; World Health Organization, "Swine Influenza," Statement by WHO Director-General Dr. Margaret Chan, 25 April 2009; Abengoa Solar, "Abengoa Solar Begins Operation of the World's Largest Solar Power Tower Plant," press release (Seville, Spain: 27 April 2009); American Lung Association, "New American Lung Association Report Finds 60 Percent of Americans Live in Areas Where Air Is Dirty Enough to Endanger Lives," press release (Washington, DC: 29 April 2009).

May 2009. Organic Trade Association, "U.S. Organic Sales Grow by a Whopping 17.1 Percent in 2008," press release (Greenfield, MA: 4 May 2009); Aiguo Dai, "Changes in Continental Freshwater Discharge from 1948 to 2004," *Journal of Climate*, May 2009, pp. 2,773–92; Conservation International, The Nature Conservancy, and WWF, "Leaders of Coral Triangle Countries Declare Action to Protect Marine Resources for People's Wellbeing," press release (Manado, Indonesia: 15 May 2009); "Sierra Leone and Liberia Create Vast Transboundary Peace Park," *Environment News Service*, 18 May 2009.

June 2009. Oli Brown and Alec Crawford, *Rising Temperatures, Rising Tensions; Climate Change and the Risk of Violent Conflict in the Middle East* (Winnipeg, Canada: International Institute for Sustainable Development, 2 June 2009); Isaac Wolf, "Recycled Radioactive Metal Contaminates Consumer Products," ScrippsNews.com, 3 June 2009; "Deadly Amazon Clashes Roil Peru," *Associated Press*, 6 June 2009; U.N. Food and Agriculture Organization (FAO), "1.02 Billion People Hungry," press release (Rome: 19 June 2009); Guus J.M. Velders et al., "The Large Contribution of Projected HFC Emissions to Future Climate Forcing," *Proceedings of the National Academy of Sciences Online*, 22 June 2009.

July 2009. Lisa M. Schloegel et al., "Magnitude of the US Trade in Amphibians and Presence of *Batrachochytrium dendrobatidis* and Ranavirus Infection in Imported North American Bullfrogs (*Rana catesbeiana*)," *Biological Conservation*, July 2009, pp. 1420–26; Office of the Mayor of San Francisco, "Executive Directive 09-03: Healthy and Sustainable Food for San Francisco" (San Francisco: 9 July 2009); Council of the European Ministers, "Council Confirms Strict Conditions for Marketing Seal Products in the EU," press release (Brussels: 27 July 2009); Boris Worm et al., "Rebuilding Global Fisheries," *Science*, 31 July 2009, pp. 578–85.

August 2009. Maplecroft, "Australia Overtakes USA as Top Polluter," press release (Bath, U.K.: 3 August 2009); Yoko Hattori et al., "The Ethylene Response Factors SNORKEL1 and SNORKEL2 Allow Rice to Adapt to Deep Water," *Nature*, 20 August 2009, pp. 1,026–30; World Agroforestry Centre, "While Farmers Frequently Blamed for Forest Loss, New Study Shows about Half of Farmlands Worldwide Have Significant Tree Cover," press release (Nairobi: 24 August 2009).

September 2009. FAO, "New Treaty Will Leave 'Fish Pirates' Without Safe Haven," press release (Rome: 1 September 2009); National Oceanic and Atmospheric Administration, "NOAA: Warmest Global Sea-Surface Temperatures for August and Summer," press release (Washington, DC: 16 September 2009); WWF, "New Species Discovered in the Greater Mekong at Risk of Extinction Due to Climate Change," press release (Gland, Switzerland: 22 September 2009); "Security Council Calls for World Free of Nuclear Weapons During Historic Summit," *UN News Service*, 24 September 2009; Group of 20, "Leaders' Statement: The Pittsburgh Summit" (Pittsburgh, PA: 24–25 September 2009).

The Rise and Fall of Consumer Cultures

1. *The Age of Stupid*, Franny Armstrong, Director, independently released, 20 March 2009.

2. World Bank, *World Development Indicators Online*, online database, at media.worldbank.org/secure/data/qquery.php, viewed 23 September 2009; vehicles from Michael Renner, "Global Auto Industry in Crisis," *Vital Signs Online*, Worldwatch Institute, 21 May 2009; "Growing World Refrigerator Market," *JARN Web Magazine*, December 2008; IDC, "PC Market Growth Evaporates in Fourth Quarter as Financial Crisis Hits Home, According to IDC," press release (Framingham, MA: 14 January 2009); IDC, "Worldwide Mobile Phone Market Declines by 12.6% in Fourth Quarter, More Challenges to Come Says IDC," press release (Framingham, MA: 4 February 2009).

3. World Bank, op. cit. note 2.

4. Gary Gardner and Payal Sampat, *Mind Over Matter: Recasting the Role of Materials in Our Lives*, Worldwatch Paper 144 (Washington, DC: Worldwatch Institute, December 1998); Michael Renner, "World Metals Production Surges," *Vital Signs Online*, Worldwatch Institute, 3 September 2009; oil and natural gas from Janet Sawin and Ishani Mukherjee, "Fossil Fuel Use Up Again," in Worldwatch Institute, *Vital Signs 2007–2008* (New York: W. W. Norton & Company, 2007); Sustainable Europe Research Institute, GLOBAL 2000, and Friends of the Earth Europe, *Overconsumption? Our Use of the World's Natural Resources* (September 2009).

5. Ecofootprint and Figure 1 from Global Footprint Network, *The Ecological Footprint Atlas 2008* (Oakland, CA: rev. ed., 16 December 2008).

6. Millennium Ecosystem Assessment (MA), *Ecosystems and Human Well-Being: Synthesis* (Washington, DC: Island Press, 2005); MA, *Living Beyond Our Means: Natural Assets and Human Well-being: Statement from the Board* (Washington, DC: World Resources Institute, 2005), p. 2.

7. C. P. McMullen and J. Jabbour, *Climate Change Science Compendium 2009* (Nairobi: United Nations Environment Programme, 2009), pp. 5, 30; Alice McKeown and Gary Gardner, "Climate Change Reference Guide and Glossary," in Worldwatch Institute, *State of the World 2009* (New York: W. W. Norton & Company, 2009), pp. 189–204; Intergovernmental Panel on Climate Change (IPCC), *Climate Change 2007: Synthesis Report* (Geneva: 2007), p. 49.

8. Andrei Sokolov et al., "Probabilistic Forecast for 21st Century Climate Based on Uncertainties in Emissions (without Policy) and Climate Parameters," *American Meteorological Society Journal of Climate*, October 2009, pp. 5,175–204; David Chandler, "Revised MIT Climate Model Sounds Alarm," *TechTalk* (Massachusetts Institute of Technology), 20 May 2009; Juliet Eilperin, "New Analysis Brings Dire Forecast of 6.3-Degree Temperature Increase," *Washington Post*, 25 September 2009; Elizabeth R. Sawin et al., "Current Emissions Reductions Proposals in the Lead-up to COP-15 Are Likely to Be Insufficient to Stabilize Atmospheric CO_2 Levels: Using C-ROADS—a Simple Computer Simulation of Climate Change—to Support Long-Term Climate Policy Development," draft presented at the Climate Change—Global Risks, Challenges, and Decisions Conference, University of Copenhagen, Denmark, 10 March 2009.

9. Ice melt from Eilperin, op. cit. note 8; refugees from International Organization for Migration, "Migration, Climate Change, and the Environment," *IOM Policy Brief* (Geneva: May 2009), p. 1; other problems from IPCC, op. cit. note 7.

10. Deforestation from Gary Gardner, "Deforestation Continues," in Worldwatch Institute, *Vital Signs 2006–2007* (New York: W. W. Norton & Company, 2006), pp. 102–03; hazardous waste from Elaine Baker et al., *Vital Waste Graphics* (Basel Convention and GRID-Arendal, 2004), pp. 34–35; for a discussion of other trends, see, for example, Peter Dauvergne, *The Shadows of Consumption: Consequences for the Global Environment* (Cambridge, MA: The MIT Press, 2008), and Gary Gardner, Erik Assadourian, and Radhika Sarin, "The State of Consumption Today," in Worldwatch Institute, *State of the World 2004* (New York: W. W. Norton & Company, 2004).

11. Stephen Pacala, "Equitable Solutions to Greenhouse Warming: On the Distribution of

Wealth, Emissions and Responsibility Within and Between Nations," presentation at International Institute for Applied Systems Analysis Global Development Conference, Vienna, Austria, 15 November 2007; Horace Herring, "Rebound Effect," in Cutler J. Cleveland, ed., *Encyclopedia of Earth* (Washington, DC: Environmental Information Coalition, National Council for Science and the Environment, rev. 18 November 2008).

12. U.S. data from World Bank, op. cit. note 2, in 2008 dollars. Table 1 from ibid., and from Global Footprint Network, op. cit. note 5; 2009 population from United Nations Population Division, *World Population Prospects, 2008 Revision* (New York: 2009).

13. G. Ananthapadmanabhan, K. Srinivas, and Vinuta Gopal, *Hiding Behind the Poor* (Bangalore: Greenpeace India Society, October 2007).

14. Ibid.

15. Ibid.; Global Footprint Network, op. cit. note 5; a third from Matthew Bentley, *Sustainable Consumption: Ethics, National Indices and International Relations*, PhD dissertation, American Graduate School of International Relations and Diplomacy, Paris, 2003, updated and revised with data from World Bank, op. cit. note 2.

16. Worldwatch calculation based on Saul Griffith, "Climate Change Recalculated," presentation at The Long Now Foundation, San Francisco, 16 January 2009; Saul Griffith, *The Game Plan: A Solution Framework for the Climate Challenge*, Slide Presentation, 13 March 2008, slides 140–47. Each of these three technologies would produce 4 terawatts of energy, producing 12 terawatts, which would replace all but 2–3 terawatts of fossil-fuel-based energy used today.

17. Population projection for 2050 from United Nations Population Division, op. cit. note 12.

18. Robert Welsch and Luis Vivanco, *Introduction to Cultural Anthropology* (McGraw-Hill Higher Education, forthcoming), Chapter 2, pp. 1–65 in draft edition.

19. Ibid., p. 9, quotation from Robert Welsch, Franklin Pierce University, discussion with author, 25 March 2009. I am indebted to Welsch and Vivanco for both this definition and the road map describing four key elements of culture.

20. Welsch and Vivanco, op. cit. note 18, p. 10; Gerrit J. van Enk and Lourens de Vries, *The Korowai of Irian Jaya: Their Language in Its Cultural Context* (New York: Oxford University Press, 1997).

21. Donella Meadows, *Leverage Points: Places to Intervene in a System* (Hartland, VT: The Sustainability Institute, 1999), pp. 17–19.

22. Paul Ekins, "The Sustainable Consumer Society: A Contradiction in Terms?" *International Environmental Affairs*, fall 1991, pp. 243–58. Box 1 from the following: Gary Gardner and Erik Assadourian, "Rethinking the Good Life," in Worldwatch Institute, *State of the World 2004*, op. cit. note 10; Daniel Kahneman, Presentation, Gallup Well-Being Forum, 2 October 2009 (note that positive affect stops increasing at higher incomes while life satisfaction levels have a logarithmic relationship, meaning it takes exponential increases in income to raise satisfaction ratings); Japan from Peter N. Stearns, *Consumerism in World History: The Global Transformation of Desire* (New York: Routledge, 2001), p. 97; United States from Rich Morin and Paul Taylor, *Luxury or Necessity? The Public Makes a U-Turn* (Washington, DC: Pew Research Center, 23 April 2009); deaths from World Health Organization (WHO), *Disease and Injury Regional Estimates for 2004*, at www.who.int/healthinfo/global_burden_disease/estimates_regional/en/index.html, and from WHO, *The World Health Report 2001* (Geneva: 2001), pp. 144–49; overweight and obese from WHO, "Obesity and Overweight," Fact Sheet No. 311 (Geneva: September 2006); life span from "Moderate Obesity Takes Years Off Life Expectancy," *Science Daily*, 20 March 2009.

23. Alan Thein Durning, *How Much Is Enough?* (New York: W. W. Norton & Company), p. 22; translations from translate.google.com, viewed 1 October 2009.

24. Leslie White as cited in Welsch and Vivanco, op. cit. note 18, p. 15; unconscious from Gráinne

M. Fitzsimons, Tanya L. Chartrand, and Gavan J. Fitzsimons, "Automatic Effects of Brand Exposure on Motivated Behavior: How Apple Makes You 'Think Different'," *Journal of Consumer Research*, June 2008, pp. 21–35; Pokémon from Andrew Balmford et al., "Why Conservationists Should Heed Pokémon," *Science*, 29 March 2002, p. 2,367; "Branded for Life? Pitching Fast Food to Kids," *Today*, NBC, 18 August 2006; "Food Fight," *Dateline*, NBC, 18 August 2006.

25. Welsch and Vivanco, op. cit. note 18, pp. 18–20; Terrence P. O'Toole et al., "Nutrition Services and Foods and Beverages Available at School: Results from the School Health Policy and Programs Study 2006," *Journal of School Health*, October 2007, pp. 500–21.

26. Welsch and Vivanco, op. cit. note 18, pp. 20–21; The Wedding Report, "Market Summary: Average Spending," at www.theweddingreport.com; Mark Harris, *Grave Matters: A Journey Through the Modern Funeral Industry to a Natural Way of Burial* (New York: Scribner, 2007).

27. Spending from Deloitte, *Savvy Consumers Demand Seasonal Savings: Annual Christmas Spending Survey 2008* (Ireland: Deloitte, 2008), p. 11; Justin McCurry, "In the Bleak Midwinter, Japanese Regain Appetite for Christmas," *Guardian* (London), 24 December 2007; 2 percent from "Japan," in Central Intelligence Agency, *The World Factbook*, at www.cia.gov/library/publications/the-world-factbook/geos/ja.html#, updated 14 July 2009; Reverend Billy from *What Would Jesus Buy?* Director, Rob VanAlkemade, 16 November 2007.

28. Welsch and Vivanco, op. cit. note 18, pp. 16–18; materialism from Tim Kasser, *The High Price of Materialism* (Cambridge: The MIT Press, 2003); students and Figure 2 from J. H. Pryor et al., *The American Freshman: National Norms for Fall 2008* (Los Angeles: Higher Education Research Institute, UCLA, 2008); Güliz Ger and Russell W. Belk, "Cross-cultural Differences in Materialism," *Journal of Economic Psychology*, no. 17 (1996), pp. 55–77.

29. Stearns, op. cit. note 22, pp. 25–36.

30. Ibid., pp. 15–24.

31. Ibid., pp. 10, 34–35.

32. Ibid., pp. 20–21.

33. Lizabeth Cohen, *A Consumer's Republic: The Politics of Mass Consumption in Postwar America* (New York: Alfred A. Knopf, 2003), pp. 123–24; Gardner, Assadourian, and Sarin, op. cit. note 10, p. 15.

34. Robert Coen, "Insider's Report: Advertising Expenditures," paper presented by Universal McCann, December 2008; Thomas N. Robinson et al., "Effects of Reducing Television Viewing on Children's Requests for Toys: A Randomized Controlled Trial," *Journal of Developmental and Behavioral Pediatrics*, June 2001, p. 179; gross domestic product from International Monetary Fund, *World Economic Outlook Database*, April 2009; Gwen B Achenreiner and Deborah R. John; "A Meaning of Brand Names to Children: A Developmental Investigation," *Journal of Consumer Psychology*, vol. 13, no. 3 (2003), pp. 205–19; Institute of Medicine, *Food Marketing to Children and Youth: Threat or Opportunity?* (Washington, DC: National Academies Press, 2006), p. 8.

35. PQ Media, *Product Placement Spending in Media: Executive Summary* (Stamford, CT: March 2005); WHO, *Smoke-free Movies: From Evidence to Action* (Geneva: 2009), p. 4.

36. PQ Media, *Word-of-Mouth Marketing Forecast: 2009–2013: Spending Trends & Analysis* (Stamford, CT: July 2009); BzzAgent from Rob Walker, "The Hidden (in Plain Sight) Persuaders," *New York Times*, 5 December 2004, from BzzAgent, "The BzzAgent Word of Mouth Network," at about.bzzagent.com/word-of-mouth/network, and from BzzAgent, "Case Studies By Campaign Objective," at about.bzzagent.com/word-of-mouth/casestudy/case-browser; Miho Inada, "Tokyo Café Targets Trend Makers," *Wall Street Journal*, 24 August 2009; Brooks Barnes, "Disney Expert Uses Science to Draw Boy Viewers," *New York Times*, 14 April 2009.

37. Victor Lebow, "Price Competition in 1955," *Journal of Retailing*, spring 1955, p. 8. Table 2 from the following: Elizabeth Royte, *Bottlemania: How Water Went on Sale and Why We Bought It* (New

York: Bloomsbury USA, 2008); 2008 data from John Rodwan, Jr., "Confronting Challenges: U.S. and International Bottled Water Developments and Statistics for 2008," *Bottled Water Reporter*, April/May 2009; 2000 data, safety, and price from Ling Li, "Bottled Water Consumption Jumps," in Worldwatch Institute, op. cit. note 4, pp. 102–03; fast-food data from First Research Industry Report, "Fast Food and Quickservice Restaurants," at www.hoovers.com/fast-food-and-quickservice-restaurants/—ID__269—/free-ind-fr-profile-basic.xhtml, viewed 28 September 2009; history from Eric Schlosser, *Fast Food Nation* (New York: Houghton Mifflin Company, 2001), pp. 197–98, advertising from "Marketer Database from Abbott to Yum," *Ad Age*, at adage.com/marketertrees09, viewed 28 September 2009; restaurants from McDonald's Corporation, *2008 Annual Report* (Oak Brook, IL: 12 March 2009); Chinese paper industry from Magdalena Kondej, "Kimberly Clark Bucks the Trend and Aims High in China," *Euromonitor International*, 15 July 2009; David W. Chen, "Shanghai Journal; A New Policy of Containment, for Baby Bottoms," *New York Times*, 5 August 2003; automobile ads from "U.S. Ad Spend Trends: 2008," *Advertising Age*, 22 June 2009; history from Peter D. Norton, *Fighting Traffic: The Dawn of the Motor Age in the American City* (Cambridge: The MIT Press, 2008), especially pp. 95–99, and from Dauvergne, op. cit. note 10, especially pp. 40–42; Center for Responsive Politics, "Lobbying Automotive, Industry Profile, 2008," online database at www.opensecrets.org/lobby/indusclient.php?lname=M02&year=2008, viewed 30 September 2009; contributions from Center for Responsive Politics, "Automotive: Long-Term Contribution Trends," at www.opensecrets.org/industries/indus.php?ind=M02, viewed 30 September 2009; $42 billion from Elizabeth Higgins, "Global Growth Trends: Sales in the Premium Segments Are Outpacing the Mid-Priced and Economy Segments," *Petfoodindustry.com*, 21 May 2007; humanization from PowerPoint presentation by Packaged Facts Pet Analyst David Lummis, *U.S. Pet Market Outlook 2009–2010: Surviving and Thriving in Challenging Times*, at www.packagedfacts.com/landing/petmarketoutlook.asp; advertising, which only includes supplies, not food, is from Packaged Facts, *Pet Supplies in the U.S.*, 7th ed. (Rockville, MD: August 2007), pp. 141–45; Robert Vale and Brenda Vale, *Time to Eat the Dog: The Real Guide to Sustainable Living* (London: Thames & Hudson, 2009), pp. 235–38.

38. Duane Elgin, interview in *Consume This Movie*, Director Gene Brockhoff, Well Crafted Films, 2008.

39. Television access from International Telecommunication Union's *World Telecommunication Development Report* database, as cited in World Bank, *World Development Indicators 2008* (Washington, DC: April 2008). Table 3 from the following: television and Internet from ibid., population and expenditures from World Bank, op. cit. note 2; Organisation for Economic Co-operation and Development, *OECD Communications Outlook 2009* (Paris: 2009), pp. 189, 199; Internet usage from Cisco, *The Connected Consumer*, slide presentation, May 2008; Online Publishers Association, "Online Publishers Association Media Usage Study Shows the Web Now Rivals TV in Reach and Extends the Impact of All Media," press release (New York: 6 June 2006); 8 billion from WHO, op. cit. note 35, p. 3.

40. Craig A. Anderson and Brad J. Bushman, "The Effects of Media Violence on Society," *Science*, 29 March 2002, pp. 2,377–79; Peter W. Vaughan, Alleyne Regis, and Edwin St. Catherine, "Effects of an Entertainment-Education Radio Soap Opera on Family Planning and HIV Prevention in St. Lucia," *International Family Planning Perspectives*, December 2000, pp. 148–57; Joan Montgomerie, "The Family Planning Soap Opera," *Peace Magazine*, October-December 2001, p. 27; WHO, op. cit. note 35; Juliet Schor, *The Overspent American: Why We Want What We Don't Need* (New York: HarperPerennial, 1999), pp. 75–83.

41. Jill Vardy and Chris Wattie, "Shopping is Patriotic, Leaders Say," *National Post* (Canada), 28 September 2001; Andrew J. Bacevich, "He Told Us to Go Shopping. Now the Bill Is Due," *Washington Post*, 5 October 2008; Norman Myers and Jennifer Kent, *Perverse Subsidies: How Tax Dollars Can Undercut the Environment and the Economy* (Washington, DC: Island Press, 2001), p. 188.

42. Political contributions from Center for Responsive Politics, "Business-Labor-Ideology Split in PAC & Individual Donations to Candidates and Parties," at www.opensecrets.org/bigpicture/blio

.php?cycle=2008, viewed 30 September 2009; lobbying dollars is a Worldwatch calculation based on Center for Responsive Politics, "Lobbying Database," at www.opensecrets.org/lobby/index.php, viewed 30 September 2009.

43. Cohen, op. cit. note 33; Lebow, op. cit. note 37, p. 7.

44. Nick Robins, Robert Clover, and Charanjit Singh, *A Climate for Recovery: The Colour of Stimulus Goes Green* (London: HSBC Global Research, 25 February 2009).

45. Inside Education, at www.insideeduca tion.ca/index.html; Division 2 materials including "The Petroleum Poster Kit," at www.insideed ucation.ca/class/div2.html, and Alberta Forest Products Association, "Teaching Materials," at www.albertaforestproducts.ca/resources/teaching _materials.aspx; Channel One News, "Frequently Asked Questions," updated 10 June 2009, at www.channelone.com/about/faq/.

46. Berry quote from interview with Caroline Webb, 2006, video at www.earth-community.org/images/BerryIV_Subtitles.mov.

47. Meadows, op. cit. note 21.

48. Ibid.

49. Richard Wilkinson and Kate Pickett, *The Spirit Level: Why More Equal Societies Almost Always Do Better* (London: Penguin Group, 2009).

50. Robert D. Putnam, *Bowling Alone: The Collapse and Revival of American Community* (New York: Simon & Schuster, 2000); Andrew Plantinga and Stephanie Bernell, "The Association Between Urban Sprawl and Obesity: Is It a Two-Way Street?" *Journal of Regional Science*, December 2007, pp. 857–79; Oregon State University, "Study Links Obesity, Urban Sprawl," *ScienceDaily*, 10 September 2005; Ohio State University, "Study Shows Urban Sprawl Continues to Gobble Up Land," *ScienceDaily*, 24 December 2007.

51. S. Kahn Ribeiro et al., "Transport and Its Infrastructure," in IPCC, *Climate Change 2007: Mitigation. Contribution of Working Group III to the Fourth Assessment Report of the Intergovernmental Panel on Climate Change* (Cambridge, U.K.: Cambridge University Press, 2007); accidents from WHO, *Disease and Injury Estimates*, op. cit. note 22; Hasselt from Roz Paterson, "Free Transport in Action," *Free Public Transport*, at www.freepublic transport.org/index.php?option=com_content&view=article&id=5&Itemid=5; Cees van Goeverden et al., "Subsidies in Public Transport," *European Transport*, no. 32 (2006), pp. 5–25.

52. William McDonough and Michael Braungart, *Cradle to Cradle: Remaking the Way We Make Things* (New York: North Point Press, 2002); Charles Moore, "Captain Charles Moore on the Seas of Plastic," *TED Talk*, Long Beach, CA, February 2009.

53. Box 2 from the following: James Davison Hunter, "To Change the World," *The Trinity Forum Briefing*, vol. 3, no. 2 (McLean, VA: 2002); for discussion of power of networks, see Nicholas A. Christakis and James H. Fowler, *Connected: The Surprising Power of Our Social Networks and How They Shape Our Lives* (New York: Little, Brown and Company, 2009); showings of *The Age of Stupid* from ageofstupid.net; 10:10 from www.1010uk.org, viewed 1 October 2009; Creel Commission, "Interview with James Lovelock," 8 August 2005, at www.creelcommission.com/interviews.php?action=show&id=3&title=James+Lovelock&date=08-08-2005.

54. "Schools Stepping Up for Active Travel with Feet First," *Environz Magazine*, March 2009; Elisabeth Rosenthal, "Students Give Up Wheels for Their Own Two Feet," *New York Times*, 26 March 2009.

55. See Johanna Mair and Kate Ganly, "Social Entrepreneurs: Innovating Toward Sustainability," in this volume; "B Corporations," at www.bcorp oration.net.

56. See Michael Maniates, "Editing Out Unsustainable Behavior," in this volume; Asamblea Constituyente, *Constitución del Ecuador*, chapter 7, article 71, at www.asambleaconstituyente.gov.ec/documentos/constitucion_de_bolsillo.pdf, p. 52.

57. Sustainable Table and Free Range Studios,

The Meatrix, 2003; Jonah Sachs, creative director, Free Range Studios, e-mail to author, 5 August 2009; Claire Atkinson, "'Desperate Housewives' Keeps Sunday Rates Competitive," *Advertising Age*, 21 September 2006; Stuart Elliott, "THE MEDIA BUSINESS: ADVERTISING; Prices for Commercials Already Vary Widely as the Season, with 38 New Series, Is Barely Under Way," *New York Times*, 2 October 2002.

58. Paul Hawken, *Blessed Unrest* (New York: Penguin Group, 2007); Paul Hawken, "Biology, Resistance, and Restoration: Sustainability as an Infinite Game," presentation at Bioneers Conference, October 2006.

59. See Gary Gardner, "Engaging Religions to Shape Worldviews," in this volume; activism from *Renewal*, Producers Marty Ostrow and Terry Kay Rockefeller, Fine Cut Productions LLC, 2007.

60. Barbara K. Rodes and Rice Odell, comp., *A Dictionary of Environmental Quotations* (New York: Simon & Schuster, 1992), p. 26.

Traditions Old and New

1. "People," in Central Intelligence Agency, *The World Factbook*, at www.cia.gov/library/publications/the-world-factbook/geos/xx.html.

2. United Nations Population Division, *World Population Prospects, 2008 Revision* (New York: 2009); Jared Diamond, *Collapse: How Societies Choose to Fail or Succeed* (New York: The Penguin Group, 2005), pp. 79–119, 286–93.

Engaging Religions in Building a Sustainable Culture

1. Stephen Scharper, "Faiths May Hold the Key to Green China," *Toronto Star*, 7 July 2009; Mary Evelyn Tucker, "Pan Yue's Vision for Ecological Civilization," blog, *Sustainable China*, at www.sustainablechina.info/2008/12/08/a-meeting-with-pan-yue.

2. "Third Taoist Ecology Forum Opens in Jurong, Jiangsu—And This One is Country-wide," at www.arcworld.org/news.asp?pageID=273; Mark Leon Goldberg, "Taoism—'The Way' for Climate Action in China?" *UN Dispatch*, at www.undispatch.com/archives/2008/11/taoism_—_guidi.php.

3. Alliance of Religions and Conservation (ARC) and the U.N. Development Programme, *Guide to Creating Your Seven-Year Plan* (Bath, U.K.: ARC, August 2008).

4. "People," in Central Intelligence Agency, *The World Factbook*, at www.cia.gov/library/publications/the-world-factbook/geos/xx.html; John A. Grim, series co-ed., *Indigenous Traditions and Ecology: The Interbeing of Cosmology and Community* (Cambridge, MA: Harvard University Center for the Study of World Religions, 2001).

5. World Values Survey, at www.worldvaluessurvey.org, viewed 4 August 2009.

6. Faith in Public Life and Public Religion Research, "Key Religious Groups Want Government to Address Climate Change and Its Impact on World's Poor," at www.faithinpubliclife.org/tools/polls/climate-change; Public Religion Research, "Climate Change & Poverty Survey, 20-27 March, 2009," e-mail to author, 16 July 2009.

7. Courses are from a survey done by Worldwatch Institute, August 2009; Parliament from Mary Evelyn Tucker, Yale University, e-mail to author, 23 August 2009. Table 4 from the following: Mary Evelyn Tucker and John Grim, series eds., *Religions of the World and Ecology*, book series (Cambridge, MA: Center for the Study of World Religions at Harvard University, 1998–2003); Bron Taylor, ed., *Encyclopedia of Religion and Nature* (London: Continuum, 2008); Willis Jenkins, ed., *The Spirit of Sustainability*, a project of the Forum on Religion and Ecology at Yale University (Great Barrington, MA: Berkshire, forthcoming); *The Green Bible* (New York: HarperOne, 2008); *Journal for the Study of Religion, Nature and Culture*, at www.religionandnature.com/journal; *Worldviews: Global Religions, Culture, and Ecology*, at www.brill.nl/m_catalogue_sub6_id9007.htm.

8. Religion, Science and the Environment, at www.rsesymposia.org.

9. Ibid.

10. Susan M. Darlington, "Practical Spirituality and Community Forests: Monks, Ritual, and Radical Conservatism in Thailand," in Anna L. Tsing and Paul Greenough, eds., *Nature in the Global South* (Durham, NC: Duke University Press, 2003); Susan Darlington, "Buddhism and Development: The Ecology Monks of Thailand," in Christopher Queen, Charles Prebish, and Damien Keown, eds., *Action Dharma: New Studies in Engaged Buddhism* (New York: Routledge Curzon, 2003).

11. The Regeneration Project, at www.theregenerationproject.org.

12. Bahá'í from www.barli.org/training-programmes.html#45; Appalachia from Renewal Project, *Renewal* (documentary film), at renewalproject.net; Sarah McFarland Taylor, *Green Sisters: A Spiritual Ecology* (Cambridge, MA: Harvard University Press, 2007).

13. Table 5 from the following: Bahá'í Reference Library, "World Peace," at reference.bahai.org/en/t/bic/SB/sb-13.html#fr5; Buddhism, Confucianism, Taoism, and Hinduism from Center for a New American Dream, "Religion and Spirituality," at www.affluenza.org/cnad/religion.html; Christianity and Judaism from *The New Jerusalem Bible* (New York: Doubleday, 1990); Islam from Islam Set, "Environmental Protection in Islam," at www.islamset.com/env/section4.html.

14. Carl Sagan, "Preserving and Cherishing the Earth—An Appeal for Joint Commitment in Science and Religion," *American Journal of Physics*, July 1990, pp. 615–17.

15. Benedict XVI, *Caritas in Veritate*, encyclical letter (Vatican City: The Vatican, 2009).

16. Nature as the book of Creation from "Benedict XVI's Very Own Shade of Green," *National Catholic Reporter*, 31 July 2009; Larry B. Stammer, "Interfaith Campaign Targets Issue of Environmental 'Sin'," *Los Angeles Times*, 29 November 1997.

17. Stephanie Kaza, "Western Buddhist Motivations for Vegetarianism," *Worldviews: Environment, Culture, Religion*, vol. 9, no. 3 (2005), pp. 385–411; 3 percent from Vegetarian Resource Group, "How Many Vegetarians are There?" at www.vrg.org/press/2009poll.htm, viewed 9 September 2009. Box 3 based on the following: Richard Sylvan and David Bennett, *The Greening of Ethics: From Human Chauvinism to Deep-Green Theory* (Cambridge, MA: White Horse Press, 1994), p. 26; distinction from enlightened human self-interest first raised by Richard Routley (later Sylvan) in "Is There a Need for a New, an Environmental Ethic?" *Proceedings of the XII World Congress of Philosophy*, No. 1 (Varna, Bulgaria: 1973), pp. 205–10.

18. International Interfaith Investment Group, at www.3ignet.org; $24 trillion is a Worldwatch calculation based on data in Social Investment Forum, *2007 Report on Socially Responsible Investing Trends in the United States* (Washington, DC: 2007); 11 percent from ibid.

19. Table 6 from the following: Buddhist economics from E. F. Schumacher, *Small is Beautiful* (New York: Harper & Row, 1973); Catholic economic teaching from United States Conference of Catholic Bishops, "Catholic Teaching on Economic Life," at www.usccb.org/jphd/economiclife; indigenous economic practices from Grim, op. cit. note 4; Islamic finance from Paul Maidmant, "A Distant Mirror," and from Elisabeth Eaves, "God and Mammon," both in *Forbes*, 21 April 2008; Sabbath economics from Ross Kinsler and Gloria Kinsler, *The Biblical Jubilee and the Struggle for Life* (Maryknoll, NY: Orbis Books, 1999).

Ritual and Taboo as Ecological Guardians

1. Rabbi Goldie Milgram, *Meaning & Mitzvah: Daily Practices for Reclaiming Judaism through Prayer, God, Torah, Hebrew, Mitzvot and Peoplehood* (Woodstock, VT: Jewish Lights Publishing, 2005).

2. Roy Rappaport, *Ritual and Religion in the Making of Humanity* (Cambridge, U.K.: Cambridge University Press, 1999), p. 24. This definition is a nontechnical adaptation of Rappaport's formal definition: "the performance of more or less invariant sequences of formal acts and utterances not entirely encoded by the performers."

3. E. N. Anderson, *Ecologies of the Heart: Emo-*

tion, Belief, and the Environment (New York: Oxford University Press, 1996); Rappaport, op. cit. note 2.

4. Anne-Christine Hornborg, *Mi'kmaq Landscapes: From Animism to Sacred Ecology* (Burlington, VT: Ashgate, 2008).

5. Ibid.; song from Maurice Bloch, "Symbols, Song, Dance and Features of Articulation," *Archives Europeénes de Sociologie*, vol. 15 (1974), pp. 51–81.

6. Roy Rappaport, *Ecology, Meaning, and Ritual* (Richmond, CA: North Atlantic Books, 1979).

7. Yaa Ntiamoa-Baidu, "Indigenous Beliefs and Biodiversity Conservation: The Effectiveness of Sacred Groves, Taboos and Totems in Ghana for Habitat and Species Conservation," *Journal for the Study of Religion, Nature, and Culture*, no. 3 (2008), p. 309.

8. Johan Colding and Carl Folke, "The Relations Among Threatened Species, Their Protection, and Taboos," *Ecology and Society*, vol. 1, no. 1 (1997).

9. The Wedding Report, "US Wedding Cost," at www.theweddingreport.com.

10. David Reay, *Climate Change Begins at Home* (New York: MacMillan, 2005); Pamela Logan, "Witness to a Tibetan Sky Burial: A Field Report for the China Exploration and Research Society," 26 September 1997, at alumnus.caltech.edu/~pamlogan/skybury.htm.

11. Central Pollution Control Board, Government of India, annual report, at cpcbenvis.nic.in/ar2000/annual_report1999-2000-34.htm.

12. Black Friday from bfads.net.

13. Centre for Natural Burial, at naturalburial.coop.

14. Box 4 from Danny Hillis, "The Millennium Clock," *Wired Magazine*, Scenarios Issue, 1995; Long Now Foundation, at www.longnow.org; Long Bets, at www.longbets.org; Rosetta Project, at www.rosettaproject.org.

15. Earth Day Network, "What is Earth Day Network?" at earthday.net/node/66.

16. Adam Vaughn, "Bishops of Liverpool and London Call for 'Carbon Fast' during Lent," *Guardian* (London), 24 February 2009; Zaher Sahloul, "Have a Blessed Green Ramadan," The Council of Islamic Organizations of Greater Chicago, at www.ciogc.org/Go.aspx?link=7654949.

17. World Car-free Day, at www.worldcarfree.net/wcfd/2008/wcfd.html.

18. Take Back Your Time, at www.timeday.org.

19. Rosita Worl, President, Sealaska Heritage Institute, Juneau, discussion with author, 13 April 2006.

20. Peter Sawtell, "Once in a Lifetime," *Eco-Justice Notes: The E-Mail Newsletter of Eco-Justice Ministries*, 7 August 2009.

Environmentally Sustainable Childbearing

1. Population data from United Nations Population Division, *World Population Prospects: The 2008 Revision Population Database*, at esa.un.org/unpp/index.asp.

2. Ibid.

3. United Nations Population Division, *World Population Policies 2007* (New York: United Nations, 2007); National Adaptation Programmes of Action, at unfccc.int/cooperation_support/least_developed_countries_portal/submitted_napas/items/4585.php.

4. United Nations Population Fund, *State of World Population 2004—The Cairo Consensus at Ten: Population, Reproductive Health and the Global Effort to End Poverty* (New York: United Nations, 2004), p. 5.

5. Ibid.

6. Robert Engelman, *More: Population, Nature, and What Women Want* (Washington, DC: Island Press, 2008).

7. Gilda Sedgh et al., "Induced Abortion: Rates

and Trends Worldwide," *The Lancet*, 13 October 2007, pp. 1,338–45; Alan Guttmacher Institute, *Sharing Responsibility: Women, Society and Abortion Worldwide* (New York: 1999); United Nations Population Division, *World Population Prospects: The 2004 Revision* (New York: United Nations, 2005); Henri Leridon, *Human Fertility: The Basic Components* (Chicago: University of Chicago Press, 1977); all cited in Guttmacher Institute, "Abortion: Worldwide Levels and Trends," PowerPoint presentation, 2007.

8. Average human fertility from United Nations Population Division, op. cit. note 1; current world replacement fertility calculated by the author based on data from this source; Malcolm Potts, "Sex and the Birth Rate: Human Biology, Demographic Change, and Access to Fertility-Regulation Methods," *Population and Development Review*, March 1997, pp. 1–39.

9. Robert Engelman, "Population & Sustainability," *Scientific American Earth 3.0*, summer 2009, pp. 22–29.

10. Anita Chandra et al., "Does Watching Sex on Television Predict Teen Pregnancy? Findings from a National Longitudinal Survey of Youth," *Pediatrics*, November 2008, pp. 1,047–54.

11. Peter W. Vaughan, Alleyne Regis, and Edwin St. Catherine, "Effects of an Entertainment-Education Radio Soap Opera on Family Planning and HIV Prevention in St. Lucia," *International Family Planning Perspectives*, December 2000, pp. 148–57; Joan Montgomerie, "The Family Planning Soap Opera," *Peace Magazine*, October-December 2001, p. 27.

12. Vaughan, Regis, and St. Catherine, op. cit. note 11.

13. Blaine Harden, "Japanese Voters Eager for Change," *Washington Post*, 27 August 2009; Bonnie Malkin (and news agencies), "Russians Told to Skip Work and Have Sex," *The Telegraph* (London), 12 September 2007.

14. Alma Cohen, Rajeev Dehejia, and Dmitri Romanov, "Do Financial Incentives Affect Fertility?" NBER Working Paper No. 13700 (Cambridge, MA: National Bureau of Economic Research, December 2007); Mikko Myrskylä, Hans-Peter Kohler, and Francesco C. Billari, "Advances in Development Reverse Fertility Declines," *Nature*, 6 August 2009, pp. 741–43; Rob Stein, "U.S. Fertility Rate Hits 35-Year High, Stabilizing Population," *Washington Post*, 21 December 2007.

15. David L. Carr, "Resource Management and Fertility in Mexico's Sian Ka'an Biosphere Reserve: Campos, Cash, and Contraception in the Lobster-Fishing Village of Punta Allen," *Population and Environment*, November 2007, pp. 83–101.

Elders: A Cultural Resource for Promoting Sustainable Development

1. Organisation for Economic Co-operation and Development, *The Well-being of Nations: The Role of Human and Social Capital* (Paris: 2001).

2. Quote from Grandmother Project and World Vision, *Report on Intergenerational Forum, Kayel Bassel, 18 December 2008* (Senegal: 2008).

3. Collins O. Airhihenbuwa, "On Being Comfortable with Being Uncomfortable: Centering an Africanist Vision in Our Gateway to Global Health," *Health Education & Behavior*, February 2007, pp. 31–42; Andreas Fuglesang, *About Understanding: Ideas and Observations on Cross-cultural Understanding* (Uppsala, Sweden: Dag Hammarskjold Foundation, 1982); Hampâté Bâ from speech at UNESCO, Paris, 1962.

4. Proverb repeated in numerous linguistic and ethnic groups in West Africa; Waly Diouf, Barry G. Sheckley, and Marijke Kehrhahn, "Adult Learning in a Non-Western Context: The Influence of Culture in a Senegalese Farming Village," *Adult Education Quarterly*, November 2000, pp. 32–44; N. K. Chadha, "Understanding Intergenerational Relationships in India," *Journal of Intergenerational Relationships*, vol. 2, issue 3/4 (2004), pp. 63–73.

5. The Elders, at www.theelders.org.

6. Ageist biases against women versus men from C. Sweetman, ed., "Editorial," *Gender and Lifecycles* (Oxford: Oxfam, 2000); biases against older women in non-western societies from Judi Aubel, Ibrahima Touré, and Mamadou Diagne, "Senegalese Grandmothers Improve Maternal and Child Nutrition Practices: The Guardians of Tradition Are Not Averse to Change," *Social Science & Medicine*, September 2004, pp. 945–59.

7. United Nations, *World Youth Report 2005* (New York: 2005), p. 76; Cara Heaven and Matthew Tubridy, "Global Youth Culture and Youth Identity," in Oxfam, *Highly Affected, Rarely Considered* (Oxford: 2008), pp. 149–60, with quote on p. 154.

8. Akopovire Oduaran, "Intergenerational Solidarity: Strengthening Economic and Social Ties," Background Paper, Expert Group Meeting, United Nations, New York, 23–25 October 2007, pp. 1–13, with quote on p. 10.

9. Haatso Youth Club of Ghana from Oxfam, op. cit. note 7, p. 157; quote from Senegalese man from Grandmother Project and World Vision, op. cit. note 2.

10. Jan Servaes and S. Lui, eds., *Moving Targets: Mapping the Paths between Communication, Technology and Social Change in Communities* (Penang, Malaysia: Southbound, 2007); quote on media role from Oxfam, op. cit. note 7, p. 154.

11. Grandmother Project field notes, Koulikoro, Mali, June 2004; maternal and child health programs from Judi Aubel, "Participatory Communication Unlocks a Powerful Cultural Resource: Grandmother Networks Promote Maternal and Child Health," *Communication for Development and Social Change*, vol. 2, no. 1 (2008), pp. 7–30.

12. Deepa Srikantaiah, "Education: Building on Indigenous Knowledge," *IK Notes* (World Bank), No. 85, 2005; Pat Pridmore and David Stephens, *Children as Partners for Health: A Critical Review of the Child-to-Child Approach* (London: Zed Books, 2000), p. 127; G. Mishra, "When Child Becomes a Teacher–The Child to Child Programme," *Indian Journal of Community Medicine*, October-December 2006, pp. 277–78.

13. Description of many intergenerational programs in North America is found on the Web site of Generations United, a U.S. nonprofit organization, at www.gu.org.

14. "Lelum'uy'lh Child Development Centre," at cowichantribes.com/memberservices/Education%20and%20Culture/Child%20Development%20Centre.

15. Judi Aubel et al., *Rapid Review of 'Time with Grandmas Initiative'* (Accra, Ghana: Government of Ghana and U.N. Population Fund, 2007).

16. Judi Aubel et al., *The "Custodians of Tradition" Promote Positive Changes for the Health of Newborns: Rapid Assessment of Ekwendeni Agogo Strategy* (Lilongwe, Malawi: Save the Children–US, 2006).

17. Ian S. McIntosh, "Nurturing Galiwin'ku Youth in Northeast Arnhem Land: Yalu Marrngikunharaw," *Cultural Survival Quarterly*, summer 2002.

18. For information on the Grandmother Project, see www.grandmotherproject.org.

19. "Enquete d'Evaluation Finale: Rapport d'Analyse," *INFO-STAT,* Bamako, Mali, 2004; Christian Children's Fund project from Aubel, Touré, and Diagne, op. cit. note 6.

20. Quote from World Vision and Grandmother Project, June 2006, Velingara, Senegal.

21. Findings from group interviews with community members from 12 villages, Girls Development Project, World Vision and Grandmother Project, Velingara, Senegal; Bam Tare local radio station, Velingara, Senegal; author's field notes during an intergenerational forum, SareFaremba village, Velingara, Senegal.

22. Helen Gould, "Culture and Social Capital," in UNESCO, *Recognising Culture* (Paris: 2001), p. 69.

From Agriculture to Permaculture

1. Jacques Cauvin, *The Birth of the Gods and the Origins of Agriculture* (Cambridge, U.K.: Cambridge University Press, 2000).

2. Ibid., pp. 51–61, and Jared Diamond, *Guns, Germs, and Steel: The Fates of Human Societies* (New York: W. W. Norton & Company, 1997); Steven Mithen, *After the Ice: A Global Human History 20,000–5,000 BC* (London: Weidenfeld & Nicolson, Ltd, 2003), p. 4.

3. A. P. Sokolov et al., "Probabilistic Forecast for 21st Century Climate Based on Uncertainties in Emissions (Without Policy) and Climate Parameters," *American Meteorological Society's Journal of Climate*, 19 May 2009; Thomas R. Karl, Jerry M. Melillo, and Thomas C. Peterson, eds., *Global Climate Change Impacts in the United States* (Cambridge, U.K.: Cambridge University Press, 2009), p. 12; J. Schahczenski and H. Hill, *Agriculture, Climate Change and Carbon Sequestration* (Fayetteville, AR: National Sustainable Agriculture Information Service, 2009), pp. 15, 104–05; Joseph Tainter, *The Collapse of Complex Societies* (Cambridge, U.K.: Cambridge University Press, 1990); Thomas Homer-Dixon, *The Upside of Down: Catastrophe, Creativity, and the Renewal of Civilization* (Washington, DC: Island Press, 2006). Box 5 from the following: Patricia Gadsby, "The Inuit Paradox," *Discover*, 1 October 2004; Committee on Food Marketing and the Diets of Children and Youth, Institute of Medicine, *Food Marketing to Children and Youth: Threat or Opportunity?* (Washington, DC: National Academy of Sciences, 2006), p. 164; World Health Organization, "Obesity and Overweight," Fact Sheet No. 311 (Geneva: September 2006); greenhouse gases from Henning Steinfeld et al., *Livestock's Long Shadow, Environmental Issues and Options* (Rome: Food and Agriculture Organization (FAO), 2006); meat consumption from Brian Halweil, "Meat Production Continues to Rise," in Worldwatch Institute, *Vital Signs 2009* (Washington, DC: 2009), pp. 15–17; long-lived cultures from John Robbins, *Healthy at 100* (New York: Random House, 2006), p. 57; U.S. consumption from FAO, "Dietary Energy, Protein and Fat Database," 7 August 2008, at www.fao.org/economic/ess/food-security-statistics/en/, viewed 22 September 2009; Michael Pollan, *In Defense of Food: An Eater's Manifesto* (New York: The Penguin Group, 2008); calorie restriction from Joseph M. Dhahbi et al., "Temporal Linkage Between the Phenotypic and Genomic Responses to Caloric Restriction," *Proceedings of the National Academy of Sciences*, 13 April 2004, pp. 5,524–29, from Michael Mason, "One for the Ages: A Prescription That May Extend Life," *New York Times*, 31 October 2006, from Ricki J. Colman et al., "Caloric Restriction Delays Disease Onset and Mortality in Rhesus Monkeys," *Science*, 10 July 2009, pp. 201–04, and from Roy B. Verdery and Roy L. Walford, "Changes in Plasma Lipids and Lipoproteins in Humans During a 2-year Period of Dietary Restriction in Biosphere 2," *Archives of Internal Medicine*, 27 April 1998, pp. 900–06; ecological benefits from Erik Assadourian, "The Living Earth Ethical Principles: Right Diet and Renewing Rituals," *World Watch*, November/December 2008, p. 32, from David Pimentel et al., "Reducing Energy Inputs in the US Food System," *Human Ecology*, August 2008, from Akifumi Ogino et al., "Evaluating Environmental Impacts of the Japanese Beef Cow-calf System by the Life Cycle Assessment Method," *Animal Science Journal*, vol. 78, issue 4 (2007), pp. 424–32, and from Daniele Fanelli, "Meat is Murder on the Environment," *New Scientist*, 18 July 2007.

4. Union of Concerned Scientists, *Industrial Agriculture: Features and Policy* (Cambridge, MA: 2007); D. Pimentel et al., "Impact of Population Growth on Food Supplies and Environment," *Population and Environment*, September 1997, pp. 9–14; A. Bouwman, *Global Estimates of Gaseous Emissions from Agricultural Land* (Rome: FAO, 2002); C. W. Rice, "Introduction to Special Section on Greenhouse Gases and Carbon Sequestration in Agriculture and Forestry," *Journal of Environmental Quality*, vol. 35 (2006), pp. 1,338–40; U.S. Environmental Protection Agency, *Global Anthropogenic Non-CO2 Greenhouse Gas Emissions: 1990–2020* (Washington, DC: 2006).

5. Secretariat of the United Nations Convention to Combat Desertification, "SLM Techniques Related to Climate Change Mitigation/Adaptation and Desertification," North American Biochar Conference, 2009.

6. F. H. King, *Farmers of Forty Centuries: or Permanent Agriculture in China, Korea and Japan* (Emmaus, PA: Rodale Press, 1990, originally published in 1911).

7. D. C. Coleman, "Through a Ped Darkly: An Ecological Assessment of Root-Soil-Microbial-Faunal Interactions," in A. H. Fitter et al., eds., *Ecological Interactions in Soil* (Cambridge, U.K.: Blackwell Scientific Publications, 1985), pp. 1–21; B. Anderson, "Soil Food Web—Opening the Lid of the Black Box," *The Permaculture Activist*, fall 2006.

8. J. Benyus, "Nature's Designs," TED Talk, February 2005; see also J. Benyus, *Biomimicry: Innovation Inspired by Nature* (New York: William Morrow, 1998).

9. D. Pimentel, "Soil Erosion: A Food and Environmental Threat," *Environment, Development and Sustainability*, February 2006, pp. 119–37; Philip H. Abelson, "A Potential Phosphate Crisis," *Science*, 26 March 1999, p. 2,015.

10. Francis Urban, "Energy, Agriculture, and the Middle East Crisis," *World Agriculture*, April 1991; D. Pfeiffer, *Eating Fossil Fuels: Oil, Food and the Coming Crisis in Agriculture* (Gabriola Island, BC: New Society Publishers, 2006).

11. For information on recalcitrant carbon, such as found in biochar or terra preta, see J. Lehmann and S. Joseph, eds., *Biochar for Environmental Management: Science and Technology* (London: Earthscan, 2009).

12. P. Hepperly, *Organic Farming Sequesters Atmospheric Carbon and Nutrients in Soils* (Emmaus PA: Rodale Institute, 2003).

13. U.S. Energy Information Administration, "Existing Electric Generating Units in the United States, (2007)," at www.eia.doe.gov/cneaf/electricity/page/capacity/capacity.html; U.S. Department of Energy, "Carbon Dioxide Emissions from the Generation of Electric Power in the United States (2000)," at www.eia.doe.gov/cneaf/electricity/page/co2_report/co2report.html.

14. E. R. Ingham, D. C. Coleman, and J. C. Moore, "An Analysis of Food-web Structure and Function in a Shortgrass Prairie, a Mountain Meadow, and a Lodgepole Pine Forest," *Biology and Fertility of Soils*, July 1989, pp. 29–37.

15. Robert Jensen, "Sustainability and Politics: An Interview with Wes Jackson," *Counterpunch*, 10 July 2003; T. X. Cox et al., "Prospects for Developing Perennial Grain Crops," *Bioscience*, August 2006, pp. 649–59.

16. M. H. Bender, *An Economic Comparison of Traditional and Conventional Agricultural Systems at a County Level* (Salina, KS: The Land Institute, 2000); J. Dewar, *Perennial Polyculture Farming: Seeds of Another Agricultural Revolution?* (Santa Monica, CA: RAND Corporation, 2007).

17. Nesbitt from A. Bates, "Going Deep in Belize," *The Permaculture Activist*, spring 2009; see also R. Nigh, "Trees, Fire And Farmers: Making Woods and Soil in The Maya Forest," *Journal Of Ethnobiology*, fall/winter 2008.

18. World Agroforestry Centre Communications Unit, "Trees on Farms Key to Climate and Food Security," press release (Nairobi: 24 July 2009); K. Trumper et al., *Natural Fix? The Role of Ecosystems in Climate Mitigation: A UNEP Rapid Response Assessment* (Cambridge, U.K.: UNEP-WCMC, 2009).

19. P. Bohlen and G. House, eds. *Sustainable Agroecosystem Management: Integrating Ecology, Economics, and Society (Advances in Agroecology)* (London: CRC Press, 2009); R. Hotinski, *Stabilization Wedges: A Concept and Game; Carbon Mitigation Initiative* (Trenton, NJ: Princeton Environmental Institute, 2007).

20. A. Lawson, "Never Let Paddocks Go Naked: Family's 'No Kill' Pasture Cropping Plan," *Meat & Livestock Australia: The Land*, 24 July 2008.

21. S. W. Duiker and J. C. Myers, *Better Soils with the No-till System* (Pennsylvania Conservation Partnership, 2002); Lawson, op. cit. note 20.

22. B. Mollison and D. Holmgren, *Permaculture One: A Perennial Agriculture for Human Settlements* (Tyalgum, NSW, Australia: Tagari Publications, 1978).

23. D. A. Perry, "Bootstrapping in Ecosystems," *BioScience*, April 1989, pp. 230–37; S. R. Gliessman, *Agroecology: Ecological Processes in Sustainable Agriculture* (Ann Arbor, MI: Ann Arbor Press, 1998).

24. For an overview of this project, see "Jordan Valley Permaculture Project," The Permaculture Research Institute of Australia, at permaculture .org.au/project_profiles/middle_east/jordan_valley_permaculture_project.htm.

25. Credits for sequestering carbon are discussed in Lehmann and Joseph, op. cit. note 11.

Education's New Assignment: Sustainability

1. UNESCO, "Educating for Sustainability," at portal.unesco.org/en/ev.php-URL_ID=1216&URL_DO=DO_TOPIC&URL_SECTION=201 .html.

Early Childhood Education to Transform Cultures for Sustainability

1. A. N. Meltzoff, A. M. Gopnik, and P. K. Kuhl, *The Scientist in the Crib: Minds, Brains, and How Children Learn* (New York: William Morrow & Company, Inc., 1999); J. P. Shonkoff and D. Phillips, eds., *From Neurons to Neighborhoods: The Science of Early Childhood Development* (Washington, DC: National Academy Press, 2000); D. Baily et al., *Critical Thinking About Critical Periods* (Baltimore: Paul H. Brooks, 2001); Organisation for Economic Co-operation and Development (OECD), *Starting Strong II: Early Childhood Education and Care* (Paris: 2006). Box 6 is based on the following: healthy and ecological development from S. Clayton and S. Opotow, eds., *Identity and the Natural Environment: The Psychological Significance of Nature* (Cambridge, MA: The MIT Press, 2003), from P. H. Kahn and S. R. Kellert, eds., *Children and Nature: Psychological, Sociocultural, and Evolutionary Investigations* (Cambridge, MA: The MIT Press, 2002), and from R. Louv, *Last Child in the Woods* (updated ed.) (Chapel Hill, NC: Algonquin Books, 2008); U.S. youth outdoor participation rates from Outdoor Foundation, *Outdoor Recreation Participation Report* 2008 (Boulder, CO: 2008); negative health effects of sedentary activities from R. R. Pate, J. R. O'Neill and F. Lobelo, "The Evolving Definition of 'sedentary': Studies of Sedentary Behavior," *Exercise and Sport Sciences Reviews*, vol. 36, no.4 (2008), pp. 173–78, and from C. Torgan, "Childhood Obesity on the Rise," *Word on Health: Consumer Health Information Based on Research from the National Institute of Health*, at www.nih.gov/news/WordonHealth/jun2002/childhoodobesity.htm; positive experiences from Sustainable Development Commission, "Outdoor Experiences," at www.sd-commission.org.uk/breakthrough.php?breakthrough=22, viewed 6 August 2009, and from Kahn and Kellert, op. cit. this note; No Child Left Inside Coalition, "About the No Child Left Inside Act," at www.cbf.org/Page.aspx?pid=948; outdoor and wilderness traditions from L. Cook, "The 1944 Education Act and Outdoor Education: From Policy to Practice," *History of Education*, vol. 28, no. 2 (1999), pp. 157–72, from B. Humberstone and K. Pedersen, "Gender, Class and Outdoor Traditions in the UK and Norway," *Sport, Education and Society*, March 2001, pp. 23–33, from P. Lynch, *Camping in the Curriculum: A History of Outdoor Education in New Zealand Schools* (Canterbury, New Zealand: PML Publications, 2006), and from R. Ramsing, "Organized Camping: A Historical Perspective," *Child and Adolescent Psychiatric Clinics of North America*, October 2007, pp. 751–54.

2. M. Woodhead, "Early Childhood and Primary Education," in M. Woodhead and P. Moss, eds., *Early Childhood and Primary Education. Early Childhood in Focus 2: Transitions in the Lives of Young Children* (Milton Keynes, U.K.: The Open University, 2007); UNICEF, *The Child Care Transition. A League Table of Early Childhood Education and Care in Economically Advanced Countries* (Florence: UNICEF Innicenti Research Centre, 2008); OECD, *Starting Strong: Early Childhood Education and Care* (Paris: 2001); UNESCO, *EFA Monitoring Report 2009. Education For All* (Oxford: Oxford University Press, 2008).

3. International Workshop, "The Role of Early Childhood Education for a Sustainable Society," Gothenburg, Sweden, 2–4 May 2007; Centre for Environment and Sustainability, *The Gothenburg Recommendations on Education for Sustainable Development* (Gothenburg, Sweden: Chalmers University of Technology and University of Gothenburg, 2008), p. 28.

4. I. Pramling Samuelsson and Y. Kaga, eds., *The Contribution of Early Childhood Education to Sustainable Society* (Paris: UNESCO, 2008); the list of 7Rs has been expanded from the 4Rs model proposed in L. Katz, "The Role of Early Childhood

Education for a Sustainable Society," prepared for International Workshop, Gothenburg, 2–4 May 2007.

5. D. Sommer, I. Pramling Samuelsson, and K. Hundheide, *Child Perspectives and Children's Perspectives in Theory and Practice* (New York: Springer, in press); Pramling Samuelsson and Kaga, op. cit. note 4.

6. E. Johansson, *Etiska Möten i Förskolebarns Världar* (Ethical Encounters in Preschool Children's Worlds), Göteborg Studies in Educational Sciences 251 (Gothenburg, Sweden: University of Gothenburg, 2007).

7. E. Johansson and I. Pramling Samuelsson, *Lek och Läroplan. Möten Mellan Barn och Lärare i Förskola och Skola* (Play and Curricula. Encounters between Children and Teachers in Preschool and School) (Gothenburg, Sweden: Acta Universitatis Gothoburgensis, 2006); E. Johansson and I. Pramling Samuelsson, "Play and Learning—An Integrated Wholeness," in R. New and M. Cochran, eds., *Early Childhood Education—An International Early Childhood Encyclopedia*, Vol. 4 (Westport, CT: Praeger Publishers, 2007), pp. 1270–73; L. Katz and S. Chard, *Engaging Children's Minds: The Project Approach* (Norwood, NJ: Ablex Publishing Corporation, 1989).

8. A. Wals, *TheEnd of ESD...The Beginning of Transformative Learning—Emphasizing the E in ESD*, presented at the Gothenburg Consultation on Sustainability in Higher Education, 2006, p. 45.

9. Pramling Samuelsson and Kaga, op. cit. note 4.

10. Y. Kaga, "The Role of Early Childhood Education in a Sustainable World," in Pramling Samuelsson and Kaga, op. cit. note 4, pp. 9–18.

11. Pramling Samuelsson and Kaga, op. cit. note 4; J. Davis, "What Might Education for Sustainability Look Like in Early Childhood? A Case for Participatory, Whole-of-Settings Approach," in ibid., pp. 18–25.

12. O. Fujii and C. Izumi, "A Silkworm is a Fascinating Insect for Children," in Pramling Samuelsson and Kaga, op. cit. note 4, pp. 87–93.

13. I. Engdahl and E. Ärlemalm-Hagsér, "Swedish Preschool Children Show Interest and Are Involved in the Future of the World—Children's Voices Must Influence Education for Sustainable Development," in Pramling Samuelsson and Kaga, op. cit. note 4, pp. 116–22; SOU, *Jämställd Förskola—Om Betydelsen av Jämställdhet och Genus i Förskolans Pedagogiska Arbetet* (An Equal Preschool—About the Importance of Equality and Gender in Preschool Practice), Slutbetänkande från delegationen för jämställdhet i förskolan (Stockholm: Fritzes, 2006), p. 75.

14. OECD, op. cit. note 1.

15. T. Herbert, "Eco-intelligent Education for a Sustainable Future Life," in Pramling Samuelsson and Kaga, op. cit. note 4, pp. 63–67.

Commercialism in Children's Lives

1. Portions of this article first appeared in Susan Linn, *The Case for Make Believe: Saving Play in a Commercialized World* (New York: The New Press, 2006); World Health Organization (WHO), *Diet, Nutrition, and the Prevention of Chronic Diseases* (Geneva: 2003); A. E. Becker et al., "Eating Behaviors and Attitudes Following Prolonged Exposure to Television Among Ethnic Fijian Adolescent Girls," *British Journal of Psychiatry*, vol. 180 (2002), pp. 509–14; American Psychological Association (APA), *Task Force on the Sexualization of Girls* (Washington, DC: 2007), p. 3; American Academy of Pediatrics, "Joint Statement on the Impact of Entertainment Violence on Children," Congressional Public Health Summit, 26 July 2000; M. Buijzen and P. M. Valkenburg, "The Effects of Television Advertising on Materialism, Parent–Child Conflict, and Unhappiness: A Review of Research," *Applied Developmental Psychology*, September 2003, pp. 437–56; U.S. Federal Trade Commission, *Self-Regulation in the Alcohol Industry: A Review of Industry Efforts to Avoid Promoting Alcohol to Underage Consumers* (Washington, DC: 1999), p. 4; National Cancer Institute, *Changing Adolescent Smoking Prevalence*, Smoking and Tobacco Control Monograph No. 14 (Washington, DC: November 2001.

2. Linn, op. cit. note 1.

3. Office of the United Nations High Commissioner for Human Rights, "Convention on the Rights of the Child: General Assembly Resolution 44/25," 20 November 1989.

4. Donald Winnicott, *Playing and Reality* (New York: Basic Books, 1971); Linn, op. cit. note 1, pp. 85–153; Angeline Lillard, "Pretend Play as Twin Earth: A Social-cognitive Analysis," *Developmental Review*, December 2001, pp. 495–531; Susan M. Burns and Charles Brainerd, "Effects of Constructive and Dramatic Play on Perspective Taking in Very Young Children," *Developmental Psychology*, September 1979, pp. 512–21; Dorothy Singer, "Team Building in the Classroom," *Early Childhood Today*, April 2002, pp. 37–41; Shirley R. Wyver and Susan H. Spence, "Play and Divergent Problem Solving: Evidence Supporting a Reciprocal Relationship," *Early Education and Development*, October 1999, pp. 419–44; Sandra Russ, Andrew L. Robins, and Beth A. Christiano, "Pretend Play: Longitudinal Prediction of Creativity and Affect in Fantasy in Children," *Creativity Research Journal*, vol. 12, no. 2 (1999), pp. 129–39; Elena Bodrova and Deborah Leong, "Self-Regulation as a Key to School Readiness: How Early Childhood Teachers Can Promote this Critical Competency," in Martha Zaslow and Ivelisse Martinez-Beck, eds., *Critical Issues in Early Childhood Professional Development* (Baltimore: Paul H. Brookes Publishing, 2006), pp. 203–24.

5. U.S. pretend play time from Sandra Hofferth, unpublished data from two Child Development Supplements to the Michigan Panel Study of Income Dynamics, 2006; Japan and France from Lego Learning Institute, *Time for Playful Learning? A Cross-cultural Study of Parental Values and Attitudes Toward Children's Time for Play* (Slough, Berks, U.K.: 2002).

6. Jill Casner-Lotto and Linda Barrington, *Are They Really Ready to Work? Employers' Perspectives on the Basic Knowledge and Applied Skills of New Entrants to the 21st Century U.S. Workforce* (New York: Conference Board, 2006).

7. Figure of $100 million from Bruce Horovitz, "Six Strategies Marketers Use to Make Kids Want Things Bad," *USA Today*, 22 November 2006; $17 billion from Juliet Schor, *Born to Buy: The Commercialized Child and the New Consumer Culture* (New York: Scribner, 2004), p. 21; Roisin Burke, "Food Giants Serve Up a €1.2bn Dish to Children," *The Sunday Independent* (Ireland), 14 June 2009. Table 7 from the following: James T. Areddy and Peter Sanders, "In China Children Learn English the Disney Way," *Wall Street Journal*, 20 April 2009; Eric Bellman, "McDonald's to Expand in India," *Wall Street Journal*, 30 June 2009; Anurag Sharma, "Cartoons—Animators Look at Licensing, Business Deals," *The Press Trust of India Limited*, 10 May 2009; MTV Networks, Africa, "SpongeBob SquarePants Surfs into Namibia," 24 June 2009.

8. Becker et al., op. cit. note 1; Israel from Daphne Lemish, "The School as a Wrestling Arena: The Modelling of a Television Series," *Communication*, vol. 22, no. 4 (1997), pp. 395–418.

9. Sam Schechner and Joseoph Pereira, "Hasbro and Discovery Form Children's TV Network," *Wall Street Journal*, 2 May 2009; Dorothy G. Singer et al., "Children's Pastimes and Play in Sixteen Nations: Is Free-Play Declining?" *American Journal of Play*, winter 2009, pp. 283–312.

10. Capitalism from Allen Kanner, "Globalization and the Commercialization of Childhood," *Tikkun*, September/October 2005, pp. 49–51; depression from Schor, op. cit. note 7; sustainable behavior from Tim Kasser, "Frugality, Generosity, and Materialism in Children and Adolescents," in Kristin Anderson Moore and Laura H. Lippman, eds., *What Do Children Need to Flourish? Conceptualizing and Measuring Indicators of Positive Development* (New York: Springer, 2005), pp. 357–74.

11. Leisure activity from Singer et al., op. cit. note 9; 40 hours from Donald F. Roberts et al., *Kids & Media @ the New Millennium* (Menlo Park, CA: Henry J. Kaiser Family Foundation, 1999), p. 78; babies from Victoria Rideout and Elizabeth Hamel, *The Media Family: Electronic Media in the Lives of Infants, Toddlers, Preschoolers, and Their Parents* (Menlo Park, CA: Henry J. Kaiser Family Foundation, May 2006), p. 18, and from Fred J. Zimmerman, Dmitri A. Christakis, and Andrew N. Meltzoff, "Television and DVD/Video Viewing in Children Younger than 2 Years," *Archives of Pediatric & Adolescent Medicine*, vol. 161, no. 5 (2007), pp. 473–79; Viet Nam and other countries from Singer et al., op. cit. note 9.

12. Less time in creative play from Elizabeth A. Vandewater, David S. Bickham, and June H. Lee, "Time Well Spent? Relating Television Use to Children's Free-Time Activities," *Pediatrics*, February 2006, pp. 181–91; some screen media encourage play from Dorothy G. Singer and Jerome L. Singer, *The House of Make-Believe: Play and the Developing Imagination* (Cambridge, MA: Harvard University Press, 1990), pp. 177–98; reading and radio from M. M. Vibbert and L. K. Meringoff, *Children's Production and Application of Story Imagery: A Cross-medium Investigation* (Cambridge, MA: Project Zero, Harvard University, 1981), and from Patti M. Valkenberg, "Television and the Child's Developing Imagination," in Dorothy G. Singer and Jerome L. Singer, eds., *Handbook of Children and the Media* (Thousand Oaks, CA: Sage Publications, 2001), pp. 121–34; boon to certain learning from Daniel R. Anderson, "Educational Television Is Not an Oxymoron," *Annals of the American Academy of Political and Social Science*, vol. 557, no. 1 (1998), pp. 24–38.

13. Box 7 based on Lucie K. Ozanne and Julie L. Ozanne, "Parental Mediation of the Market's Influence on their Children: Toy Libraries as Safe Havens," paper presented at The Academy of Marketing Conference, Leeds, U.K., 7–9 July 2009.

14. Estimate of $6.2 billion from Anita Frazier, Toy Industry Analyst, NPD Group, New York, discussion with author, 11 September 2009.

15. Quebec, Norway, Sweden, and Greece from Corinna Hawkes, *Marketing Food to Children: The Global Regulatory Environment* (Geneva: WHO, 2004), p. 19; France from "Shows Aimed at Toddlers Banned," *The Independent* (London), 21 August 2008.

16. Box 8 based on the following: data on environmental benefits from Arup (consulting firm), "California Academy of Sciences," at www.arup.com/Projects/California_Academy_of_Sciences.aspx; Rana Creek Living Architecture, "California Academy of Sciences, The Osher Living Roof," at www.greenroofs.com/projects/pview.php?id=509; California Academy of Sciences, "New California Academy of Sciences Receives Highest Possible Rating from U.S. Green Building Council: LEED Platinum," press release (San Francisco: 8 October 2008); for more information on the Academy and its programs, see www.calacademy.org.

17. Campaign for a Commercial-Free Childhood, "TV Cultura Goes Commercial-Free," *News*, October 2008.

18. American Academy of Pediatrics, "News Briefs," 3 October 2005; Brian L. Wilcox et al., *Report of the APA Task Force on Advertising and Children* (Washington, DC: APA, 20 February 2004).

19. Christopher Morgan, "Archbishop Warns of Dysfunctional 'Infant Adults,'" *Sunday Times* (London), 17 September 2006; Zoe Williams, *The Commercialisation of Childhood* (London: Compass, December 2006).

20. Green space from Mary Ann Kirkby, "Nature as a Refuge in Children's Environments," *Children's Environments Quarterly*, spring 1989, pp. 7–12; Children & Nature Network from Richard Louv, *Last Child in the Woods: Saving Our Children from Nature-Deficit Disorder* (New York: Algonquin Publishing, 2005); No Child Left Inside Act and Netherlands from Cheryl Charles et al., *Children and Nature 2008: A Report on the Movement to Reconnect Children to the Natural World* (Santa Fe, NM: Children & Nature Network, January 2008), pp. 10, 38; Waldkindergärtens from Harry de Quetteville, "Waldkindergärten: The Forest Nurseries Where Children Learn in Nature's Classroom," *Daily Telegraph* (London), 18 October 2008.

Rethinking School Food: The Power of the Public Plate

1. K. Morgan and R. Sonnino, *The School Food Revolution: Public Food and the Challenge of Sustainable Development* (London: Earthscan, 2008).

2. R. Sonnino, "Quality Food, Public Procurement and Sustainable Development: The School Meal Revolution in Rome," *Environment and Planning A*, vol. 41, no. 2 (2009), pp. 425–40.

3. Morgan and Sonnino, op. cit. note 1, pp. 137–64, 177.

4. K. Morgan, "Greening the Realm: Sustainable Food Chains and the Public Plate," *Regional Studies*, November 2008, pp. 1,237–50.

5. K. J. Morgan and R. Sonnino, "Empowering Consumers: Creative Procurement and School Meals in Italy and the UK," *International Journal of Consumer Studies*, vol. 31, no. 1 (2007), pp. 19–25.

6. S. Garnett, *School Districts and Federal Procurement Regulations* (Alexandria, VA: Food and Nutrition Service, U.S. Department of Agriculture, 2007).

7. *Jamie's School Dinners*, Channel 4, at www.channel4.com/life/microsites/J/jamies_school_dinners; investment and reform description from Scottish Executive, *Hungry for Success: A Whole School Approach to School Meals in Scotland* (Edinburgh: The Stationery Office, 2002).

8. R. Sonnino, "Escaping the Local Trap: Insights on Re-Localization from School Food Reform," *Journal of Environmental Policy and Planning*, forthcoming.

9. R. Sonnino and K. Morgan, "Localizing the Economy: The Untapped Potential of Green Procurement," in A. Cumbers and G. Whittam, eds., *Reclaiming the Economy: Alternatives to Market Fundamentalism in Scotland and Beyond* (Biggar, U.K.: Scottish Left Review Press, 2007), pp. 127–40.

10. R. Gourlay, "Sustainable School Meals: Local and Organic Produce," in V. Wheelock, ed., *Healthy Eating in Schools: A Handbook of Practical Case Studies* (Skipton, U.K.: Verner Wheelock Associates, 2007); C. Bowden, M. Holmes, and H. MacKenzie, *Evaluation of a Pilot Scheme to Encourage Local Suppliers to Supply Food to Schools* (Edinburgh: Scottish Executive, Environment and Rural Affairs Division, 2006).

11. Soil Association, *Food for Life: Healthy, Local, Organic School Meals* (Bristol, U.K.: 2003).

12. Morgan and Sonnino, op. cit. note 1.

13. Sonnino, op. cit. note 2, p. 432.

14. Ibid.

15. A. Dobson, *Citizenship and the Environment* (Oxford: Oxford University Press, 2003); J. Meadowcroft, "Who Is in Charge Here? Governance for Sustainable Development in a Complex World," *Journal of Environmental Policy & Planning*, September 2007, pp. 299–314; L. B. DeLind, "Of Bodies, Place, and Culture: Re-situating Local Food," *Journal of Agricultural and Environmental Ethics*, April 2006, pp. 121–46.

16. K. Morgan and R. Sonnino, "The Urban Foodscape: World Cities and the New Food Equation," *Cambridge Journal of Regions, Economy and Society*, forthcoming; R. Sonnino, "Feeding the City: Towards a New Research and Planning Agenda," *International Planning Studies*, forthcoming.

What Is Higher Education for Now?

1. "The Tbilisi Declaration—Intergovernmental Conference on Environmental Education: October 14–26, 1977," at www.cnr.uidaho.edu/css487/The_Tbilisi_Declaration.pdf.

2. Chet Bowers, *Education, Cultural Myths, and the Ecological Crisis* (Albany: State University of New York Press, 1993); Chet Bowers, *Educating for an Ecologically Sustainable Culture* (Albany: State University of New York Press, 1995); Stephen Sterling, *Sustainable Education* (Dartington, U.K.: Green Books, 2001); John Huckle and Stephen Sterling, eds., *Education for Sustainability* (London: Earthscan, 1996); John Blewitt and Cedric Cullingford, eds., *The Sustainability Curriculum* (London: Earthscan, 2004). Box 9 from the following: Herman Kahn, William Brown, and Leon Martel, *The Next 200 Years* (New York: William Morrow, 1976), pp. 163–80; for a discussion of what is "natural," see Lawrence Krieger, "What's Wrong with Plastic Trees," *Science*, 2 February 1973.

3. Box 10 from the following: Winnie Carruth, Administration Manager, IUCN Academy of Environmental Law, e-mail to Stefanie Bowles, Worldwatch Institute, 16 September 2009; Elizabeth Redden, "Green Revolution" *Inside Higher Ed*, 23 April 2009; Harvard Medical School Green Program, at www.greencampus.harvard.edu/hms/

green-program; "'Green' Initiatives Move Medical School toward Sustainability," *Medicine@Yale*, January/February 2009, p. 7; Francesca Di Meglio, "MBA Programs Go Green," *Business Week*, 19 January 2009; Presidio School of Management, at www.presidioedu.org; Bainbridge Graduate Institute, at www.bgiedu.org; Aspen Institute Centre for Business Education, *Beyond Grey Pinstripes 2007–2008: Preparing MBA's for Social and Environmental Stewardship*, at www.beyondgrey pinstripes.org; "The MBA Oath: Responsible Value Creation," at www.mbaoath.com; "Forswearing Greed," *The Economist*, 6 June 2009, pp. 66–68; Anne VanderMey, "Harvard's MBA Oath Goes Viral," *Business Week*, 11 June 2009; Brenda Kiefer, "Magnify Your Impact," Net Impact Media Kit (San Francisco: Net Impact, 2009); Net Impact, "Curriculum Change" and "Service Corps," at www.netimpact.org.

4. Peter Corcoran and Arjen Wals, *Higher Education and the Challenge of Sustainability* (Dordrecht, Netherlands: Kluwer, 2004); Association of University Leaders for a Sustainable Future, "Talloires Declaration," at www.ulsf.org/programs _talloires.html.

5. Karin E. Karlfeldt and Jennica M. Kjällstrand, "Campus Greening at Chalmers University of Technology," prepared for Environmental Management for Sustainable Universities Fifth International Conference, Barcelona, 15–17 October 2008; D. Ferrer-Balas et al., "An International Comparative Analysis of Sustainability Transformation across Seven Universities," *International Journal of Sustainability in Higher Education*, forthcoming; UNESCO chairs from Ariana Stahmer, UNESCO Division for Higher Education, e-mail to Erik Assadourian, 16 September 2009.

6. Ferrer-Balas et al., op. cit. note 5.

7. National Wildlife Federation, *Campus Environmental Report 2008* (Washington, DC: 2008).

8. April Smith, "In Our Backyard," Master's Thesis in Urban Planning, University of California–Los Angeles, unpublished, 1989; Sam Passmore, "A Study of Hendrix College Food System," unpublished, 1989.

9. David W. Orr, "The Liberal Arts, the Campus, and the Biosphere," *Harvard Educational Review*, summer 1990; Sarah Hammond Creighton, *Greening the Ivory Tower* (Cambridge, MA: The MIT Press, 1998); Peggy Barlett and Geoffrey Chase, eds., *Sustainability on Campus* (Cambridge: The MIT Press, 2004).

10. David W. Orr, *Design on the Edge* (Cambridge: The MIT Press, 2006).

11. For details of the four reports, see Intergovernmental Panel on Climate Change, at www.ipcc.ch.

12. David W. Orr, "2020: A Proposal," *Conservation Biology*, April 2000, reprinted in the *Chronicle of Higher Education*, 2000; Ann Rappaport and Sarah Creighton, *Degrees that Matter* (Cambridge, MA: The MIT Press, 2007); American College & University Presidents' Climate Commitment, at www.presidentsclimatecommitment.org/signatories; Architecture2020.org, "Think You're Making a Difference? Think Again," ad in *The New Yorker*.

13. Box 11 from the following: For the possibilities of collapse, see, for example, P. R. Ehrlich and A. H. Ehrlich, *The Dominant Animal: Human Evolution and the Environment* (Washington, DC: Island Press, 2009); for evolution of norms and cultures, see P. Ehrlich and S. Levin, "The Evolution of Norms," *Public Library of Science*, June 2005, pp. 943–48, and D. Rogers and P. Ehrlich, "Natural Selection and Cultural Rates of Change. *Proceedings of the National Academy of Sciences*, 4 March 2008, pp. 3,416–20, and references therein; for more on the concept of the MAHB, see P. R. Ehrlich and A. H. Ehrlich, *One with Nineveh: Politics, Consumption, and the Human Future* (Washington, DC: Island Press, 2004), pp. 282–85, and P. R. Ehrlich and D. Kennedy, "Millennium Assessment of Human Behavior: A Challenge to Scientists," *Science*, 22 July 2005, pp. 562–63.

14. M. Rubio, C. Hidalgo, and P. Ysern, "Collaboraton between Universities and Local Administrations to Promote Sustainability through Greening Events: A Case Study," Environment Office, Universitat Autònoma de Barcelona, prepared for Environmental Management for Sustainable Universities Fifth International Conference,

Barcelona, 15–17 October 2008; Judith Rodin, *The University and Urban Revival: Out of the Ivory Tower and Into the Streets* (Philadelphia: University of Pennsylvania Press, 2007).

15. Ronald Coleman, founder, Genuine Progress Indicator Atlantic, discussion with author, December 2008.

Business and Economy: Management Priorities

1. Paul Hawken, "Commencement Address to the Class of 2009," University of Portland, Maine, 3 May 2009.

2. Gross world product from International Monetary Fund, *World Economic Outlook Database*, April 2009; transnational corporations from U.N. Conference on Trade and Development, *World Investment Report 2008* (New York: 2008), pp. 26–30.

Adapting Institutions for Life in a Full World

1. This article is based in part on a longer version by Rachael Beddoe et al., "Overcoming Systemic Roadblocks to Sustainability: The Evolutionary Redesign of Worldviews, Institutions and Technologies," *Proceedings of the National Academy of Sciences*, 24 February 2009, pp. 2,483–89. We thank the co-authors of the longer version and two anonymous reviewers for helpful comments on earlier versions. Herman E. Daly and Joshua Farley, *Ecological Economics: Principles and Applications* (Washington, DC: Island Press, 2005); Robert Costanza, "Stewardship for a 'Full' World," *Current History*, vol. 107 (2008), pp. 30–35.

2. Thomas R. Malthus, *An Essay on the Principle of Population* (Oxford, U.K.: Oxford World's Classics reprint, 1999); Joseph A. Tainter, *The Collapse of Complex Societies* (Cambridge, U.K.: Cambridge University Press, 1988); Joseph A. Tainter, "Problem Solving: Complexity, History, Sustainability," *Population and Environment*, September 2000, pp. 3–41; Joseph A. Tainter, "Social Complexity and Sustainability," *Ecological Complexity*, June 2006, pp. 91–103; P. O'Sullivan, "The 'Collapse' of Civilizations: What Palaeoenvironmental Reconstruction Cannot Tell Us, But Anthropology Can," *The Holocene*, vol. 18, no. 1 (2008), pp. 45–55.

3. Pew Campaign for Responsible Mining, "Waiting for Mining Reform," www.pewminingreform.org/137years.html, 6 May 2009.

4. Herman E. Daly, "On Economics as a Life Science," *Journal of Political Economy*, vol. 76 (1968), pp. 392–406; Robert Costanza and Herman E. Daly, "Natural Capital and Sustainable Development," *Conservation Biology*, March 1992, pp. 37–46. Box 12 based on the following: "A Blueprint for Survival," *The Ecologist*, January 1972; Jay Forrester, *World Dynamics* (Cambridge, MA: Wright Allen Press, 1970); World Wide Fund for Nature (WWF), Zoological Society of London, and Global Footprint Network, *Living Planet Report 2006* (Gland, Switzerland: WWF, 2006); John Stuart Mill, *Principles of Political Economy, Vol. II* (London: John W Parker, 1857); Kenneth Boulding, "Environment and Economics," in William W. Murdoch, ed., *Environment: Resources, Pollution & Society* (Stamford, CT: Sinauer Associates, 1971), pp. 359–67; Paul Hawken, Amory Lovins, and L. Hunter Lovins, *Natural Capitalism: Creating the Next Industrial Revolution* (New York: Little, Brown and Company, 1999); 1958 car fleet from American Automobile Manufacturers Association, *World Motor Vehicle Data*, 1998 edition (Washington, DC: 1998); 2008 fleet from Michael Renner, "Global Auto Industry in Crisis," *Vital Signs Online*, Worldwatch Institute, 18 May 2009; Zoë Chafe, "Air Travel Reaches New Heights," in Worldwatch Institute, *Vital Signs 2007–2008* (New York: W. W. Norton & Company, 2007), pp. 70–71.

5. John Talberth, Clifford Cobb, and Noah Slattery, *The Genuine Progress Indicator 2006: A Tool for Sustainable Development* (Oakland, CA: Redefining Progress, 2007); Figure 3 from Redefining Progress Web site, www.rprogress.org.

6. Richard Layard, *Happiness: Lessons from a New Science* (New York: Penguin Press, 2005); Richard Easterlin, "Explaining Happiness," *Proceedings of the National Academy of Sciences*, 4 September 2003, pp. 11,176–83.

7. Robert Costanza et al., "The Value of the World's Ecosystem Services and Natural Capital," *Nature*, 15 May 1997, pp. 253–60.

8. Figure of $2 trillion from Norman Myers and

Jennifer Kent, *The Citizen is Willing but Society Won't Deliver: The Problem of Institutional Roadblocks* (Winnipeg, MB: International Institute for Sustainable Development, 2008); commons from Peter Barnes, *Capitalism 3.0* (San Francisco: Berrett-Koehler Publishers, Inc., 2006); Peter G. Brown, *The Commonwealth of Life: New Environmental Economics—A Treatise on Stewardship* (Montreal: Black Rose Books, 2007); Peter R. Barnes et al., "Creating an Earth Atmospheric Trust," *Science*, 8 February 2008, p. 724; Herman Daly, *Ecological Economics and Sustainable Development, Selected Essays of Herman Daly* (Northampton, MA: Edward Elgar, 2007).

9. Robert Costanza et al., *Beyond GDP: The Need for New Measures of Progress*, The Pardee Papers No. 4 (Boston: The Frederick S. Pardee Center for the Study of the Longer-Range Future, Boston University, 2009).

10. Steve Bernow et al., "Ecological Tax Reform," *BioScience*, vol. 48 (1998), pp. 193–96.

11. Harvey Weiss and Raymond S. Bradley, "What Drives Societal Collapse?" *Science*, 26 January 2001, pp. 609–10; Robert Boyd and Peter J. Richerson, *The Origin and Evolution of Cultures* (New York: Oxford University Press, 2005); Tainter, "Problem Solving: Complexity, History, Sustainability," op. cit. note 2; Tainter, "Social Complexity and Sustainability," op. cit. note 2; Jared Diamond, *Collapse: How Societies Choose to Fail or Succeed* (New York: Viking, 2005); Robert Costanza, Lisa J. Graumlich, and Will L. Steffen, eds., *Sustainability or Collapse? An Integrated History and Future of People on Earth* (Cambridge, MA: The MIT Press, 2007).

12. Tainter, *The Collapse of Complex Societies*, op. cit. note 2.

13. Peter Barnes, *Who Owns the Sky? Our Common Assets and the Future of Capitalism* (Washington, DC: Island Press, 2003).

14. Aaron Smith, *The Internet's Role in Campaign 2008* (Washington, DC: Pew Internet & American Life Project, April 2009).

Sustainable Work Schedules for All

1. Annual hours from Lawrence Mishel, Jared Bernstein, and Heidi Shierholz, *The State of Working America* (Ithaca, NY: Cornell University Press, 2009), Table 3.2; employment to population ratio from Council of Economic Advisers, *Economic Report of the President* (Washington, DC: Government Printing Office, 2009), pp. 326–27.

2. Decline of 600 hours from Angus Maddison, "Growth and Slowdown in Advanced Capitalist Economies: Techniques of Quantitative Assessment," *Journal of Economic Literature*, vol. 25, no. 2 (1987), pp. 649–98; A. Maddison, *The World Economy: A Millennial Perspective* (Paris: Organisation for Economic Co-operation and Development, 2001); 400 hours reduction calculated from Conference Board, *Groningen Total Economy Data Base 2008*, at www.conference-board.org/economics/database.cfm, viewed 15 March 2009.

3. Figure 4 from Conference Board, op. cit. note 2.

4. David Rosnick and Mark Weisbrot, *Are Shorter Work Hours Good for the Environment? A Comparison of U.S. and European Energy Consumption*, Working Paper (Washington, DC: Center for Economic and Policy Research, 2006).

5. Ibid.

6. Juliet B. Schor, *Plenitude: The New Economics of True Wealth* (New York: The Penguin Press, forthcoming), Chapter 4.

7. Benjamin Hunnicutt, *Work Without End* (Philadelphia: Temple University Press, 1990); Juliet B. Schor, *The Overworked American: The Unexpected Decline of Leisure* (New York: Basic Books, 1992); Gary Cross, *Of Time and Money* (London: Routledge, 1993).

8. Schor, op. cit. note 6; Christopher Nyland, *Reduced Working Time and the Management of Production* (Cambridge, U.K.: Cambridge University Press, 1991).

9. Juliet B. Schor, *The Overspent American* (New

York: Basic Books, 1997), Chapter 4. A 2004 poll found that 48 percent of a national sample of adults said that within the last five years they had made lifestyle changes that entailed earning less money; Center for a New American Dream, *More of What Matters Poll*, 2005, at newdream.org/about/polls.php.

10. Center for a New American Dream, op. cit. note 9.

11. Elizabeth D. Elmerm, Jeffrey R. Cohen, and Louise E. Single, "Is it the Kids or the Schedule? The Incremental Effect of Families and Flexible Scheduling on Perceived Career Success," *Journal of Business Ethics*, vol. 54, no. 1 (2004), pp. 51–65.

12. Average time worked in 2007 (33.9 hours) from "Employment and Earnings, Average Weekly Hours, Establishment Data, by Major Industry Sector, 1964 to Date," at ftp://ftp.bls.gov/pub/suppl/empsit.ceseeb2.txt; time worked in July 2009 (33.1 hours) from Bureau of Labor Statistics, "Economic News Release," 7 August 2009, at www.bls.gov/news.release/empsit.t15.htm.

13. Hewitt survey from "Survey Highlights: Cost Reduction & Engagement Survey, 2009," at www.hewittassociates.com/_MetaBasicCMAsset-Cache_/Assets/Articles/2009/Hewitt_Survey_Highlights_Cost_Reduction_and_Engagement_042009.pdf; Towers Perrin survey from: "Cross Cutting Strategies in the Downturn—A Balancing Act," at www.towersperrin.com/.../2009/200906/cost-cutting_strategies_pulse-svy_6-5-09.pdf; Jim McNett, "High-tech Companies Use Furloughs to Weather Recession," *The Oregonian*, 14 March 2009.

14. Utah experience from Bryan Walsh, "The Four Day Workweek is Winning Fans," *Time*, 7 September 2009.

15. Atlanta and California examples from Shaila Dewan, "A Slowdown that May Slow Us Down," *New York Times*, 1 March 2009; Office of the President, "Furlough Program Begins across UC," press release (Oakland, CA: University of California, 9 March 2009).

Changing Business Cultures from Within

1. Paul Hawken, *Blessed Unrest* (New York: Penguin Group, 2007); Paul H. Ray and Sherry Ruth Anderson, *The Cultural Creatives* (New York: Three Rivers Press, 2000); Mary Jo Hatch and Majken Schultz, "Dynamics of Organizational Identity," *Human Relations*, vol. 55, no. 8 (2002), pp. 989–1,018; Pasquale Gagliardi, "The Creation of Change of Organizational Cultures: A Conceptual Framework," *Organizational Studies*, January 1986, pp. 117–34; Edgar H. Schein, "Coming to New Awareness of Organizational Culture," *Sloan Management Review*, 1985; Jean M. Bartunek and Michael K. Mock, "First-Order, Second-Order, and Third-Order Change and Organizational Development Interventions: A Cognitive Approach," *Journal of Applied Behavioral Science*, December 1987, pp. 483–500; Terry E. Deal and Allan A. Kennedy, *Corporate Culture* (Reading, PA: Addison-Wesley, 1982).

2. Lovins quoted in Carl Fussman, "The Energizer," *Discover*, February 2006.

3. Romona A. Amodeo, "Becoming Sustainable at Interface: A Study of Identity Dynamics Within Transformational Culture Change," unpublished doctoral dissertation, Benedictine University, 2005; Romona A. Amodeo, "Interface Inc.'s Journey to Sustainability," in Peter Docherty, Jan Forslin, and A. B. Shani, eds., *Creating Sustainable Work Systems*, 2nd ed. (London: Routledge, 2008); Mona Amodeo and C. K. Cox, "Systemic Sustainability: Moving Ideas to Action," in William Rothwell, Roland Sullivan, and J. Stravos, eds., *Practicing Organization Development*, 3rd ed. (San Francisco: Josey Bass, 2009); Romona A. Amodeo and Jim Hartzfeld, "The Next Ascent Using Appreciative Inquiry to Support Interface's Continuing Sustainability Journey," *AI Practitioner*, August 2008, pp. 6–13.

4. Amodeo and Cox, op. cit. note 3.

5. A deeper survey of Interface's "EcoMetrics" can be found at Interface, Inc., "Metrics: What Gets Measured Gets Managed," at www.interfaceglobal.com/Media-Center/Ecometrics.aspx.

6. Globescan, "Companies and Governments Lag NGOs in Driving Sustainability but New Corporate Leaders Emerging, According to Experts," press release (London: 22 July 2009); Amodeo, "Becoming Sustainable at Interface," op. cit. note 3.

7. Nike from David Vogel, *The Market for Virtue* (Washington, DC: Brookings Institution Press, 2005), pp. 77–82; Electrolux from Samuel O. Idowu and Walter Leal Filho, *Global Practices of Corporate Social Responsibility* (New York: Springer, 2009); Tachi Kiuchi and William Shireman, *What We Learned in the Rainforest: Business Lessons from Nature* (San Francisco: Berrett-Koehler, 2002).

8. Paul Hawken, *The Ecology of Commerce* (New York: Harper Business, 1993); KB Home, *KB Home Sustainability Report* (Los Angeles: July 2008).

9. Wal-Mart Stores, Inc., "Lee Scott's 21st Century Leadership Speech" (Bentonville, AR: 24 October 2005).

10. Marvin Ross Weisbord and Sandra Janoff, *Future Search* (San Francisco: Berrett-Koehler, 1995); Peter M. Senge, *The Fifth Discipline: The Art and Practice of the Learning Organization* (New York: Doubleday Currency, 1990); quote from Amodeo and Cox, op. cit. note 3, p. 413.

11. Vogel, op. cit. note 7.

12. Wal-Mart Stores, Inc., "Wal-Mart CEO Leads Quarterly Sustainability Network Meeting," press release (Bentonville, AR: 12 July 2006); Amanda Little, "Al Gore Takes His Green Message to Wal-Mart Headquarters," *Grist Magazine*, 19 July 2006.

13. Bradley K. Googins, Philip H. Mirvis, and Steven A. Rochlin, *Beyond Good Company: Next Generation Corporate Citizenship* (New York: Palgrave Macmillan, 2007); Paul Kielstra, *Doing Good: Business and the Sustainability Challenge* (London: Economist Intelligence Unit, 2008).

14. Walmart Stores, Inc., "Walmart Associates Develop Personal Sustainability Projects," fact sheet (Bentonville, AK: 1 September 2009); Wal-Mart Stores, Inc., "Wal-Mart Announces Initial Results of Packaging Scorecard," press release (Bentonville, AR: 12 March 2007).

15. InterfaceRAISE Web site, at www.interfaceraise
.com.

16. Box 13 from the following: "B Corporations," at www.bcorporation.net; Richard Stengel, "For American Consumers, a Responsibility Revolution," *Time Magazine*, 10 September 2009; World Bank, *World Development Indicators 2008* (Washington, DC: 2008); Hannah Clark, "A New Kind of Company," *Inc.*, 1 July 2007.

Social Entrepreneurs: Innovating Toward Sustainability

1. Office of the Press Secretary, The White House, "President Obama to Request $50 Million to Identify and Expand Effective, Innovative Non-Profits," press release (Washington, DC: 5 May 2009).

2. J. Defourny and M. Nyssens, eds., *Social Enterprise in Europe: Recent Trends and Developments*, Working Paper 08/01 (Liège, Belgium: EMES European Research Network, 2008); information on La Fageda from Schwab Foundation profile, at www.schwabfound.org, viewed 16 July 2009; information on community interest companies from Social Enterprise Coalition (UK), at www.socialenterprise.org.uk.

3. Celia W. Dugger, "Peace Prize to Pioneer of Loans to Poor No Bank Would Touch," *New York Times*, 14 October 2006; information on Grameen Bank available at www.grameen-info.org.

4. C. Borzaga, G. Galera, and R. Nogales, eds., *Social Enterprise: A New Model for Poverty Reduction and Employment Generation* (Bratislava, Slovak Republic: UNDP Regional Bureau for Europe and the CIS, 2008); Office of the Third Sector–Cabinet Office, *Social Enterprise Action Plan: Scaling New Heights* (London: U.K. Government, 2006); R. Harding, *GEM UK: Social Entrepreneurs Specialist Summary* (London: Global Entrepreneurship Monitor, 2006); Y. Inoue, D. Hirose, and M. Nakayama, *Framework for Venture Philanthropy Country Market Studies: Japan Briefing Study* (Asia Venture Philanthropy Network, 2009).

5. Origin of the phrase from P. Light, "Social Entrepreneurship Revisited," *Stanford Social Innovation Review*, summer 2009, pp. 21–22; information on Ashoka at www.ashoka.org, on the Schwab Foundation at www.schwabfound.org, and on the Skoll Foundation at www.skollfoundation.org.

6. Information on Global Giving at www.globalgiving.com; U.S. examples of venture philanthropy organizations include the Acumen Fund, Good Capital, and Social Investment Forum, while European ones include Triodos Bank, Bonventure, and LGT Venture Philanthropy.

7. J. Mair and S. Seelos, *The Sekem Initiative*, IESE Case Study DG-1466-E (Barcelona: IESE Business School, 2004); Sekem Web site, at www.sekem.com/english/default.aspx; I. Abouleish and H. Abouleish, "Garden in the Desert: Sekem Makes Comprehensive Sustainable Development a Reality in Egypt," *Innovations*, summer 2008. Box 14 based on William McDonough and Michael Braungart, *Cradle to Cradle: Remaking the Way We Make Things* (New York: North Point Press, 2002).

8. J. Mair and J. Mitchell, *Waste Concern*, IESE Case Study DG-1502-E (Barcelona: IESE Business School, 2006); Waste Concern's impact from the Dhaka Case Study at the 2007 C40 Large Cities Climate Summit, New York, at www.nycclimatesummit.com/casestudies_waste.html, viewed 15 July 2009.

9. J. Mair and J. Shortall, *PDA* (Barcelona: IESE Business School, in press).

10. J. Mair and C. Seelos, "The Sekem Initiative: A Holistic Vision to Develop People," in F. Perrini, ed., *New Social Entrepreneurship: What Awaits Social Entrepreneurship Ventures?* (Cheltenham, U.K.: Edward Elgar, 2006), pp. 210–23; Waste Concern as best practice for Clean Development Mechanism projects from UNESCAP, *Economic and Social Survey of Asia and the Pacific*, 2008, at www.unescap.org/survey2008/download/index.asp. Box 15 based on the following: World Federation of Exchanges, "The World Federation of Exchanges," at www.world-exchanges.org/about-wfe, viewed 7 August 2009; Bank for International Settlements, "Statistical Annex," in *BIS Quarterly Review* (Basel, Switzerland: December 2008); Standard & Poor's, "S&P U.S. Carbon Efficient Index," fact sheet, at www2.standardandpoors.com, viewed 7 August 2009; Robert Kropp, "S&P Adds Carbon Index to Its Family of Environmental Indices," *SocialFunds.com*, 16 March 2009; STOXX Indexes, at www.stoxx.com/index.html, viewed 8 August 2009; U.S. Environmental Protection Agency, "Greenhouse Gas Reporting Rule," at www.epa.gov/climatechange/emissions/ghgrulemaking.html, viewed 9 August 2009; BM&FBOVESPA, at www.bmfbovespa.com.br; Eduardo Athayde, "Principio do Preservador-Pagador," *O Estado de São Paulo* (Brazil), 4 June 2009.

11. Transfair USA, at www.transfairusa.org; Rugmark, at www.rugmark.org/home.php.

12. For information on El Poder de Consumidor and Interrupcion, see Alejandro Cavillo Unna and Diego Carvajal profiles, at www.ashoka.org/fellows; for the Akatu Institute for Conscious Consumption, see Mattar Helio profile, at www.schwabfound.org/sf/SocialEntrepreneurs/Profiles/index.htm; for Polish organizations addressing the problems caused by mass consumption, see Zdzislaw Nitak, Jacek Schindler, and Ewa Smuk Stratenwerth profiles, at www.ashoka.org/fellows.

13. O. Sulla, "Philanthropic Foundations and Multilateral Aid Institutions like the World Bank: Increased Opportunities for Collaboration in ACP Agriculture," presented at 6th Brussels Development Briefing, Brussels, 2 July 2008; M. Jarvis and J. Goldberg, *Business and Philanthropy: The Blurring of Boundaries*, Business and Development Discussion Paper No. 09 (Washington, DC: World Bank Institute, fall 2008).

Relocalizing Business

1. Natural Resources Defense Council, at smartercities.nrdc.org/rankings/small; Sustainable Connections, at www.sustainableconnections.org.

2. Applied Research Northwest survey from New Rules Project, "Study Finds More People Shopping Locally Thanks to 'Think Local First'," 3 December 2006, at www.newrules.org; Michelle Long, executive director, Sustainable Connections, e-mail to author, 14 August 2009.

3. Numbers of communities affiliated with U.S. business alliances available at Web sites of BALLE (www.livingeconomies.org), AMIBA (www.amiba.net), and the Post-Carbon Institute (www.postcarbon.org); Michael Brownlee, U.S. director, Transition Towns, e-mail to author, 8 August 2009.

4. U.S. Census Bureau, *Statistical Abstract of the United States: 2009* (Washington, DC: U.S. Government Printing Office, 2009), Table 738, p. 493.

5. Sustainability definition from World Commission on Environment and Development, *Our Common Future* (Oxford: Oxford University Press, 1987), p. 8.

6. Pete Hurrey, "Chicken Poop a Real Concern for Chesapeake Bay Waters," 24 April 2009, at www.thebaynet.com/news/index.cfm/fa/viewStory/story_ID/12976.

7. Stacy Mitchell, "Will Wal-Mart Eat Britain," speech to the New Economics Foundation, 25 May 2005.

8. National Federation of Independent Business, "Charitable Contributions Comparison," January 2003.

9. See, for example, Annelies Van Hauwermeiren et al., "Energy Lifecycle Inputs in Food Systems: A Comparison of Local versus Mainstream Cases," *Journal of Environmental Policy & Planning*, March 2007, pp. 31–51.

10. James McWilliams, "Food That Travels Well," *New York Times*, 6 August 2007; for a critique, see Michael H. Shuman, "On the Lamb," *The Ethicurian*, 10 August 2007.

11. Oklahoma Food Coop, at www.oklahomafood.coop; David Shapero, managing director, Future Energy Pty. Ltd., "Going Local" Workshop, Melbourne, Australia, 24 June 2009.

12. Sherri Buri McDonald and Christian Wihtol, "Small Businesses: The Success Story," *The Register-Guard* (Eugene, OR), 10 August 2003; Michael H. Shuman, "Go Local and Prosper," *Eugene Weekly*, 8 January 2004.

13. Austin study in Civic Economics, *Economic Impact Analysis: A Case Study* (Austin, TX: December 2002).

14. For other studies on this point, see Institute for Local Self-Reliance, *The Economic Impact of Locally Owned Businesses vs. Chains: A Case Study in Midcoast Maine* (Minneapolis, MN: September 2003); David Morris, *The New City-States* (Washington, DC: Institute for Local Self-Reliance, 1982), p. 6; Christopher Gunn and Hazel Dayton Gunn, *Reclaiming Capital: Democratic Initiatives and Community Control* (Ithaca, NY: Cornell University Press, 1991); Gbenga Ajilore, "Toledo-Lucas County Merchant Study," Urban Affairs Center, Toledo, OH, 21 June 2004; and Justin Sachs, *The Money Trail* (London: New Economics Foundation, 2002), whose multiplier methodology has been used in dozens of small U.K. communities

15. Richard Florida, *The Rise of the Creative Class* (New York: Basic Books, 2002).

16. Stewart Smith, e-mail to author, 2 December 2005, updating Stewart Smith, "Sustainable Agriculture and Public Policy," *Maine Policy Review*, April 1993, pp. 68–78.

17. See Michael H. Shuman, *The Small-Mart Revolution: How Local Businesses Are Beating the Global Competition* (San Francisco: Berrett-Koehler, 2006), pp. 65–67.

18. Descriptions of these tools can be found in Shuman, op. cit. note 17, and in Stacy Mitchell, *The Big Box Swindle: The True Cost of Mega Retailers and the Fight for America's Independent Businesses* (Boston: Beacon Press, 2006).

19. Sustainable Business Network of Greater Philadelphia, at www.sbnphiladelphia.org; Tucson Originals, at www.tucsonoriginals.com.

20. Calculated from Australian Bureau of Statistics, "Counts of Australian Businesses, Including Entries and Exits," Report 8165, 14 December 2007.

21. Lisa Lerer, "Chamber Under Fire on Warming," *The Politico*, 5 May 2009, p. 1.

22. See, for example, the unsuccessful initiative in

Austin, Texas, to eliminate all city "economic development" subsidies for chain and other nonlocal stores; Leigh McIlvaine, "State and Local Ballot Initiative Round-Up, 7 November 2007, at www.clawback.org/2008/11/07/state-and-local-ballot-initiative-round-up.

Government's Role in Design

1. Rwanda Environment Management Authority, "FAQs," at www.rema.gov.rw/index.php?option=com_content&view=article&id=93&Itemid=41&lang=en; Karen Ann Gajewski, "Nations Set Goals to Phase Out the Use and Sale of Incandescent Light Bulbs," *The Humanist*, July-August 2007, p. 48; Gwladys Fouché, "Sweden's Carbon-Tax Solution to Climate Change Puts It Top of the Green List," *Guardian* (London), 29 April 2008; Jim Bai and Leonora Walet, "China Offers Big Solar Subsidy, Shares Up," *Reuters*, 21 July 2009.

2. Sam Perlo-Freeman, "Military Expenditure," in Stockholm International Peace Research Institute, *SIPRI Yearbook 2009. Armaments, Disarmament and International Security* (Oxford: Oxford University Press, 2009), p. 179.

Editing Out Unsustainable Behavior

1. Karen Ann Gajewski, "Nations Set Goals to Phase Out the Use and Sale of Incandescent Light Bulbs," *The Humanist*, July-August 2007, p. 48; Alexander Jung, "Getting Around the EU Ban: Germans Hoarding Traditional Light Bulbs," *Seigel Online International*, 27 July 2009.

2. Lester R. Brown, "Ban the Bulb: Worldwide Shift from Incandescents to Compact Fluorescents Could Close 270 Coal-Fired Power Plants," *Earth Policy Update* (Washington, DC: Earth Policy Institute, 9 May 2007); Jung, op. cit. note 1; Warna Oosterbaan, "Good Light Bulbs are Hard on the Eyes," NRC Handelsblad, 19 January 2009; "The Rise of the Light Bulb Fascist," 27 July 2009, at freestudents.blogspot.com/2009/07/rise-of-light-bulb-fascist.html; "Liberal Fascism," *TheAmericanScene.com*, 6 February 2008.

3. Sustainable Consumption Roundtable, *Looking Back, Looking Forward: Lessons in Choice Editing for Sustainability* (London: Sustainable Development Commission, May 2006). Box 16 from the following: signatory countries committed to the 10-year framework at www.un.org/esa/sustdev/documents/WSSD_POI_PD/English/WSSD_PlanImpl.pdf; U.N. Department of Economic and Social Affairs (UNDESA) and U.N. Environment Programme (UNEP), *Proposed Input to CSD 18 and 19 on a 10 Year Framework of Programmes on Sustainable Consumption and Production (10YFP on SCP): Third Public Draft* (2 September 2009), at esa.un.org/marrakechprocess/pdf/Draft3_10yfpniputtoCSD2Sep09.pdf; overview of the work of the Marrakech Task Force on Sustainable Lifestyles, at www.unep.fr/scp/marrakech/taskforces/lifestyles.htm; UNDESA/UNEP, Marrakech Task Force on Cooperation with Africa, at esa.un.org/marrakechprocess/tfcooperationafrica.shtml; UNDESA/UNEP, Marrakech Task Force on Sustainable Public Procurement, at esa.un.org/marrakechprocess/tfsuspubproc.shtml; International Task Force on Sustainable Products, at www.itfsp.org; Marrakech Task Force on Sustainable Tourism, Green Passport Program, at www.unep.fr/greenpassport; UNEP, *Sowing the Seeds of Change: An Environmental and Sustainable Tourism Teaching Pack for the Hospitality Industry* (Nairobi: 2008); Task Force on Sustainable Tourism Development, at www.veilleinfotourisme.fr/taskforce, Marrakech Task Force on Sustainable Buildings and Construction, at www.environment.fi/default.asp?contentid=328751&lan=EN, Marrakech Task Force on Education for Sustainable Consumption, at esa.un.org/marrakechprocess/tfedususconsump.shtml; UNESCO Associated Schools Network, at portal.unesco.org/education/en/ev.php-URL_ID=7366&URL_DO=DO_TOPIC&URL_SECTION=201.html.

4. Marla Cone, "Barbecue Ruling Adopted to Take a Bite Out of Smog," *Los Angeles Times*, 6 October 1990; Bob Pool, "Fanning the Flames," *Los Angeles Times*, 10 March 1991; shift from leaded to unleaded petrol from Sustainable Consumption Roundtable, op. cit. note 3, and from U.S. Environmental Protection Agency, "EPA Takes Final Step in Phaseout of Leaded Gasoline," press release (Washington, DC: 29 January 1996); Frank Convery, Simon McDonnell, and Susana Ferreira, "The Most Popular Tax in Europe? Lessons from the Irish Plastic Bags Levy," *Environmental and Resource Economics*, September 2007, pp. 1–12;

Leo Hickman, "Should You Have the Choice to Choose?" *Guardian* (London), 7 September 2007.

5. Lizabeth Cohen, *A Consumers' Republic: The Politics of Mass Consumption in Postwar America* (New York: Vintage Books, 2004), p. 7; David St. Clair, *The Motorization of American Cities* (New York: Praeger, 1986); Jim Klein and Martha Olson, *Taken for a Ride* (videorecording) (Hohokus, NJ: New Day Films, 1996).

6. Sustainable Consumption Roundtable, op. cit. note 3, p. 2.

7. James Maxwell and Sanford Weiner, "Green Consciousness or Dollar Diplomacy?" *International Environmental Affairs*, winter 1993, p. 36.

8. Lang quote from Leo Hickman, "Does the Consumer Really Know Best?" *Guardian* (London), 25 October 2007; Paul Hawken, *The Ecology of Commerce: A Declaration of Sustainability* (New York: HarperBusiness, 1994).

9. Ralph Horne, "Limits to Labels: The Role of Eco-Labels in the Assessment of Product Sustainability and Routes to Sustainable Consumption," *International Journal of Consumer Studies*, March 2009; Isabelle Szmigin, Marylyn Carrigan, and Morven G. McEachern, "The Conscious Consumer: Taking a Flexible Approach to Ethical Behaviour," *International Journal of Consumer Studies*, March 2009.

10. Marion Nestle, *What to Eat: An Aisle-by-Aisle Guide to Savvy Food Choices and Good Eating* (New York: North Point Press, 2007); Sustainable Consumption Roundtable, op. cit. note 3, p. 3.

11. For sustainability initiatives in higher education in the United States, see Association for the Advancement of Sustainability in Higher Education, at www.aashe.org.

12. "Renewable & Alternative Energy Portfolio Standards," Pew Center on Global Climate Change, at www.pewclimate.org/what_s_being_done/in_the_states/rps.cfm.

13. Todd Litman, *London Congestion Pricing: Implications for Other Cities* (Victoria, BC: Victoria Transport Policy Institute, 2004); Santosh A. Jalihal and T. S. Reddy, "Assessment of the Impact of Improvement Measures on Air Quality: Case Study of Delhi," *Journal of Transportation Engineering*, June 2006.

14. Home Depot "Wood Purchasing Policy," at corporate.homedepot.com/wps/portal/Wood_Purchasing.

15. Knight quote from Michael Jenkins and Emily Smith, *The Business of Sustainable Forestry* (Washington DC: Island Press, 1999), p. 75.

16. "Wal-Mart Stores Inc., Introduces New Label to Distinguish Sustainable Seafood," press release (Bentonville, AK: 31 August 2006).

17. "Hannaford Supermarkets to License Guiding Stars," press release (Portland, ME: 29 November 2007); Dan Goleman, "Look to the Future, Not the Past," *Greenbiz.com*, 17 June 2009; Andrew Martin, "Store Chain's Test Concludes That Nutrition Sells," *New York Times*, 6 September 2007.

18. "Hannaford Supermarkets," op. cit. note 17.

19. Thomas Princen, "Consumer Sovereignty, Heroic Sacrifice: Two Insidious Concepts in an Endlessly Expansionist Economy," in Michael Maniates and John M. Meyer, eds., *The Environmental Politics of Sacrifice* (Cambridge, Mass: The MIT Press, forthcoming).

20. Maike Bunse et al., *Top Runner Approach* (Wuppertal, Germany: UNEP–Wuppertal Institute Collaborating Center on Sustainable Consumption and Production, September 2007); Joakim Nordqvist, "The Top Runner Policy Concept: Pass it Down?" *Proceedings of the European Council for an Energy Efficient Economy (ECEEE) 2007 Summer Study* (Stockholm: 2007), pp. 1209–14; Ben Block, "Wal-Mart Scrutinizes Supply-Chain Sustainability," *Eye on Earth* (Worldwatch Institute), 20 July 2009.

21. See John de Graaf, "Reducing Work Time as a Path to Sustainability" in this volume; see also Anders Hayden, *Sharing the Work, Sparing the Planet* (Ontario, Canada: Zed Books, 1999).

22. Robert H. Frank, "Just What This Downturn Demands: A Consumption Tax," *New York Times*, 8 November 2008; Robert H. Frank, *Luxury Fever* (Princeton, NJ: Princeton University Press, 1999).

23. Richard Thaler and Cass Sunstein, *Nudge: Improving Decisions About Health, Wealth, and Happiness* (New York: Penguin, 2008).

Broadening the Understanding of Security

1. Daniel Deudney, "Forging Missiles into Spaceships," *World Policy Journal*, spring 1985, p. 273.

2. Michael Klare, *Rising Powers, Shrinking Planet: The New Geopolitics of Energy* (New York: Macmillan, 2008).

3. Water scarcity from Wissenschaftlicher Beirat Globale Umweltveränderungen der Bundesregierung (WBGU, German Advisory Council for Global Change), *Climate Change as a Security Risk* (London: Earthscan, 2008), pp. 64–65; food security study from Ian Sample, "Billions Face Food Shortages, Study Warns," *Guardian* (London), 9 January 2009.

4. Disaster trends from Centre for Research on the Epidemiology of Disasters, Université Catholique de Louvain, Belgium, "EM-DAT Emergency Events Database," at www.emdat.be/Database/AdvanceSearch/advsearch.php, viewed 7 August 2009; cases of unrest and conflict from WBGU, op. cit. note 3, pp. 31–33, and from Michael Renner and Zoë Chafe, *Beyond Disasters: Creating Opportunities for Peace* (Washington, DC: Worldwatch Institute, 2007).

5. International Labour Organization, *Global Employment Trends Update, May 2009*, at www.ilo.org/public/libdoc/ilo/P/09332/09332(2009-May).pdf.

6. Refugees and internally displaced from U.N. High Commissioner for Refugees (UNHCR), *2008 Global Trends: Refugees, Asylum-seekers, Returnees, Internally-Displaced and Stateless Persons* (Geneva: June 2009), p. 3; disaster-uprooted people from UNHCR, *2007 Global Trends: Refugees, Asylum-seekers, Returnees, Internally-Displaced and Stateless Persons* (Geneva: June 2009), p. 2; number displaced by development projects from Christian Aid, *Human Tide: The Real Migration Crisis* (London: May 2007), p. 5; 2050 projections from International Organization for Migration, "Migration, Climate Change, and the Environment," *IOM Policy Brief* (Geneva: May 2009), p. 1; unrest in host areas from Deutsche Gesellschaft für Technische Zusammenarbeit, *Climate Change and Security. Challenges for German Development Cooperation* (Eschborn, Germany: April 2008), p. 23, and from WBGU, op. cit. note 3, pp. 124–25.

7. Human Security Network, at www.humansecuritynetwork.org/network-e.php; Institute for Environmental Security, *Inventory of Environment and Security Policies and Practices* (The Hague, October 2006); United Nations Security Council, "Security Council Holds First-ever Debate on Impact of Climate Change on Peace, Security, Hearing over 50 Speakers," press release (New York: 17 April 2007).

8. See, for instance, CNA Corporation, *National Security and the Threat of Climate Change* (Alexandria, VA: 2007); John M. Broder, "Climate Change Seen as Threat to U.S. Security," *New York Times*, 9 August 2009.

9. Sam Perlo-Freeman, "Military Expenditure," in Stockholm International Peace Research Institute, *SIPRI Yearbook 2009. Armaments, Disarmament and International Security* (Oxford: Oxford University Press, 2009), p. 179; Organisation for Economic Co-operation and Development, "International Development Statistics," online database, at www.oecd.org/dac/stats/idsonline, viewed 14 August 2009.

10. U.S. military and climate budget ratio from Miriam Pemberton, *Military vs. Climate Security. Mapping the Shift from the Bush Years to the Obama Era* (Washington, DC: Institute for Policy Studies, July 2009); nuclear weapons and renewable energy/energy efficiency budget from U.S. Department of Energy, *FY 2010 Congressional Budget Request. Budget Highlights* (Washington, DC: May 2009), pp. 24, 63; Germany from Presse- und Informationsamt der Bundesregierung (Press and Information Office of the German Federal Government), "Bundesregierung beschließt Energie- und Klimaprogramm," press release (Berlin: 5

December 2007), and from Bundesfinanzministerium, "Bundeshaushaltsplan 2008," at www.bundesfinanzministerium.de/bundeshaushalt2008/html/vsp2i-e.html; Japan from Shigeru Sato and Yuji Okada, "Japan Plans 27% Increase in Budget to Cut Emissions," *Bloomberg*, 27 August 2009; Ministry of Finance Japan, "Budget," at www.mof.go.jp/english/budget/budget.htm.

11. Funds for climate mitigation and adaptation based on the following: Gareth Porter et al., *New Finance for Climate Change and the Environment* (Washington, DC: WWF and Heinrich Böll Stiftung, July 2008), pp. 24–25; U.N. Framework Convention on Climate Change, *Investment and Financial Flows to Address Climate Change* (Bonn: 2007); Manish Bapna and Heather McGray, *Financing Adaptation: Opportunities for Innovation and Experimentation* (Washington, DC: World Resources Institute, 2008); and World Bank, "Climate Investment Funds (CIF)," at www.worldbank.org/cif. U.S. military aid from Pemberton, op. cit. note 10, p. 26. Arms transfers to developing countries from Richard F. Grimmett, *Conventional Arms Transfers to Developing Nations, 2000–2007*, CRS Report (Washington, DC: Congressional Research Service, October 2008), p. 45.

12. U.N. Department of Economic and Social Affairs, Statistics Division, "Millennium Development Goals: 2009 Progress Chart," at mdgs.un.org/unsd/mdg/Resources/Static/Products/Progress2009/MDG_Report_2009_Progress_Chart_En.pdf.

13. International Renewable Energy Agency, at www.irena.org/foundingcon.htm; Ottmar Edenhofer and Lord Nicholas Stern, *Towards a Green Recovery: Recommendations for Immediate G20 Action* (Berlin and London: Potsdam Institute for Climate Impacts Research and Grantham Research Institute on Climate Change and the Environment, April 2009), pp. 37–38.

14. For examples of transparency initiatives, see the Kimberley Process, at www.kimberleyprocess.com, the Extractive Industries Transparency Initiative, at eitransparency.org, and Forest Law Enforcement, Governance and Trade Regulation, at ec.europa.eu/environment/forests/flegt.htm.

15. Ken Conca et al., "Building Peace Through Environmental Cooperation," in Worldwatch Institute, *State of the World 2005* (New York: W. W. Norton & Company, 2005), pp. 144–57.

16. Aaron T. Wolf et al., "Managing Water Conflict and Cooperation," in Worldwatch Institute, op. cit. note 15, pp. 80–95.

17. Saleem H. Ali, ed., *Peace Parks. Conservation and Conflict Resolution* (Cambridge, MA: The MIT Press, 2007).

18. United Nations News Centre, "Blue Helmets Planting Trees in Bid to 'Green' Planet," press release (New York: 22 July 2009); Nathanial Gronewold, "Environmental Demands Grow for U.N. Peacekeeping Troops," *New York Times*, 11 August 2009.

19. U.N. Environment Programme (UNEP), "Disasters and Conflicts," at www.unep.org/conflictsanddisasters/Home/tabid/146/language/en-US/Default.aspx.

20. Renner and Chafe, op. cit. note 4.

21. Ibid.

22. Thomas Novotny and Vincanne Adams, *Global Health Diplomacy*, Global Health Sciences Working Paper (La Jolla, CA: Institute on Global Conflict and Cooperation, University of California, 16 January 2007); "Global Health Diplomacy," Institute on Global Conflict and Cooperation, University of California, at igcc.ucsd.edu/research/globalhealth/index.php.

23. Margaret Blunden, "South-South Development Cooperation: Cuba's Health Programmes In Africa," *The International Journal of Cuban Studies*, June 2008; C. William Keck, "Cuba's Contribution to Global Health Diplomacy," Global Health Diplomacy Workshop, Institute on Global Conflict and Cooperation, University of California, 12 March 2007.

24. Green New Deal Group, *A Green New Deal* (London: New Economics Foundation, July 2008); UNEP, *Global Green New Deal: Policy Brief* (Nairobi: March 2009); Maikel R. Lieuw-Kie-

Song, *Green Jobs for the Poor: A Public Employment Approach*, Poverty Reduction Discussion Paper (New York: U.N. Development Programme, April 2009).

Building the Cities of the Future

1. J. Scheurer and P. Newman, "Vauban: A European Model Bridging the Brown and Green Agendas," case study prepared for UN Habitat, *Global Report on Human Settlements 2009* (Nairobi: 2009); City of Hanover, *Hannover-Kronsberg: Model of a Sustainable Community* (Hanover, Germany: 1998); City of Hanover, *CO_2 Audit 1991–2001* (Hanover, Germany: 2003).

2. P. Newman and I. Jennings, *Cities as Sustainable Ecosystems* (Washington DC: Island Press, 2008).

3. P. Newman, T. Beatley, and H. Boyer, *Resilient Cities: Responding to Peak Oil and Climate Change* (Washington DC: Island Press, 2009).

4. WWF, Zoological Society of London, and Global Footprint Network, *Living Planet Report 2008* (Gland, Switzerland: WWF, 2008), p. 32.

5. Jan Scheurer, "Urban Ecology," PhD Thesis, Institute for Sustainability and Technology Policy, Murdoch University, 2002.

6. P. Newman and J. Kenworthy, *Sustainability and Cities* (Washington, DC: Island Press, 1999).

7. Perth data from Public Transport Authority of Western Australian Government; Copenhagen data from P. Newman and J. Kenworthy, "Greening Urban Transportation," in Worldwatch Institute, *State of the World 2007* (New York: W. W. Norton & Company, 2007), p. 81.

8. Center for Transit-Oriented Development, *Hidden in Plain Sight: Capturing the Demand for Housing Near Transit* (Oakland, CA: 2004).

9. G. P. Metschies, *Prices and Vehicle Taxation* (Germany: Deutsche Geslleschaft fur Technische Zusammenarbeit GmbH, 2001); R. Porter, *Economics at the Wheel: The Costs of Cars and Drivers* (London: Academic Press, 1999).

10. See R. Salzman, "TravelSmart: A Marketing Program Empowers Citizens to be a Part of the Solution in Improving the Environment," *Mass Transit: Sustainability Concepts*, April 2008, pp. 8–11; R. Salzman, "Now That's What I Call Intelligent Transport...SmartTravel," *Thinking Highways*, March 2008, pp. 51–58.

11. Department of Transport, Government of Western Australia, "Publication and Maps," at www.dpi.wa.gov.au/travelsmart; R. Salzman and P. Newman, *Kicking the Car Habit* (New York: New Society Press, forthcoming).

12. C. Ashton-Graham, *TravelSmart + TOD = Sustainability and Synergy*, Transit-Oriented Development Conference Fremantle, at: www.transport.wa.gov.au/mediaFils/ts_tod.pdf.

13. I. Ker, *North Brisbane Household TravelSmart: Peer Review and Evaluation*, for Brisbane City Council, Queensland Transport, and Australian Greenhouse Office (Brisbane, Australia: February 2008).

14. Socialdata Australia, *TravelSmart Household Final Evaluation Report Murdoch Station Catchment (City of Melville 2007)*, Department of Transport (forthcoming); see www.transport.wa.gov.au/travelsmart; Public Transport Authority, Western Australian Government.

15. Department of Transport, Government of Western Australia, "Living Smart: Acting on Climate Change," at www.dpi.wa.gov.au/livingsmart.

16. C. Ashton-Graham, *Garnaut Climate Change Review: TravelSmart and LivingSmart Case Study—Western Australia*, at www.garnautreview.org.au/CA25734E0016A131/WebObj/Casestudy-TravelSmartandLivingSmart-WesternAustralia/$File/Case%20study%20-%20TravelSmart%20and%20LivingSmart%20-%20Western%20Australia.pdf; Socialdata Australia, *LivingSmart Delivery Report*, Department of Transport (forthcoming), see www.transport.wa.gov.au/livingsmart.

17. Synovate, *LivingSmart Quality Survey*, Department of Transport (forthcoming), see www.transport.wa.gov.au/livingsmart.

18. Cascadia Green Building Council, "The Living Building Challenge," at ilbi.org/stuff/lbc-web-brochure.pdf.

Reinventing Health Care: From Panacea to Hygeia

1. Theoi Greek Mythology, at www.theoi.com.

2. A. Behbehani, "The Smallpox Story: Life and Death of an Old Disease," *Microbiological Reviews*, December 1983, pp. 455–509; World Health Organization (WHO), "Poliomyelitis," fact sheet, at www.who.int/mediacentre/factsheets/fs114/en/index.html; life expectancy from T. McKeown, *Origins of Human Disease* (Oxford: Basil Blackwell, 1988), p. 76.

3. *Food, Inc.*, documentary distributed by Magnolia and directed by Robert Kenner, 2009; D. Kessler, *The End of Overeating* (Emmaus, PA: Rodale Press, 2009); C. Newman, "The Heavy Cost of Fat," *National Geographic*, August 2004, pp. 48–61.

4. "Years of life lost" takes into account the age at which deaths occur by giving greater weight to deaths occurring at younger ages and lower weight to deaths occurring at older ages; WHO, *World Health Report 2002*, Statistical Annex #2 (Geneva: 2002), Table 16; Ali H. Mokdad et al., "Actual Causes of Death in the United States, 2000," *Journal of the American Medical Association*, 10 March 2004, pp. 1,238–45.

5. John P. Holdren, "Science and Technology for Sustainable Well Being," *Science*, 25 January 2008, pp. 424–34; A. M. Prentice and S. A. Jebb, "Obesity in Britain: Gluttony or Sloth?" *British Medical Journal*, 12 August 1995, pp. 437–39.

6. H. J. Aaron and W. B. Schwartz, eds., *Coping with Methuselah* (Washington, DC: Brookings Institution Press, 2004).

7. WHO, *World Health Report 2005* (Geneva: 2005), Annex #6: Selected National Healthcare Indicators: Measured Levels of Expenditures on Health 1996–2002; Uwe E. Reinhardt, Peter S. Hussey, and Gerard E. Anderson, "American Healthcare Spending in an International Context," *Health Affairs*, vol. 23, no. 3 (2004), pp. 10–25; high payments from A. S. Relman, "The New Medical-Industrial Complex," *New England Journal of Medicine*, 23 October 1980, pp. 963–70; ranking from WHO, *World Health Report 2000* (Geneva: 2000); infant mortality from UNICEF, "Childinfo: Monitoring the Situation of Children and Women," at www.childinfo.org; P. Starr, *The Social Transformation of American Medicine* (New York: Basic Books, 1982); S. J. Olshansky et al., "A Potential Decline in Life Expectancy in the United States in the 21st Century," *New England Journal of Medicine*, 17 March 2005, pp. 1,138–45; M. Ezzati et al., "The Reversal of Fortunes: Trends in County Mortality and Cross County Mortality Disparities in the United States," *PLoS Medicine*, April 2008. Table 9 from WHO, *Statistical Information System Database*, at www.who.int/whosis.en, viewed 25 September 2009, and ranking from WHO, *World Health Report 2000*, op. cit. this note. All data are from 2006, except healthy active life expectancy from 2003.

8. Kenneth J. Arrow, "Uncertainty and the Welfare Economics of Medical Care," *American Economic Review*, December 1963, pp. 941–69.

9. J. M. McGinnis and W. H. Foege, "Actual Causes of Death," *Journal of the American Medical Association*, November 1993, pp. 2,207–12. Box 17 from the following: Organisation for Economic Co-operation and Development (OECD), "OECD Stat Extracts," at stats.oecd.org/Index.aspx?datasetcode=SOCX_AGG; Gosta Esping-Anderson, "After the Welfare State," *Public Welfare*, winter 1983, p. 28; Francesco di Iacovo, "Social Farming: Dealing with Communities Rebuilding Local Economy," presentation at *Rural Futures Conference*, University of Plymouth, U.K., 1–4 April 2008; Laurene Mainguy, "Cell Block Green," *EJ Magazine*, spring 2008, pp. 20–21; Nilsen Arne Kvernvik, Bastøy Prison warden, discussion with Erik Assadourian, 3 November 2008; Kathy Lindert, "Brazil: Bolsa Familia Program—Scaling-up Cash Transfers for the Poor," in *Managing for Development Results, Principles in Action: Sourcebook of Emerging Good Practices*, 3rd ed. (Paris: OECD, 2008), pp. 67–74; Gustavo Nigenda and Luz María González-Robledo, *Lessons Offered by Latin American Cash Transfer Programmes, Mexico's Oportunidades and Nicaragua's SPN: Implications for Africa* (London: DFID Health Systems Resource Centre, 2005).

10. P. Puska et al., "Cardiovascular Risk Factor Changes in a Three Year Followup of a Cohort in Connection with a Community Program (the North Karelia Project)," *Acta Medica Scandinavica*, vol. 204 (1976), pp. 381–88.

11. WHO, *World Health Report 2008* (Geneva: 2008); WHO, "The French Country Doctor: Caring for the Sick Through the Centuries," *WHO Bulletin*, October 2008, pp. 743–44.

12. Medical Education Cooperation with Cuba, *Salud!*, documentary, 2006.

13. WHO, *World Health Report 2002* (Geneva: 2002), pp. 188, 224–27.

14. Health Care Without Harm, at www.noharm.org, viewed 16 September 2009.

15. Remarks of Governor Christine Todd Whitman, Administrator of the U.S. Environmental Protection Agency, at the Department of Veterans Affairs, Washington, D.C., 14 May 2003, at yosemite.epa.gov/opa/admpress.nsf/a162fa4bfc0fd2ef8525701a004f20d7/361ee2b093512fe88525701a0052e4c0!OpenDocument; Susan Germain, "The Ecological Footprint of Lions Gate Hospital," *Hospital Quarterly*, winter 2001/2002, pp. 61–66.

16. Health Care Without Harm, op. cit. note 14; Trevor Hancock, *Doing Less Harm: Assessing and Reducing the Environmental and Health Impact of Canada's Health Care System* (Canadian Coalition for Green Health Care, 2001).

Earth Jurisprucence: From Colonization to Participation

1. Address to the Rights of Nature Conference, 24–26 November 2008, Quito, Ecuador, organized by the Fundación Pachamama; see pachamama.org.ec.

2. For information on the Community Environmental Defense Fund, see www.celdf.org.

3. Cormac Cullinan, *Wild Law: A Manifesto for Earth Justice* (Dartington, U.K.: Green Books, 2003).

4. Christopher D. Stone, "Should Trees Have Standing? Towards Legal Rights for Natural Objects," *Southern California Law Review*, vol. 45 (1972), p. 450.

5. Godofredo Stutzin, "Nature's Rights," *Resurgence*, January-February 2002, pp. 24–26.

6. Thomas Berry, *The Great Work: Our Way Into the Future* (New York: Bell Tower, 1999), p. 161.

7. Cullinan, op. cit. note 3; Mike Bell, "Thomas Berry and an Earth Jurisprudence: An Exploratory Essay," *The Trumpeter*, vol. 19, no. 1 (2003).

8. For further information regarding the ongoing development of Earth jurisprudence, see the online community wildfrontiers.ning.com.

9. The ecosystem approach is described in Decision V/6 of the Fifth Meeting of the Conference of the Parties to the Convention on Biological Diversity.

10. Ben Price, CELDF director, discussion with author, 25 August 2009.

11. For information on the Democracy School, see www.celdf.org/DemocracySchool/tabid/60/Default.aspx.

12. Price, op. cit. note 10.

13. Ibid.

14. Act 24 of 2008, see www.info.gov.za/view/DynamicAction?pageid=623&myID=183965.

15. For information on the Center for Earth Jurisprudence, see www.earthjuris.org; for information on the activities of UK Environmental Law Association, see www.ukela.org.

16. Mellese Damtie, unpublished paper presented at Earth Jurisprudence Course at Schumacher College, Devon, U.K., October 2009.

17. "Speech of President Morales before the UN General Assembly on April 22nd, International Mother Earth Day," at www.boliviaun.org/cms/?cat=7&paged=2.

18. For information on Navdanya, see www.navdanya/earthdcracy/index.htm; for a synopsis of this approach, see Vandana Shiva, "Paradigm Shift: Earth Democracy. Rebuilding True Security in an Age of Insecurity," *Resurgence*, September/October 2002.

19. "Community Ecological Governance Global Alliance," Gaia Foundation, at www.gaiafoundation.org/areas/community.php.

Media: Broadcasting Sustainability

1. Rosario G. Gómez, "El Fin de la Publicidad en TVE Revoluciona el Sector Audiovisual," *El Pais*, 9 May 2009; Rosario G. Gómez, "TVE dará Salida Hasta Fin de Año a sus Contratos Publicitarios," *El Pais*, 30 July 2009.

2. Chris Jordan, "Running the Numbers II: Portraits of Global Mass Culture," at www.chrisjordan.com/current_set2.php?id=9

From Selling Soap to Selling Sustainability: Social Marketing

1. Andrea Fuhrel-Forbis, P. Gayle Nadorff, and Leslie Snyder, "Analysis of Public Service Announcements on National Television, 2001–2006," *Social Marketing Quarterly*, March 2009, pp. 49–69; Robert Coen, "Insider's Report: Advertising Expenditures," paper presented by Universal McCann, December 2008, p. 2.

2. Table 10 from the following: ICMR Center for Management Research, "The Marlboro Story," at www.icmrindia.org/casestudies/catalogue/Marketing/The%20Marlboro%20Story.htm; Beetle from Brian Akre, "VW Beetle Led the First U.S. Import Invasion," *Associated Press,* 6 August 1997, and from Thomas Frank, *The Conquest of Cool: Business Culture, Counterculture, and the Rise of Hip Consumerism* (Chicago: University of Chicago Press, 1998), pp. 59–73; seatbelt use from Ad Council, "Safety Belt Education (1985–Present)," at www.adcouncil.org/default.aspx?id=138, from N. Russell, P. Dreyfuss, and M. Cosgrove, "Legislative History of Recent Primary Safety Belt Laws," *National Highway Traffic and Safety Administration* (Washington, DC: January 1999), and from R. J. Arnould and H. Grabowski, "Auto Safety Regulation: An Analysis of Market Failure," *The Bell Journal of Economics*, spring 1981, pp. 27–48; Apple from Juliann Sivulka, *Soap, Sex, and Cigarettes: A Cultural History of American Advertising* (Florence, KY: Wadsworth Publishing, 1997), pp. 353–56, and from Jeremy Reimer, "Total Share: 30 Years of Personal Computer Market Share Figures," at arstechnica.com/old/content/2005/12/totalshare.ars; "The Story of Stuff International," at www.storyofstuff.com/international; Suemedha Sood, "Weighing the Impact of 'Super Size Me'," *Wiretapmag.org*, 29 June 2004.

3. Joseph Campbell, *The Hero With a Thousand Faces* (Princeton, NJ: Princeton University Press, 1973).

4. Ibid.

5. Joseph Campbell with Bill Moyers, *The Power of Myth* (New York: Anchor Books, 1988), p. 48.

6. "How Can Entertainment-Education Influence Behavior?" in *INFO Reports: Entertainment-Education for Better Health* (The INFO Project, Johns Hopkins Bloomberg School of Public Health), February 2008.

7. *INFO Reports*, op. cit. note 6.

8. Based on a survey of the climate "landing pages" on the Web sites of Environmental Defense Fund, Greenpeace, National Audubon Society, Natural Resources Defense Council, Sierra Club, The Nature Conservancy, World Resources Institute, and World Wildlife Fund; Simon Retallack, "Ankelohe and Beyond: Communicating Climate Change," *opendemocracy.net*, 17 May 2006.

9. Edward Maibach, Connie Roser-Renouf, and Anthony Leiserowitz, *Global Warming's Six Americas 2009: An Audience Segmentation Analysis* (New Haven, CT, and Fairfax, VA: Yale Project on Climate Change and George Mason University Center for Climate Change Communication, May 2009).

10. Geller cited in D. McKenzie-Mohr and W. Smith, *Fostering Sustainable Behavior: An Introduction to Community-Based Social Marketing* (Gabriola Island, BC: New Society Publishers, 1999), p. 9.

11. McKibben quoted in Elizabeth Kolbert, "Profiles: The Catastrophist," *New Yorker*, 29 June 2009, p. 39.

12. "Understanding 350," at www.350.org/understanding-350.

13. Robert Scoble, "What is Social Media?" 16 February 2007, at www.scobleizer.com/02/16/what-is-social-media.

14. Lee Rainie et al., *The Strength of Internet Ties* (Washington, DC: Pew Internet & American Life Project, 2006); "Facebook Statistics," at www.facebook.com/press/info.php?statistics#/press/info.php?statistics.

15. Rainie et al., op. cit. note 14.

16. "Global Advertising: Consumers Trust Real Friends and Virtual Strangers the Most," blog, *Nielsen Wire*, 7 July 2009.

17. "TV Gala Helps Raise 1.5 bln Yuan for Earthquake-hit Areas," *China.com*, 19 May 2008.

18. "Protests Worldwide Call for End to Iranian Rights Abuses," *USA Today*, 26 July 2009.

19. Rainie et al., op. cit. note 14.

20. "350 Home Page," at www.350.org.

Media Literacy, Citizenship, and Sustainability

1. Terry Richardson/Marcel Paris, "Diesel Global Warming Ready," 2007, advertisment.

2. Diane Farsetta, "Video News Releases: A Hidden Epidemic of Fake TV News," in Robin Andersen and Jonathan Gray, eds., *Battleground: The Media* (Westport, CT: Greenwood, 2008), pp. 542–49.

3. Jim Hansen, "The Threat to the Planet," *New York Review of Books*, 13 July 2006; Jules Boykoff and Maxwell Boykoff, "Journalistic Balance as Global Warming Bias," *EXTRA!* November/December 2004; George Monbiot, "The Denial Industry," (London) *Guardian*, 19 September 2006.

4. Alliance for a Media Literate America, *What is Media Literacy? AMLA's Short Answer and a Longer Thought* (Center for Media Literacy Reading Room, 2001); Action Coalition for Media Education, *About ACME*, at www.acmecoalition.org/about_acme; Bill Yousman, "Media Literacy: Creating Better Citizens or Better Consumers?" in Andersen and Gray, op. cit. note 2, pp. 238–47.

5. Yousman, op. cit. note 4, p. 244.

6. Ibid., p. 238.

7. Abdul Waheed Khan, "Foreword: UNESCO," in UN-Alliance of Civilizations in cooperation with Grupo Comunicar, *Mapping Media Education Policies in the World: Visions, Programmes and Challenges* (New York and Huelva, Spain: 2009), p. 9. Table 11 is based on the following: Argentina from Roxana Morduchowicz, "When Media Education is State Policy," in UN-Alliance of Civilizations in cooperation with Grupo Comunicar, op. cit. this note, pp. 177–87; Australian Communications and Media Authority, *The ACMA Digital Media Literacy Research Program*, at www.acma.gov.au/WEB/STANDARD/pc=PC_311472; Austria from Maria Koller, Astrid Haider, and Elke Dall, *Case Studies of Conditions and Success Criteria in Media Literacy Education* (Vienna: Centre for Social Innovation); Carolyn Wilson and Barry Duncan, "Implementing Mandates in Media Education: the Ontario Experience," in UN-Alliance of Civilizations in cooperation with Grupo Comunicar, op. cit. this note, pp. 127–40; Finland from Sirkku Kotilainen, "Promoting Youth Civic Participation with Media Production: The Case of Youth Voice Editorial Board," in UN-Alliance of Civilizations in cooperation with Grupo Comunicar, op. cit. this note, pp. 243–56; France from Centre for Liaison between Teaching and Information Media, "The CLEMI at a Glance," at www.clemi.org/fr/anglais; C. K. Cheung, "Education Reform as an Agent of Change: The Development of Media Literacy in Hong Kong During the Last Decade," in UN-Alliance of Civilizations in cooperation with Grupo Comunicar, op. cit. this note, pp. 95–109; Alexander Fedorov, "Media Education in Russia: A Brief History," in Marcus Leaning, ed., *Issues in Information and Media Literacy: Criticism, History and Policy* (Santa Rosa, CA: Informing Science Press, 2009), pp. 167–88; Hyeon-Seon Jeong et al., "History, Pol-

icy and Practices of Media Education in South Korea," in UN-Alliance of Civilizations in cooperation with Grupo Comunicar, op. cit. this note, pp. 111–25; Sweden from Nordicom, "The International Clearinghouse on Children, Youth and Media," at www.nordicom.gu.se/clearinghouse.php; Turkey from E. Nezih Orhon, "Media Education in Turkey: Toward a Multi-Stakeholder Framework," in UN-Alliance of Civilizations in cooperation with Grupo Comunicar, op. cit. this note, pp. 211–24; United Kingdom from Office of Communications, at www.ofcom.org.uk.

8. UNESCO from Khan, op. cit. note 7, pp. 9–10.

9. Divina Frau-Meigs and Jordi Torrent, "Media Education Policy: Toward a Global Rationale," in UN-Alliance of Civilizations in cooperation with Grupo Comunicar, op. cit. note 7, pp. 15–21.

10. Costas Criticos, "Media Education for a Critical Citizenry in South Africa," in Robert Kubey, ed., *Media Literacy in the Information Age* (New Brunswick, NJ: Transaction, 2001), pp. 229–40.

11. Fackson Banda, "Exploring Media Education as Civic Praxis in Africa," in UN-Alliance of Civilizations in cooperation with Grupo Comunicar, op. cit. note 7, pp. 225–42.

12. Ibid., p. 235.

13. International Telecommunication Union's World Telecommunication Development Report database, as cited in World Bank, *World Development Indicators Database 2008* (Washington, DC: April 2008).

14. Roxana Morduchowicz, "When Media Education is State Policy," in UN-Alliance of Civilizations in cooperation with Grupo Comunicar, op. cit. note 7, pp. 177–87; India Resource Center, *Campaign to Hold Coca-Cola Accountable*, at www.indiaresource.org/campaigns/coke; Frau-Meigs and Torrent, op. cit. note 9, p. 19.

15. Robin Blake, "An International Model for Media Literacy," International Media Literacy Research Forum (London), 15 May 2008; DeeDee Halleck, *Waves of Change Blog*, at www.deepdish wavesofchange.blogspot.com.

16. David Gauntlett, *Video Critical: Children, the Environment and Media Power* (Bedfordshire, U.K.: University of Luton Press, 2005).

17. Ibid.; DeeDee Halleck, *Handheld Visions: The Impossible Possibilities of Community Media* (New York: Fordham University Press, 2002).

Music: Using Education and Entertainment to Motivate Change

1. Peter Marler and Hans Willem Slabbekoorn, *Nature's Music: The Science of Birdsong* (San Diego, CA: Elsevier Academic Press, 2004), p. 386.

2. Neil Edmunds, *Soviet Music and Society under Lenin and Stalin: The Baton and Sickle* (London: Routledge, 2004), p. 182; Ron Eyerman and Andrew Jamison, *Music and Social Movements: Mobilizing Traditions in the Twentieth Century* (Cambridge, U.K.: Cambridge University Press, 1998), p. 3.

3. U.S. Environmental Protection Agency, "Administrator Lisa P. Jackson, Remarks to the National Association of Black Journalists, as Prepared," 7 August 2009.

4. Box 20 from the following: *Home* from www.home-2009.com; earnings of *An Inconvenient Truth* (2006) from Box Office Mojo, at www.boxofficemojo.com/movies/?id=inconvenienttruth.htm; *The Age of Stupid* (2009) from www.ageofstupid.net. Box 21 from the following: Beuys cited in Ken Hopper and William Hopper, *The Puritan Gift: Triumph, Collapse and Revival of an American Dream* (New York: St Martin's Press, 2007), p. 141; Coomaraswamy cited in David Nicholls, *The Cambridge Companion to John Cage* (Cambridge, U.K.: Cambridge University Press, 2002), p. 46; William Morris, *Hopes and Fears for Art* (Boston: Robert's Brothers, 1882), pp. 71–113.

5. Frances H. Rauscher, Gordon L. Shaw, and Katherine N. Ky, "Music and Spatial Task Performance," *Nature*, 14 October 1993, p. 611; Anne Mitchell and Judy David, eds., *Explorations with Young Children: A Curriculum Guide from Bank Street College of Education* (Beltsville, MD: Gryphon House, 1992), p. 218; *Shinichi Suzuki: His Speeches and Essays* (Princeton, NJ: Suzuki Method International, 1993); Albert L. Blackwell, *The Sacred in Music* (Louisville, KY: Westminster John Knox Press, 1999), p. 170.

6. Ecogainder at www.kids-station.com/mini site/ecogainder/about/index.html; RaffiNews Web site, at www.raffinews.com; Raffi News, "BG Feedback," at www.raffinews.com/beluga-grads/feedback.

7. Penguins on Thin Ice Web site, at www.pen guinsonthinice.com.

8. Massukos Web site, at www.massukos.org; The Goldman Environmental Prize, "Feliciano dos Santos," at www.goldmanprize.org/2008/africa.

9. "Live Aid," *Wikipedia*, at en.wikipedia.org/wiki/Live_Aid; Michael Scott and Mutombo Mpanya, *We Are the World: An Evaluation of Pop Aid for Africa* (Washington, DC: InterAction, 1994), p. 3; "We Are the World," *Wikipedia*, at en.wikipedia .org/wiki/We_Are_the_World; $63 million from Cindy Clark et al., "Moments of Sex, Drugs and Rock 'n' Roll," *USA Today*, 28 July 2006.

10. LiveEarth Web site, at liveearth.org/en/liveearth.

11. "Josh Tyrangiel, "Bono's Mission," *Time Magazine*, 23 February 2002; ONE Web site, at www.one.org.

12. Sean Michaels, "U2 Criticised for World Tour Carbon Footprint," *Guardian* (London), 10 July 2009.

13. Roskilde Festival Web site, at www.roskilde-fes tival.dk/uk/about_the_festival/green_footsteps; Glastonbury Festival of Contemporary Performing Arts Web site, at www.glastonburyfestivals .co.uk/information/green-glastonbury/our-green -policies.

14. Glastonbury Festival, op. cit. note 13; Bumbershoot 2009, "Greening Bumbershoot," at www.bumbershoot.org/fresh/green; High Sierra, "Greening," at www.highsierramusic.com/event-info/greening; Leave No Trace Center for Outdoor Ethics, "About Us," at www.lnt.org/aboutUs/index.php.

15. Ojai Music Festival, "Ojai Music Festival Goes Greener," *Shuman Associates News*, 30 April 2008; Sarah Van Schagen, "Dave Matthews Band Offers Free Music Downloads for Eco-pledges," *Grist Magazine*, 21 August 2009; So Much to Save Web site, at www.somuchtosave.org.

16. "Big Yellow Taxi," *Wikipedia*, at en.wiki pedia.org/wiki/Big_Yellow_Taxi; "Tracy Chapman: The Rape of the World," *Wikia*, at lyrics.wikia.com/Tracy_Chapman:The_Rape_Of_The_World; CultureChange, "Depavers: Eco-Rock," at www.culturechange.org/cms/index .php?option=com_content&task=view&id=225&Itemid=53.

17. Bonnie Raitt Web site, at www.bon nieraitt.com/bio.php; "Bonnie Raitt," *Wikipedia*, at en.wikipedia.org/wiki/Bonnie_Raitt.

18. "Willie Nelson," *Wikipedia*, at en.wikipedia .org/wiki/Willie_Nelson; Farm Aid Web site, at www.farmaid.org/site/c.qlI5IhNVJsE/b.2723609/k.C8F1/About_Us.htm.

19. Tipping Point Art & Climate Change Web site, at www.tippingpoint.org.uk/index.htm.

20. Judith Marcuse Projects ICASC Web site, at www.icasc.ca/jmp.

21. The Climate Group, "Live Earth: When the Music Stops, We Must Start," press release (London: 9 July 2007).

The Power of Social Movements

1. Paul Hawken, "Biology, Resistance, and Restoration: Sustainability as an Infinite Game," presentation at Bioneers Conference, October 2006.

2. Erik Assadourian, "Engaging Communities for a Sustainable World," in Worldwatch Institute, *State of the World 2008* (New York: W. W. Norton & Company, 2008), pp. 151–65.

Reducing Work Time as a Path to Sustainability

1. Ruhm cited in Drake Bennett, "The Good Recession," *Boston Globe*, 23 March 2009; Joe Rojas-Burke, "Does Our Health Actually Get Better in Some Ways During a Down Economy?" *The Oregonian*, 22 April 2009.

2. Rojas-Burke, op. cit. note 1; see also Stephen Bezruchka, "The Effect of Economic Recessions on Population Health," *Canadian Medical Association Journal*, in press.

3. Global Footprint Network, "Earth Overshoot Day," at www.footprintnetwork.org/en/index.php/GFN/page/earth_overshoot_day, updated 16 July 2009.

4. James Gustave Speth, *The Bridge at the Edge of the World* (New Haven, CT: Yale University Press, 2009), p. 51.

5. Ibid.

6. Juliet Schor, "The Even More Overworked American," in John de Graaf, *Take Back Your Time* (San Francisco: Berrett-Koehler, 2003), p. 10.

7. Author's analysis of country-by-country working hours at Organisation for Economic Co-operation and Development, 2007, at statlinks.oecdcode.org/302009011P1T082.XLS.

8. U.S. life expectancy from Central Intelligence Agency, *The CIA World Factbook, 2009* (New York: Skyhorse Publishing, 2008); Lisa Girion, "Europe Healthier than U.S.—Older Americans Have Higher Rates of Serious Diseases than Aging Europeans, A Study Says," *Los Angeles Times*, 2 October 2007.

9. Lauren Sherman, "Europe's Happiest Places," *Forbes*, 12 August 2009.

10. Center for Economic and Policy Research, "Long U.S. Work Hours Are Bad for the Environment, Study Shows," press release (Washington, DC: 20 December 2006).

11. Quote from Donald Worster, *A Passion for Nature: The Life of John Muir* (Oxford, U.K.: Oxford University Press, 2009), p. 225.

12. Paid vacations from Jody Heymann and Allison Earle, *Raising the Global Floor* (Stanford, CA: Stanford University Press, 2009); U.S. median vacation time from "Results of Take Back Your Time's Right2Vacation Poll," *right2vacation.com*, at www.timeday.org/right2vacation/poll_results.asp; Scott Gediman, media relations director, Yosemite National Park, discussion with author, August 2008.

13. Take Back Your Time, at www.timeday.org.

14. Human Resources and Skills Development Canada, "The Netherlands: Improving Work-Life Balance—What Are Other Countries Doing?" 24 November 2004.

15. Heymann and Earle, op. cit. note 12.

16. See www.momsrising.org.

17. Take Back Your Time, op. cit. note 13; John de Graaf, "H.R. 2564: The Paid Vacation Act of 2009—Rebutting the Opposition," Take Back Your Time, Seattle, WA, undated.

18. The White House, "President Barack Obama's Inaugural Address," Washington, DC, 21 January 2009; Dean Baker, "When Less Is More," *Guardian* (London), 26 January 2009.

Inspiring People to See That Less Is More

1. Goldian Bandenbroeck, *Less is More: The Art of Voluntary Poverty: An Anthology of Ancient and Modern Voices Raised in Praise of Simplicity* (Rochester, VT: Inner Traditions International, 1991); David E. Shi, *The Simple Life: Plain Living and High Thinking in American Culture* (Oxford: Oxford University Press, 1985).

2. Abraham Joshua Heschel and Susannah Heschel, *Moral Grandeur and Spiritual Audacity: Essays* (New York: Farrar, Straus and Giroux, 1997), p. 31.

3. Tim Kasser, *The High Price of Materialism* (Cambridge, MA: The MIT Press, 2002).

4. Richard Wilkinson, *The Impact of Inequality: How to Make Sick Societies Healthier* (New York: The New Press, 2005); Richard Wilkinson and Kate Pickett, *The Spirit Level: Why More Equal Societies Almost Always Do Better* (London: Penguin Group, 2009).

5. George Lakoff, *Thinking Points: Communi-*

cating Our American Values and Vision (New York: Farrar, Straus and Giroux, 2006).

6. Bill McKibben, "Together, We Save the Planet," *The Nation*, 23 March 2009.

7. Myles Horton and Paulo Freire, *We Make the Road by Walking: Conversations on Education and Social Change* (Philadelphia: Temple University Press, 1990); Paulo Friere, *The Pedagogy of the Oppressed*, anniversary ed. (London: Continuum, 2000).

8. "What Is a Transition Town (or village/city/forest/island)?" Transition Towns WIKI, at www.transitiontowns.org. Box 22 from the following: degrowth in France from the Network of Growth Objectors for Post-Development ROCADe, at www.apres-developpement.org, from Institute for Social and Economic studies for Sustainable Degrowth, at www.decroissance.org, from the French Degrowth Party, at www.partipourladecroissance.net, and from *La Decroissance*, at www.ladecroissance.net; degrowth in Italy from the Italian Degrowth Movement, at www.decrescita.it, from Italian Movement for Happy Degrowth, at www.decrescitafelice.it, and from the Italian Degrowth Party, at www.partitoperladecrescita.it; degrowth in Spain from Spanish Degrowth Movement, at www.decrecimiento.info, from "Diario del Decrecimiento," Spanish Degrowth News, at www.decrecimiento.es, from Catalan network for Degrowth, at www.decreixement.info, and from Temps de Re-voltes campaign, at www.temsdere-voltes.cat; Transition Towns and local currencies from the Totnes Pound Project, at www.totnes.transitionnetwork.org/totnespound.home, from Ithaca Hours Online, at www.ithacahours.com, from Transition Towns Wiki, at www.transitiontowns.org, from Transition Town Kinsale, at www.transitiontownkinsale.org, and from Transition Town Totnes, at www.totnes.transitionnetwork.org.

9. Rob Hopkins, *The Transition Handbook: From Oil Dependency to Local Resilience* (Totnes, Devon, U.K.: Green Books Ltd, 2008).

10. Carl Honore, *In Praise of Slowness: How a Worldwide Movement Is Challenging the Cult of Speed* (San Francisco: HarperSanFrancisco, 2004). Box 23 from the following: Carlo Petrini, *Slow Food: The Case for Taste* (New York: Columbia University Press, 2003); Slow Food International, at www.slowfood.com and www.slowfood.it; Slow Food Foundation for Biodiversity, at www.slowfoodfoundation.org; information on the Ark of Taste from ibid.; Slow Food Earthmarkets, at www.earthmarkets.net; Slow Fish Exhibition 2009, at www.slowfish.it/welcome_eng.lasso; Slow Food USA, "Time for Lunch" campaign, at www.slowfoodusa.org/index.php/campaign/time_for_lunch/about.

11. Take Back Your Time is a U.S. and Canadian initiative to challenge overwork, overscheduling, and time famine; see www.timeday.org.

12. Robert Wuthnow, *American Mythos: Why Our Best Efforts to Be a Better Nation Fall Short* (Princeton, NJ: Princeton University Press, 2006).

13. Robert Putnam, *Bowling Alone* (New York: Simon & Schuster, 2000).

14. Tim Kasser and Tom Crompton, *Meeting Environmental Challenges: The Role of Human Identity* (Godalming, Surrey, U.K.: WWF-UK, 2009).

15. "Open Space Planning," Show 413, June 2008, and "Simplicity & Spirituality," Show 106, August 2004, on *Simple Living with Wanda Urbanska*.

16. "The Thing That Refused to Die," Show 108, September 2004, on *Simple Living with Wanda Urbanska*.

17. For information on the Buy Nothing Day campaign, see www.adbusters.org/campaigns.bnd.

18. For The Compact, see sfcompact.blogspot.com; freeganism information available at freegan.info/?page_id=2; World Wide Opportunities on Organic Farms, at www.wwoof.org.

19. Cecile Andrews, *Slow is Beautiful: New Visions of Community, Leisure, and Joie de Vivre* (Gabriola Island, B.C., Canada: New Society Publishers, 2006); "Barefoot College," at www.barefootcollege.org; Alan Weisman, *Gaviotas: A Village to Reinvent the World* (White River Junction, VT:

Chelsea Green, 1999); "Friends of Gaviotas," at www.friendsofgaviotas.org/Friends_of_Gaviotas/Home.html; Jan Gehl and Lars Gemzoe, *Public Spaces, Public Life* (Copenhagen: Danish Architectural Press, 2004); "Jan Gehl," Project for Public Spaces, at www.pps.org/info/placemakingtools/placemakers/jgehl; "The Vision of City Repair," at cityrepair.org.

Ecovillages and the Transformation of Values

1. Claus Schenk, "Paradise With Side Effects," Capricorn Film, International Society for Ecology and Culture, Ladakh, India, 2004.

2. Global Ecovillage Network, at gen.ecovillage.org.

3. Robert Gilman, " *In Context*, summer 1991, p. 10.

4. Information on both communes from Kommune Niederkaufungen, Gemeinschaftlich Nachhaltig, at www.usf.uni-kassel.de/glww/ziele.htm.

5. Jason R. Brown, *Comparative Analysis of Energy Consumption Trends in Cohousing and Alternate Housing Arrangements*, Department of Civil and Environmental Engineering, Massachusetts Institute of Technology, 2004, unpublished thesis; Jonathan Dawson, "Findhorn's Incredible Shrinking Footprint," *Communities*, summer 2009.

6. Kenneth Mulder, Robert Costanza, and Jon Erickson, "The Contribution of Built, Human, Social and Natural Capital to Quality of Life in Intentional and Unintentional Communities," *Ecological Economics*, August 2006, pp. 18–19.

7. Ibid., p. 20.

8. "Work Areas," More About Twin Oaks, at www.twinoaks.org/community/index.html.

9. Tency Baetens, "The Use of Horizontal Planted Filters for Decentralised Wastewater Treatment in Auroville, An Overview and Description," prepared for Conference on Constructed Wetlands for Wastewater Treatment in Tropical and Subtropical Regions, December 2000; Raven Le Fay, "From Dust to Dawn," *Permaculture Magazine*, No. 45, pp. 39–42; Sólheimar, at solheimar.hlutverk.is/page.asp?Id=834; The Farm, at www.thefarm.org.

10. Russian ecovillage Grishino, at www.grishino.ecology.net.ru/en/index.htm; Findhorn Foundation, at www.findhorn.org/index.php?tz=240.

11. Sarvodaya Empowerment Programmes, at www.sarvodaya.org/about/empowerment-programmes.

12. The Ladakh Project, International Society for Ecology and Culture, at www.isec.org.uk/pages/ladakh.html#womensallianceofladakh.

13. For more on systems thinking, see David W. Orr, *Earth in Mind* (Washington, DC: Island Press, rev. 2004), and Fritjof Capra, "Ecoliteracy: The Challenge for Education in the Next Century," Liverpool Schumacher Lectures, Center for Ecoliteracy, Berkeley, CA, 20 March 1999.

14. The Farm Ecovillage Training Center, at www.thefarm.org/etc; Lotan Center for Creative Ecology, at www.kibbutzlotan.com/creativeEcology; EcoCentre, Ecological Solutions, at www.ecologicalsolutions.com.au/venue/ecocentre.html.

15. CIFAL Findhorn, at www.cifalfindhorn.org.

16. Findhorn Foundation College, at www.findhorncollege.org/index.php; Sustainable Community Design, Heriot-Watt University, at www.postgraduate.hw.ac.uk/course/327.

17. Gaia Education, at www.gaiaeducation.org.

18. Living Routes, at www.livingroutes.org; "Partnerships for Sustainability Education," Ithaca College, at www.ithaca.edu/hs/science_in_the_community.

Index

D

E

F

H

I